Honda CBR900RR Owners Workshop Manual

by Penny Cox and Matthew Coombs

Models covered:
Honda CBR900RR FireBlade. 893cc. UK 1992 through 1995
Honda CBR900RR. 893cc. US 1993 through 1995

ABCDE
FGHIJ
KLMNO
PQRS

Haynes Publishing
Sparkford Nr Yeovil
Somerset BA22 7JJ England

Haynes North America, Inc
861 Lawrence Drive
Newbury Park
California 91320 USA

Acknowledgements

Our thanks are due to Paul Branson Motorcycles of Yeovil, Somerset, who supplied the machine featured in the photographs throughout this manual and to Melvin Rawlings of MHR Engineering who carried out the mechanical work. Thanks are also due to NGK Spark Plugs (UK) Ltd who supplied the color spark plug condition photos and the Avon Rubber Company who supplied information on tire fitting.

© **Haynes Publishing 1995**

A book in the Haynes Owners Workshop Manual Series

Printed by J. H. Haynes & Co. Ltd., Sparkford, Nr Yeovil, Somerset, BA22 7JJ, England

ISBN 1 56392 161 8

Library of Congress Catalog Card Number 95-80720

British Library Cataloguing in Publication Data
A catalogue record for this book is available from the British Library

We take great pride in the accuracy of information given in this manual, but motorcycle manufacturers make alterations and design changes during the production run of a particular motorcycle of which they do not inform us. No liability can be accepted by the authors or publishers for loss, damage or injury caused by any errors in, or omissions from, the information given.

Contents

1
2
3
4
5
6
7
8
9

IND

The UK CBR900RR-N model

The UK CBR900RR-S model

About this manual

Its purpose

The purpose of this manual is to help you get the best value from your motorcycle. It can do so in several ways. It can help you decide what work must be done, even if you choose to have it done by a dealer service department or a repair shop; it provides information and procedures for routine maintenance and servicing; and it offers diagnostic and repair procedures to follow when trouble occurs.

We hope you use the manual to tackle the work yourself. For many simpler jobs, doing it yourself may be quicker than arranging an appointment to get the vehicle into a shop and making the trips to leave it and pick it up. More importantly, a lot of money can be saved by avoiding the expense the shop must pass on to you to cover its labour and overhead costs. An added benefit is the sense of satisfaction and accomplishment that you feel after doing the job yourself.

Using the manual

The manual is divided into Chapters. Each Chapter is divided into numbered Sections, which are headed in bold type between horizontal lines. Each Section consists of consecutively numbered paragraphs or steps.

At the beginning of each numbered Section you will be referred to any illustrations which apply to the procedures in that Section. The reference numbers used in illustration captions pinpoint the pertinent Section and the Step within that Section. That is, illustration 3.2 means the illustration refers to Section 3 and Step (or paragraph) 2 within that Section.

Procedures, once described in the text, are not normally repeated. When it's necessary to refer to another Chapter, the reference will be given as Chapter and Section number. Cross references given without use of the word 'Chapter' apply to Sections and/or paragraphs in the same Chapter. For example, 'see Section 8' means in the same Chapter.

References to the left or right side of the vehicle assume you are sitting on the seat, facing forward.

Motorcycle manufacturers continually make changes to specifications and recommendations, and these, when notified, are incorporated into our manuals at the earliest opportunity.

Even though we have prepared this manual with extreme care, neither the publisher nor the author can accept responsibility for any errors in, or omissions from, the information given.

NOTE

A **Note** provides information necessary to properly complete a procedure or information which will make the procedure easier to understand.

CAUTION

A **Caution** provides a special procedure or special steps which must be taken while completing the procedure where the Caution is found. Not heeding a Caution can result in damage to the assembly being worked on.

WARNING

A **Warning** provides a special procedure or special steps which must be taken while completing the procedure where the Warning is found. Not heeding a Warning can result in personal injury.

Introduction to the Honda CBR900RR FireBlade

The CBR900RR has been designed for ultimate lightness and high performance. Its dry weight is around 185 kg and engine power output around 121 bhp, depending on the market.

Its liquid-cooled in-line four cylinder engine is derived from the CBR600F2, in which the cylinder block is integral with the upper crankcase. A wet multi-plate clutch delivers power to the 6-speed transmission. Final drive to the rear wheel is by chain.

The fuel system comprises four 38 mm Keihin flat-slide CV carburettors. Ignition is digital.

The twin-spar frame is made of lightweight aluminium and uses the engine as a stressed member. Suspension is by Showa. The front forks are of conventional design, although lower leg design gives the impression of an inverted fork. Honda's "Pro-link" rear suspension is used with an extruded box section aluminium swingarm.

Wheel sizes are 16 inch front, and 17 inch rear. Four piston opposed twin brake calipers are used at the front, with a single piston caliper at the rear.

Identification numbers

Frame and engine numbers

The frame serial number is stamped into the right side of the steering head and also appears on the identification plate attached to the right side of the frame. The engine number is stamped into the right upper side of the crankcase, directly above the clutch unit. Both of these numbers should be recorded and kept in a safe place so they can be furnished to law enforcement officials in the event of a theft.

The frame serial number, engine serial number and carburetor identification number should also be kept in a handy place (such as with your driver's license) so they are always available when purchasing or ordering parts for your machine.

Model	Initial engine number
UK CBR900RR-N (1992)	SC28E-2000064-2001053
UK CBR900RR-P (1993)	SC28E-2105049-2115746
UK CBR900RR-R (1994)	SC28E-2250388 on
UK CBR900RR-S (1995)	SC28E-2350001 on
US CBR900RR 1993 - California	SC28E-2100012 on
US CBR900RR 1993 - except California	SC28E-2100007 on
US CBR900RR 1994 - California	SC28E-2200001 on
US CBR900RR 1994 - except California	SC28E-2200001 on
US CBR900RR 1995 - California	SC28E-2300001 on
US CBR900RR 1995 - except California	SC28E-2300001 on

Identifying codes

The procedures in this manual identify the bikes by model code for UK models and production year for US models. The model code (eg CBR900RR-S) or production year is printed on the color code label, which is stuck to the top of the frame, next to the fuel tank rear mounting.

The model code or production year can also be determined from the engine and frame serial numbers as follows:

Initial frame number
SC28-2000043-2009388
SC28-2100006-2108740
SC28-2250391 on
SC28-2350001 on
SC281*PM100004 on
SC280*PM100002 on
SC281*RM200001 on
SC280*RM200001 on
SC281*SM300001 on
SC280*SM300001 on

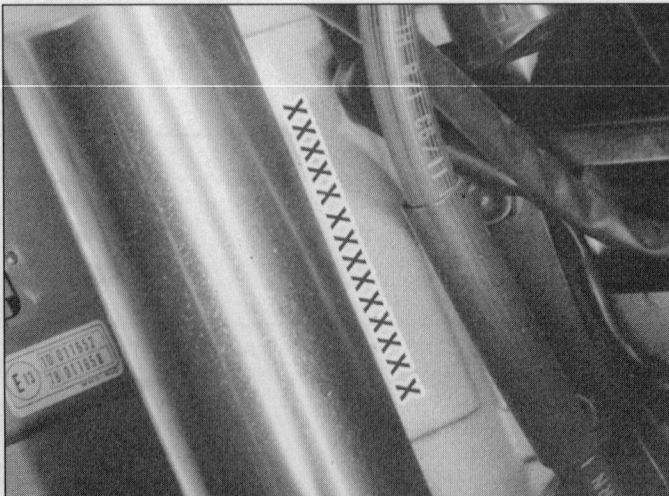

Frame number is stamped on the right side of the steering head

Engine number is stamped on the right side of the crankcase

Buying parts

Once you have found all the identification numbers, record them for reference when buying parts. Since the manufacturers change specifications, parts and vendors (companies that manufacture various components on the machine), providing the ID numbers is the only way to be reasonably sure that you are buying the correct parts.

Whenever possible, take the worn part to the dealer so direct comparison with the new component can be made. Along the trail from the manufacturer to the parts shelf, there are numerous places that the part can end up with the wrong number or be listed incorrectly.

The two places to purchase new parts for your motorcycle - the accessory store and the franchised dealer - differ in the type of parts they carry. While dealers can obtain virtually every part for your motorcycle, the accessory dealer is usually limited to normal high wear items such as shock absorbers, tune-up parts, various engine gaskets, cables, chains, brake parts, etc. Rarely will an accessory outlet have major suspension components, cylinders, transmission gears, or cases.

Used parts can be obtained for roughly half the price of new ones, but you can't always be sure of what you're getting. Once again, take your worn part to the wrecking yard (breaker) for direct comparison.

Whether buying new, used or rebuilt parts, the best course is to deal directly with someone who specializes in parts for your particular make.

General specifications

Wheelbase	
US models	1400 mm (55.1 in)
UK models	1405 mm (55.3 in)
Overall length	
US models	2055 mm (81.0 in)
UK models	2030 mm (80.0 in)
Overall width	685 mm (27.0 in)
Overall height	
US 1993 and 1994 models, UK CBR900RR-N and RR-P	1115 mm (44.0 in)
US 1995 model, UK CBR900RR-R and RR-S	1130 mm (44.5 in)
Seat height	800 mm (31.5 in)
Ground clearance	130 mm (5.1 in)
Weight (dry)	
California models	187 kg (412 lbs)
UK models and US models (except California)	185 kg (408 lbs)
Weight (with oil and full fuel tank)	
California models	208 kg (459 lbs)
UK models and US models (except California)	206 kg (454 lbs)

Maintenance techniques, tools and working facilities

Basic maintenance techniques

There are a number of techniques involved in maintenance and repair that will be referred to throughout this manual. Application of these techniques will enable the amateur mechanic to be more efficient, better organized and capable of performing the various tasks properly, which will ensure that the repair job is thorough and complete.

Fastening systems

Fasteners, basically, are nuts, bolts and screws used to hold two or more parts together. There are a few things to keep in mind when working with fasteners. Almost all of them use a locking device of some type (either a lock washer, locknut, locking tab or thread adhesive). All threaded fasteners should be clean, straight, have undamaged threads and undamaged corners on the hex head where the wrench fits. Develop the habit of replacing all damaged nuts and bolts with new ones.

Rusted nuts and bolts should be treated with a penetrating oil to ease removal and prevent breakage. Some mechanics use turpentine in a spout type oil can, which works quite well. After applying the rust penetrant, let it -work for a few minutes before trying to loosen the nut or bolt. Badly rusted fasteners may have to be chiseled off or removed with a special nut breaker, available at tool stores.

If a bolt or stud breaks off in an assembly, it can be drilled out and removed with a special tool called an E-Z out (or screw extractor). Most dealer service departments and motorcycle repair shops can perform this task, as well as others (such as the repair of threaded holes that have been stripped out).

Flat washers and lock washers, when removed from an assembly, should always be replaced exactly as removed. Replace any damaged washers with new ones. Always use a flat washer between a lock washer and any soft metal surface (such as aluminum), thin sheet metal or plastic. Special locknuts can only be used once or twice before they lose their locking ability and must be replaced.

Tightening sequences and procedures

When threaded fasteners are tightened, they are often tightened to a specific torque value (torque is basically a twisting force). Over-tightening the fastener can weaken it and cause it to break, while under-tightening can cause it to eventually come loose. Each bolt, depending on the material it's made of, the diameter of its shank and the material it is threaded into, has a specific torque value, which is noted in the Specifications. Be sure to follow the torque recommendations closely.

Fasteners laid out in a pattern (i.e. cylinder head bolts, engine case bolts, etc.) must be loosened or tightened in a sequence to avoid warping the component. Initially, the bolts/nuts should go on finger tight only. Next, they should be tightened one full turn each, in a criss-cross or diagonal pattern. After each one has been tightened one full turn, return to the first one tightened and tighten them all one half turn, following the same pattern. Finally, tighten each of them one quarter turn at a time until each fastener has been tightened to the proper torque. To loosen and remove the fasteners the procedure would be reversed.

Disassembly sequence

Component disassembly should be done with care and purpose to help ensure that the parts go back together properly during reassembly. Always keep track of the sequence in which parts are removed. Take note of special characteristics or marks on parts that can be installed more than one way (such as a grooved thrust washer on a shaft). It's a good idea to lay the disassembled parts out on a clean surface in the order that they were removed. It may also be helpful to make sketches or take instant photos of components before removal.

When removing fasteners from a component, keep track of their locations. Sometimes threading a bolt back in a part, or putting the washers and nut back on a stud, can prevent mixups later. If nuts and bolts can't be returned to their original locations, they should be kept in a compartmented box or a series of small boxes. A cupcake or muffin tin is ideal for this purpose, since each cavity can hold the bolts and nuts from a particular area (i.e. engine case bolts, valve cover bolts, engine mount bolts, etc.). A pan of this type is especially helpful when working on assemblies with very small parts (such as the carburetors and the valve train). The cavities can be marked with paint or tape to identify the contents.

Whenever wiring looms, harnesses or connectors are separated, it's a good idea to identify the two halves with numbered pieces of masking tape so they can be easily reconnected.

Gasket sealing surfaces

Throughout any motorcycle, gaskets are used to seal the mating surfaces between components and keep lubricants, fluids, vacuum or pressure contained in an assembly.

Many times these gaskets are coated with a liquid or paste type gasket sealing compound before assembly. Age, heat and pressure can sometimes cause the two parts to stick together so tightly that they are very difficult to separate. In most cases, the part can be loosened by striking it with a soft-faced hammer near the mating surfaces. A regular hammer can be used if a block of wood is placed between the hammer and the part. Do not hammer on cast parts or parts that could be easily damaged. With any particularly stubborn part, always recheck to make sure that every fastener has been removed.

Avoid using a screwdriver or bar to pry apart components, as they can easily mar the gasket sealing surfaces of the parts (which must remain smooth). If prying is absolutely necessary, use a piece of wood, but keep in mind that extra clean-up will be necessary if the wood splinters.

After the parts are separated, the old gasket must be carefully scraped off and the gasket surfaces cleaned. Stubborn gasket material can be soaked with a gasket remover (available in aerosol cans) to soften it so it can be easily scraped off. A scraper can be fashioned from a piece of copper tubing by flattening and sharpening one end. Copper is recommended because it is usually softer than the surfaces to be scraped, which reduces the chance of gouging the part. Some gaskets can be removed with a wire brush, but regardless of the method used, the mating surfaces must be left clean and smooth. If for some reason the gasket surface is gouged, then a gasket sealer thick enough to fill scratches will have to be used during reassembly of the components. For most applications, a non-drying (or semi-drying) gasket sealer is best.

Hose removal tips

Hose removal precautions closely parallel gasket removal precautions. Avoid scratching or gouging the surface that the hose mates against or the connection may leak. Because of various chemical reactions, the rubber in hoses can bond itself to the metal spigot that the hose fits over. To remove a hose, first loosen the hose clamps that secure it to the spigot. Then, with slip joint pliers, grab the hose at the clamp and rotate it around the spigot. Work it back and forth until it is completely free, then pull it off (silicone or other lubricants will ease removal if they can be applied between the hose and the outside of the spigot). Apply the same lubricant to the inside of the hose and the outside of the spigot to simplify installation.

If a hose clamp is broken or damaged, do not reuse it. Also, do not reuse hoses that are cracked, split or torn.

Spark plug gap adjusting tool

Feeler gauge set

Tools

A selection of good tools is a basic requirement for anyone who plans to maintain and repair a motorcycle. For the owner who has few tools, if any, the initial investment might seem high, but when compared to the spiraling costs of routine maintenance and repair, it is a wise one.

To help the owner decide which tools are needed to perform the tasks detailed in this manual, the following tool lists are offered: Maintenance and minor repair, Repair and overhaul and Special. The newcomer to practical mechanics should start off with the Maintenance and minor repair tool kit, which is adequate for the simpler jobs. Then, as confidence and experience grow, the owner can tackle more difficult tasks, buying additional tools as they are needed. Eventually the basic kit will be built into the Repair and overhaul tool set. Over a period of time, the experienced do-it-yourselfer will assemble a tool set complete enough for most repair and overhaul procedures and will add tools from the Special category when it is felt that the expense is justified by the frequency of use.

Maintenance and minor repair tool kit

The tools in this list should be considered the minimum required for performance of routine maintenance, servicing and minor repair work. We recommend the purchase of combination wrenches (box end and

Control cable pressure luber

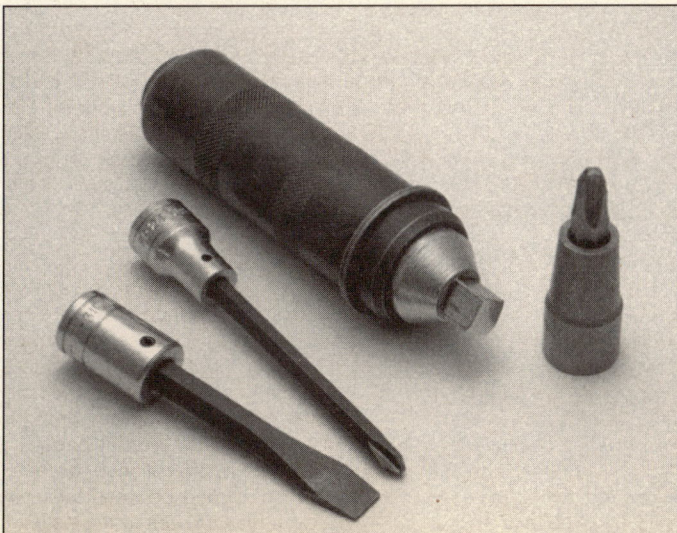

Hand impact screwdriver and bits

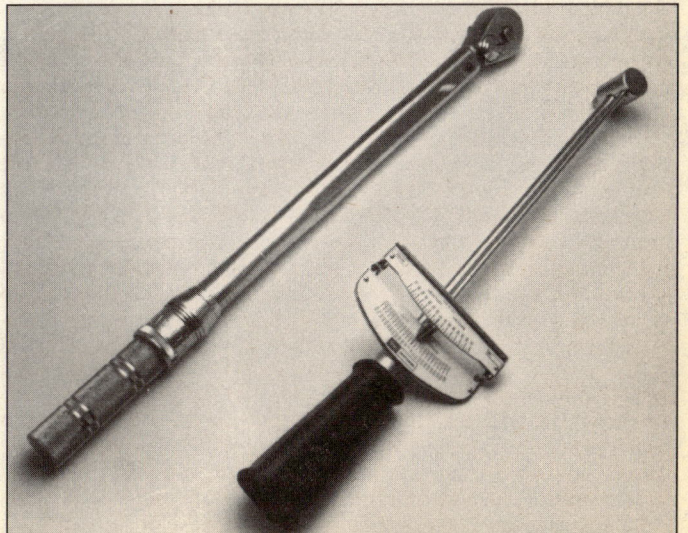

Torque wrenches (left — click; right — beam type)

Snap-ring pliers (top - external; bottom - internal)

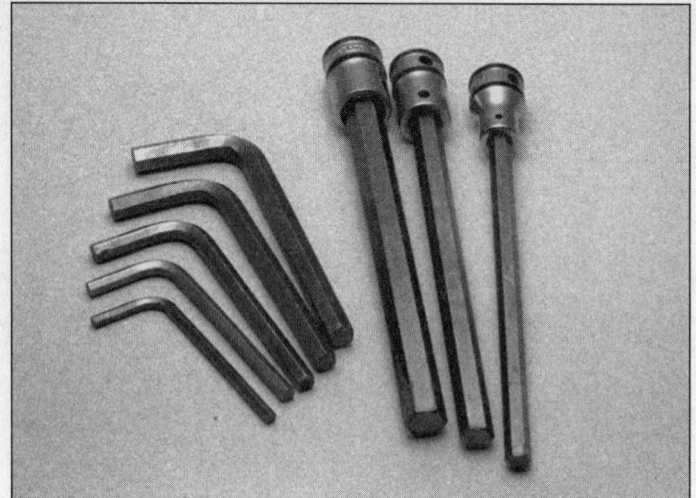

Allen wrenches (left), and Allen head sockets (right)

open end combined in one wrench); while more expensive than open-ended ones, they offer the advantages of both types of wrench.

Combination wrench set (6 mm to 22 mm)
Adjustable wrench - 8 in
Spark plug socket (with rubber insert)
Spark plug gap adjusting tool
Feeler gauge set
Standard screwdriver (5/16 in x 6 in)
Phillips screwdriver (No. 2 x 6 in)
Allen (hex) wrench set (4 mm to 12 mm)
Combination (slip-joint) pliers - 6 in
Hacksaw and assortment of blades
Tire pressure gauge
Control cable pressure luber
Grease gun
Oil can
Fine emery cloth
Wire brush
Hand impact screwdriver and bits
Funnel (medium size)
Safety goggles
Drain pan
Work light with extension cord

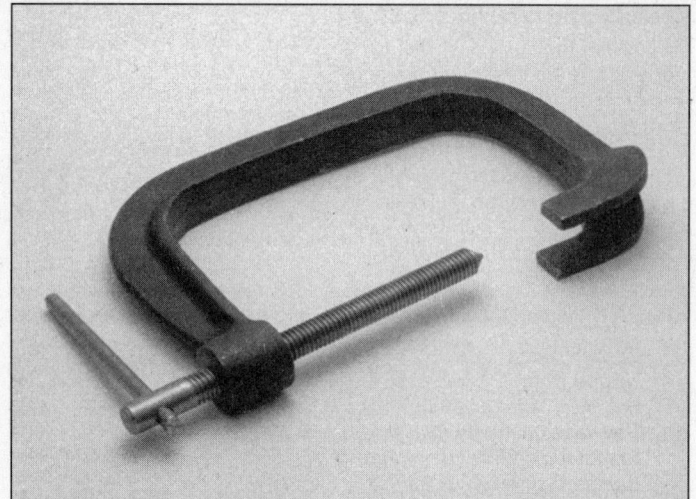

Valve spring compressor

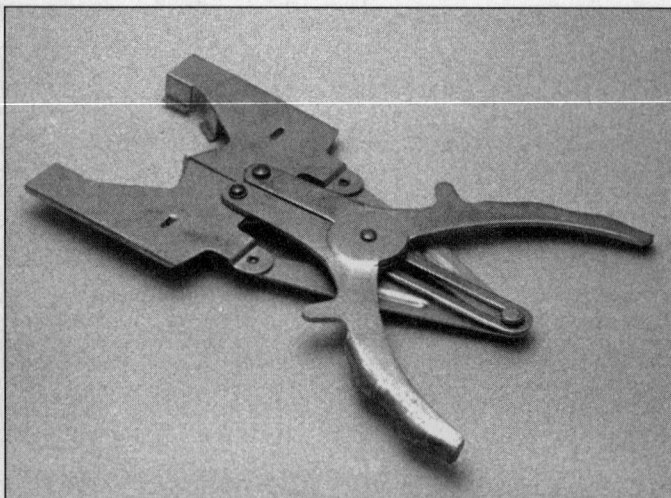

Piston ring removal/installation tool

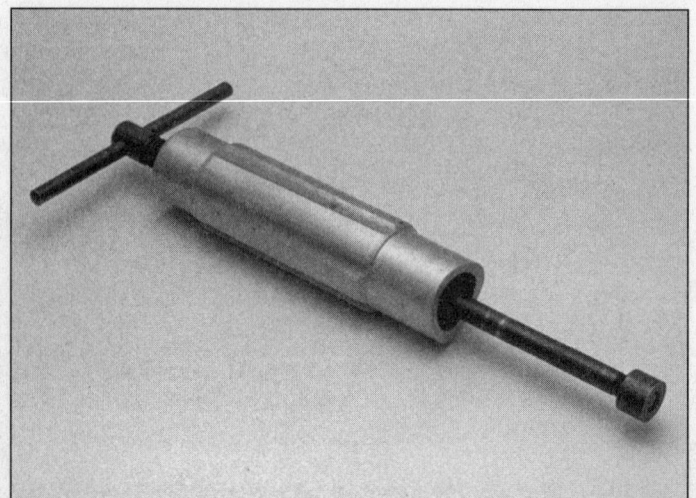

Piston pin puller

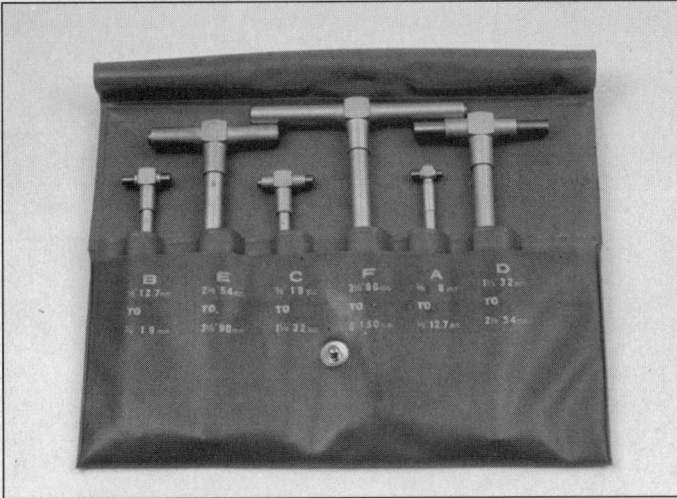

Telescoping gauges

0-to-1 inch micrometer

Repair and overhaul tool set

These tools are essential for anyone who plans to perform major repairs and are intended to supplement those in the Maintenance and minor repair tool kit. Included is a comprehensive set of sockets which, though expensive, are invaluable because of their versatility (especially when various extensions and drives are available). We recommend the 3/8 inch drive over the 1/2 inch drive for general motorcycle maintenance and repair (ideally, the mechanic would have a 3/8 inch drive set and a 1/2 inch drive set).

Alternator rotor removal tool
Socket set(s)
Reversible ratchet
Extension - 6 in
Universal joint
Torque wrench (same size drive as sockets)
Ball pein hammer - 8 oz
Soft-faced hammer (plastic/rubber)
Standard screwdriver (1/4 in x 6 in)
Standard screwdriver (stubby - 5/16 in)
Phillips screwdriver (No. 3 x 8 in)
Phillips screwdriver (stubby - No. 2)
Pliers - locking
Pliers - lineman's
Pliers - needle nose
Pliers - snap-ring (internal and external)
Cold chisel - 1/2 in
Scriber
Scraper (made from flattened copper tubing)
Center punch
Pin punches (1/16, 1/8, 3/16 in)
Steel rule/straightedge - 12 in
Pin-type spanner wrench
A selection of files
Wire brush (large)

Note: *Another tool which is often useful is an electric drill with a chuck capacity of 3/8 inch (and a set of good quality drill bits).*

Special tools

The tools in this list include those which are not used regularly, are expensive to buy, or which need to be used in accordance with their manufacturer's instructions. Unless these tools will be used frequently, it is not very economical to purchase many of them. A consideration would be to split the cost and use between yourself and a friend or friends (i.e. members of a motorcycle club).

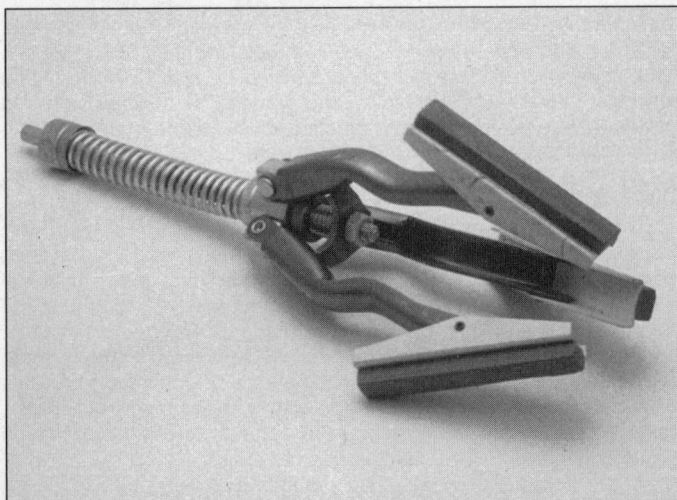

Cylinder surfacing hone

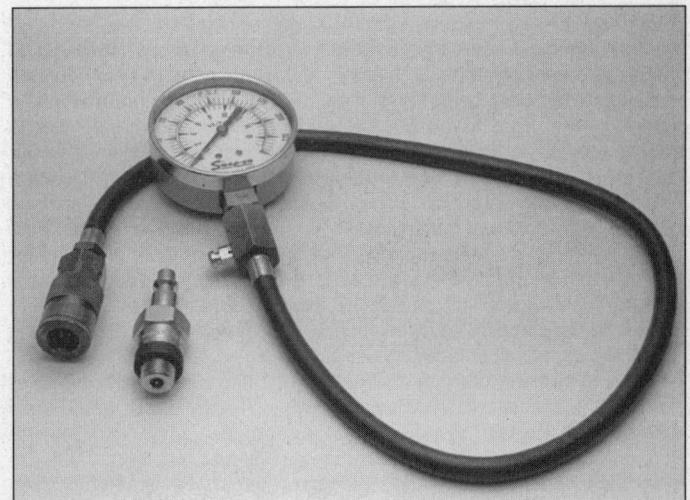

Cylinder compression gauge

Dial indicator set

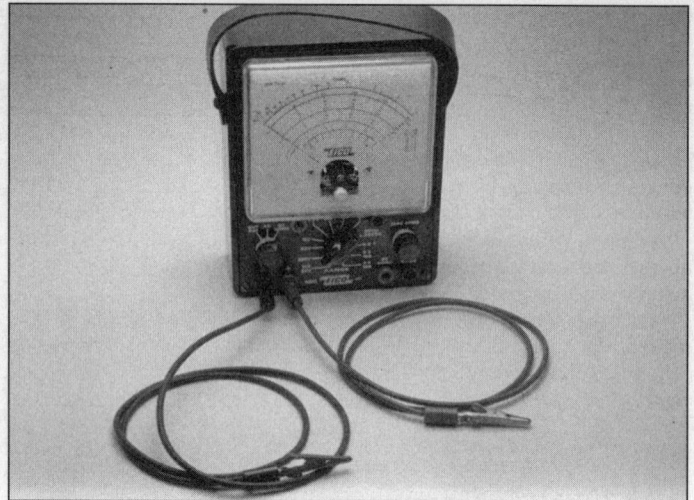

Multimeter (volt/ohm/ammeter)

This list primarily contains tools and instruments widely available to the public, as well as some special tools produced by the vehicle manufacturer for distribution to dealer service departments. As a result, references to the manufacturer's special tools are occasionally included in the text of this manual. Generally, an alternative method of doing the job without the special tool is offered. However, sometimes there is no alternative to their use. Where this is the case, and the tool can't be purchased or borrowed, the work should be turned over to the dealer service department or a motorcycle repair shop.

Valve spring compressor
Piston ring removal and installation tool
Piston pin puller
Telescoping gauges
Micrometer(s) and/or dial/Vernier calipers
Cylinder surfacing hone
Cylinder compression gauge
Dial indicator set
Multimeter
Adjustable spanner
Manometer or vacuum gauge set
Small air compressor with blow gun and tire chuck

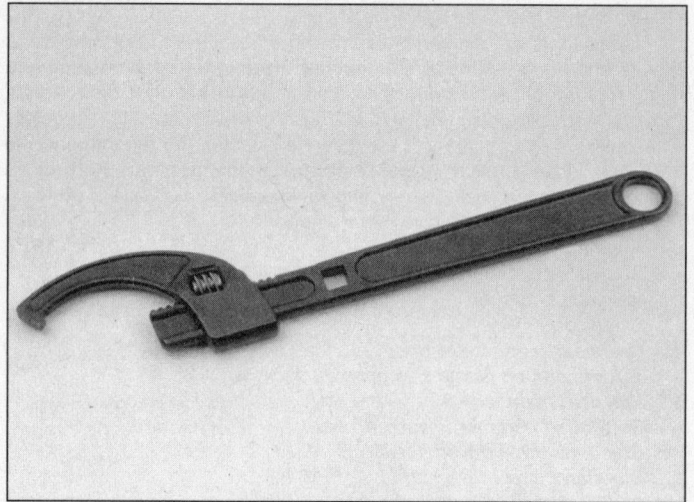

Adjustable spanner

Buying tools

For the do-it-yourselfer who is just starting to get involved in motorcycle maintenance and repair, there are a number of options available when purchasing tools. If maintenance and minor repair is the extent of the work to be done, the purchase of individual tools is satisfactory. If, on the other hand, extensive work is planned, it would be a good idea to purchase a modest tool set from one of the large retail chain stores. A set can usually be bought at a substantial savings over the individual tool prices (and they often come with a tool box). As additional tools are needed, add-on sets, individual tools and a larger tool box can be purchased to expand the tool selection. Building a tool set gradually allows the cost of the tools to be spread over a longer period of time and gives the mechanic the freedom to choose only those tools that will actually be used.

Tool stores and motorcycle dealers will often be the only source of some of the special tools that are needed, but regardless of where tools are bought, try to avoid cheap ones (especially when buying screwdrivers and sockets) because they won't last very long. There are plenty of tools around at reasonable prices, but always aim to purchase items which meet the relevant national safety standards. The expense involved in replacing cheap tools will eventually be greater than the initial cost of quality tools.

It is obviously not possible to cover the subject of tools fully here. For those who wish to learn more about tools and their use, there is a book entitled *Motorcycle Workshop Practice Manual* (Book no. 1454) available from the publishers of this manual. It also provides an introduction to basic workshop practice which will be of interest to a home mechanic working on any type of motorcycle.

Care and maintenance of tools

Good tools are expensive, so it makes sense to treat them with respect. Keep them clean and in usable condition and store them properly when not in use. Always wipe off any dirt, grease or metal chips before putting them away. Never leave tools lying around in the work area.

Some tools, such as screwdrivers, pliers, wrenches and sockets, can be hung on a panel mounted on the garage or workshop wall, while others should be kept in a tool box or tray. Measuring instruments, gauges, meters, etc. must be carefully stored where they can't be damaged by weather or impact from other tools.

When tools are used with care and stored properly, they will last a very long time. Even with the best of care, tools will wear out if used frequently. When a tool is damaged or worn out, replace it; subsequent jobs will be safer and more enjoyable if you do.

Working facilities

Not to be overlooked when discussing tools is the workshop. If anything more than routine maintenance is to be carried out, some sort of suitable work area is essential.

It is understood, and appreciated, that many home mechanics do not have a good workshop or garage available and end up removing an engine or doing major repairs outside (it is recommended, however, that the overhaul or repair be completed under the cover of a roof).

A clean, flat workbench or table of comfortable working height is an absolute necessity. The workbench should be equipped with a vise that has a jaw opening of at least four inches.

As mentioned previously, some clean, dry storage space is also required for tools, as well as the lubricants, fluids, cleaning solvents, etc. which soon become necessary.

Sometimes waste oil and fluids, drained from the engine or cooling system during normal maintenance or repairs, present a disposal problem. To avoid pouring them on the ground or into a sewage system, simply pour the used fluids into large containers, seal them with caps and take them to an authorized disposal site or service station. Plastic jugs (such as old antifreeze containers) are ideal for this purpose.

Always keep a supply of old newspapers and clean rags available. Old towels are excellent for mopping up spills. Many mechanics use rolls of paper towels for most work because they are readily available and disposable. To help keep the area under the motorcycle clean, a large cardboard box can be cut open and flattened to protect the garage or shop floor.

Whenever working over a painted surface (such as the fuel tank) cover it with an old blanket or bedspread to protect the finish.

Safety first!

Professional mechanics are trained in safe working procedures. However enthusiastic you may be about getting on with the job at hand, take the time to ensure that your safety is not put at risk. A moment's lack of attention can result in an accident, as can failure to observe simple precautions.

There will always be new ways of having accidents, and the following is not a comprehensive list of all dangers; it is intended rather to make you aware of the risks and to encourage a safe approach to all work you carry out on your bike.

Essential DOs and DON'Ts

DON'T start the engine without first ascertaining that the transmission is in neutral.

DON'T suddenly remove the pressure cap from a hot cooling system - cover it with a cloth and release the pressure gradually first, or you may get scalded by escaping coolant.

DON'T attempt to drain oil until you are sure it has cooled sufficiently to avoid scalding you.

DON'T grasp any part of the engine or exhaust system without first ascertaining that it is cool enough not to burn you.

DON'T allow brake fluid or antifreeze to contact the machine's paint work or plastic components.

DON'T siphon toxic liquids such as fuel, hydraulic fluid or antifreeze by mouth, or allow them to remain on your skin.

DON'T inhale dust - it may be injurious to health (see *Asbestos* heading).

DON'T allow any spilled oil or grease to remain on the floor - wipe it up right away, before someone slips on it.

DON'T use ill fitting wrenches or other tools which may slip and cause injury.

DON'T attempt to lift a heavy component which may be beyond your capability - get assistance.

DON'T rush to finish a job or take unverified short cuts.

DON'T allow children or animals in or around an unattended vehicle.

DON'T inflate a tire to a pressure above the recommended maximum. Apart from over stressing the carcase and wheel rim, in extreme cases the tire may blow off forcibly.

DO ensure that the machine is supported securely at all times. This is especially important when the machine is blocked up to aid wheel or fork removal.

DO take care when attempting to loosen a stubborn nut or bolt. It is generally better to pull on a wrench, rather than push, so that if you slip, you fall away from the machine rather than onto it.

DO wear eye protection when using power tools such as drill, sander, bench grinder etc.

DO use a barrier cream on your hands prior to undertaking dirty jobs - it will protect your skin from infection as well as making the dirt easier to remove afterwards; but make sure your hands aren't left slippery. Note that long-term contact with used engine oil can be a health hazard.

DO keep loose clothing (cuffs, ties etc. and long hair) well out of the way of moving mechanical parts.

DO remove rings, wristwatch etc., before working on the vehicle - especially the electrical system.

DO keep your work area tidy - it is only too easy to fall over articles left lying around.

DO exercise caution when compressing springs for removal or installation. Ensure that the tension is applied and released in a controlled manner, using suitable tools which preclude the possibility of the spring escaping violently.

DO ensure that any lifting tackle used has a safe working load rating adequate for the job.

DO get someone to check periodically that all is well, when working alone on the vehicle.

DO carry out work in a logical sequence and check that everything is correctly assembled and tightened afterwards.

DO remember that your vehicle's safety affects that of yourself and others. If in doubt on any point, get professional advice.

IF, in spite of following these precautions, you are unfortunate enough to injure yourself, seek medical attention as soon as possible.

Asbestos

Certain friction, insulating, sealing and other products - such as brake pads, clutch linings, gaskets, etc. - contain asbestos. *Extreme care must be taken to avoid inhalation of dust from such products since it is hazardous to health*. If in doubt, assume that they *do* contain asbestos.

Fire

Remember at all times that gasoline (petrol) is highly flammable. Never smoke or have any kind of naked flame around, when working on the vehicle. But the risk does not end there - a spark caused by an electrical short-circuit, by two metal surfaces contacting each other, by careless use of tools, or even by static electricity built up in your body under certain conditions, can ignite gasoline (petrol) vapor, which in a confined space is highly explosive. Never use gasoline (petrol) as a cleaning solvent. Use an approved safety solvent.

Always disconnect the battery ground (earth) terminal before working on any part of the fuel or electrical system, and never risk spilling fuel on to a hot engine or exhaust.

It is recommended that a fire extinguisher of a type suitable for fuel and electrical fires is kept handy in the garage or workplace at all times. Never try to extinguish a fuel or electrical fire with water.

Fumes

Certain fumes are highly toxic and can quickly cause unconsciousness and even death if inhaled to any extent. Gasoline (petrol) vapor comes into this category, as do the vapors from certain solvents such as trichloroethylene. Any draining or pouring of such volatile fluids should be done in a well ventilated area.

When using cleaning fluids and solvents, read the instructions carefully. Never use materials from unmarked containers - they may give off poisonous vapors.

Never run the engine of a motor vehicle in an enclosed space such as a garage. Exhaust fumes contain carbon monoxide which is extremely poisonous; if you need to run the engine, always do so in the open air or at least have the rear of the vehicle outside the workplace.

The battery

Never cause a spark, or allow a naked light near the vehicle's battery. It will normally be giving off a certain amount of hydrogen gas, which is highly explosive.

Always disconnect the battery ground (earth) terminal before working on the fuel or electrical systems (except where noted).

If possible, loosen the filler plugs or cover when charging the battery from an external source. Do not charge at an excessive rate or the battery may burst.

Take care when topping up, cleaning or carrying the battery. The acid electrolyte, even when diluted, is very corrosive and should not be allowed to contact the eyes or skin. Always wear rubber gloves and goggles or a face shield. If you ever need to prepare electrolyte yourself, always add the acid slowly to the water; never add the water to the acid.

Electricity

When using an electric power tool, inspection light etc., always ensure that the appliance is correctly connected to its plug and that, where necessary, it is properly grounded (earthed). Do not use such appliances in damp conditions and, again, beware of creating a spark or applying excessive heat in the vicinity of fuel or fuel vapor. Also ensure that the appliances meet national safety standards.

A severe electric shock can result from touching certain parts of the electrical system, such as the spark plug wires (HT leads), when the engine is running or being cranked, particularly if components are damp or the insulation is defective. Where an electronic ignition system is used, the secondary (HT) voltage is much higher and could prove fatal.

Motorcycle chemicals and lubricants

A number of chemicals and lubricants are available for use in motorcycle maintenance and repair. They include a wide variety of products ranging from cleaning solvents and degreasers to lubricants and protective sprays for rubber, plastic and vinyl.

Contact point/spark plug cleaner is a solvent used to clean oily film and dirt from points, grime from electrical connectors and oil deposits from spark plugs. It is oil free and leaves no residue. It is also used to remove gum and varnish from carburetor jets and other orifices.

Carburetor cleaner is similar to contact point/spark plug cleaner but it usually has a stronger solvent and may leave a slight oily reside. It is not recommended for cleaning electrical components or connections.

Brake system cleaner is used to remove grease or brake fluid from brake system components (where clean surfaces are absolutely necessary and petroleum-based solvents cannot be used); it also leaves no residue.

Silicone-based lubricants are used to protect rubber parts such as hoses and grommets, and are used as lubricants for hinges and locks.

Multi-purpose grease is an all purpose lubricant used wherever grease is more practical than a liquid lubricant such as oil. Some multi-purpose grease is colored white and specially formulated to be more resistant to water than ordinary grease.

Gear oil (sometimes called gear lube) is a specially designed oil used in transmissions and final drive units, a s well as other areas where high friction, high temperature lubrication is required. It is available in a number of viscosities (weights) for various applications.

Motor oil, of course, is the lubricant specially formulated for use in the engine. It normally contains a wide variety of additives to prevent corrosion and reduce foaming and wear. Motor oil comes in various weights (viscosity ratings) of from 5 to 80. The recommended weight of the oil depends on the seasonal temperature and the demands on the engine. Light oil is used in cold climates and under light load conditions; heavy oil is used in hot climates and where high loads are encountered. Multi-viscosity oils are designed to have characteristics of both light and heavy oils and are available in a number of weights from 5W-20 to 20W-50.

Gas (petrol) additives perform several functions, depending on their chemical makeup. They usually contain solvents that help dissolve gum and varnish that build up on carburetor and intake parts. They also serve to break down carbon deposits that form on the inside surfaces of the combustion chambers. Some additives contain upper cylinder lubricants for valves and piston rings.

Brake fluid is a specially formulated hydraulic fluid that can withstand the heat and pressure encountered in brake systems. Care must be taken that this fluid does not come in contact with painted surfaces or plastics. An opened container should always be resealed to prevent contamination by water or dirt.

Chain lubricants are formulated especially for use on motorcycle final drive chains. A good chain lube should adhere well and have good penetrating qualities to be effective as a lubricant inside the chain and on the side plates, pins and rollers. Most chain lubes are either the foaming type or quick drying type and are usually marketed as sprays.

Degreasers are heavy duty solvents used to remove grease and grime that may accumulate on engine and frame components. They can be sprayed or brushed on and, depending on the type, are rinsed with either water or solvent.

Solvents are used alone or in combination with degreasers to clean parts and assemblies during repair and overhaul. The home mechanic should use only solvents that are non-flammable and that do not produce irritating fumes.

Gasket sealing compounds may be used in conjunction with gaskets, to improve their sealing capabilities, or alone, to seal metal-to-metal joints. Many gasket sealers can withstand extreme heat, some are impervious to gasoline and lubricants, while others are capable of filling and sealing large cavities. Depending on the intended use, gasket sealers either dry hard or stay relatively soft and pliable. They are usually applied by hand, with a brush, or are sprayed on the gasket sealing surfaces.

Thread cement is an adhesive locking compound that prevents threaded fasteners from loosening because of vibration. It is available in a variety of types for different applications.

Moisture dispersants are usually sprays that can be used to dry out electrical components such as the fuse block and wiring connectors. Some types can also be used as treatment for rubber and as a lubricant for hinges, cables and locks.

Waxes and polishes are used to help protect painted and plated surfaces from the weather. Different types of paint may require the use of different types of wax polish. Some polishes utilize a chemical or abrasive cleaner to help remove the top layer of oxidized (dull) paint on older vehicles. In recent years, many non-wax polishes (that contain a wide variety of chemicals such as polymers and silicones) have been introduced. These non-wax polishes are usually easier to apply and last longer than conventional waxes and polishes.

Troubleshooting

Contents

Engine doesn't start or is difficult to start

1 Starter motor doesn't rotate

1 Engine kill switch OFF.
2 Fuse blown. Check main fuse and starter circuit fuse (Chapter 9).
3 Battery voltage low. Check and recharge battery (Chapter 9).
4 Starter motor defective. Make sure the wiring to the starter is secure. Make sure the starter relay clicks when the start button is pushed. If the relay clicks, then the fault is in the wiring or motor.
5 Starter relay faulty. Check it with reference to Chapter 9.
6 Starter switch not contacting. The contacts could be wet, corroded or dirty. Disassemble and clean the switch (Chapter 9).
7 Wiring open or shorted. Check all wiring connections and harnesses to make sure that they are dry, tight and not corroded. Also check for broken or frayed wires that can cause a short to ground (earth) (see wiring diagram, Chapter 9).
8 Ignition (main) switch defective. Check the switch according to the procedure in Chapter 9. Replace the switch with a new one if it is defective.
9 Engine kill switch defective. Check for wet, dirty or corroded contacts. Clean or replace the switch as necessary (Chapter 9).
10 Faulty neutral, side stand or clutch switch. Check the wiring to each switch and the switch itself, referring Chapter 9.

2 Starter motor rotates but engine does not turn over

1 Starter motor clutch defective. Inspect and repair or replace (Chapter 2).
2 Damaged idler or starter gears. Inspect and replace the damaged parts (Chapter 2).

3 Starter works but engine won't turn over (seized)

1   Seized engine caused by one or more internally damaged components. Failure due to wear, abuse or lack of lubrication. Damage can include seized valves, followers, camshafts, pistons, crankshaft, connecting rod bearings, or transmission gears or bearings. Refer to Chapter 2 for engine disassembly.

4 No fuel flow

1   No fuel in tank.
2   Fuel pump failure or in-line filter blockage (see Chapters 1 and 9 respectively).
3   Fuel tank breather hose obstructed (not California models).
4   Fuel tap filter clogged. Remove the tap and clean it and the filter (Chapter 1).
5   Fuel line clogged. Pull the fuel line loose and carefully blow through it.
6   Float needle valve clogged. For all of the valves to be clogged, either a very bad batch of fuel with an unusual additive has been used, or some other foreign material has entered the tank. Many times after a machine has been stored for many months without running, the fuel turns to a varnish-like liquid and forms deposits on the inlet needle valves and jets. The carburetors should be removed and overhauled if draining the float chambers doesn't solve the problem.

5 Engine flooded

1   Float height too high. Check as described in Chapter 4.
2   Float needle valve worn or stuck open. A piece of dirt, rust or other debris can cause the valve to seat improperly, causing excess fuel to be admitted to the float chamber. In this case, the float chamber should be cleaned and the needle valve and seat inspected. If the needle and seat are worn, then the leaking will persist and the parts should be replaced with new ones (Chapter 4).
3   Starting technique incorrect. Under normal circumstances (ie, if all the carburetor functions are sound) the machine should start with little or no throttle. When the engine is cold, the choke should be operated and the engine started without opening the throttle. When the engine is at operating temperature, only a very slight amount of throttle should be necessary. If the engine is flooded, turn the fuel tap OFF and hold the throttle open while cranking the engine. This will allow additional air to reach the cylinders. Remember to turn the fuel tap back ON after the engine starts.

6 No spark or weak spark

1   Ignition switch OFF.
2   Engine kill switch turned to the OFF position.
3   Battery voltage low. Check and recharge the battery as necessary (Chapter 9).
4   Spark plugs dirty, defective or worn out. Locate reason for fouled plugs using spark plug condition chart and follow the plug maintenance procedures (Chapter 1).
5   Spark plug caps or secondary (HT) wiring faulty. Check condition. Replace either or both components if cracks or deterioration are evident (Chapter 5).
6   Spark plug caps not making good contact. Make sure that the plug caps fit snugly over the plug ends.
7   Ignition control module defective. Check the module, referring to Chapter 5 for details.
8   Pulse generator defective. Check the unit with reference to Chapter 5.
9   Ignition HT coils defective. Check the coils, referring to Chapter 5.
10   Ignition or kill switch shorted. This is usually caused by water, corrosion, damage or excessive wear. The switches can be disassembled and cleaned with electrical contact cleaner. If cleaning does not help, replace the switches (Chapter 9).

11   Wiring shorted or broken between:
  a)   *Ignition (main) switch and engine kill switch (or blown fuse)*
  b)   *Ignition control module and engine kill switch*
  c)   *Ignition control module and ignition HT coils*
  d)   *Ignition HT coils and spark plugs*
  e)   *Ignition control module and pulse generator*
12   Make sure that all wiring connections are clean, dry and tight. Look for chafed and broken wires (Chapters 5 and 9).

7 Compression low

1   Spark plugs loose. Remove the plugs and inspect their threads. Reinstall and tighten to the specified torque (Chapter 1).
2   Cylinder head not sufficiently tightened down. If the cylinder head is suspected of being loose, then there's a chance that the gasket or head is damaged if the problem has persisted for any length of time. The head bolts should be tightened to the proper torque in the correct sequence (Chapter 2).
3   Improper valve clearance. This means that the valve is not closing completely and compression pressure is leaking past the valve. Check and adjust the valve clearances (Chapter 1).
4   Cylinder and/or piston worn. Excessive wear will cause compression pressure to leak past the rings. This is usually accompanied by worn rings as well. A top-end overhaul is necessary (Chapter 2).
5   Piston rings worn, weak, broken, or sticking. Broken or sticking piston rings usually indicate a lubrication or carburation problem that causes excess carbon deposits or seizures to form on the pistons and rings. Top-end overhaul is necessary (Chapter 2).
6   Piston ring-to-groove clearance excessive. This is caused by excessive wear of the piston ring lands. Piston replacement is necessary (Chapter 2).
7   Cylinder head gasket damaged. If the head is allowed to become loose, or if excessive carbon build-up on the piston crown and combustion chamber causes extremely high compression, the head gasket may leak. Retorquing the head is not always sufficient to restore the seal, so gasket replacement is necessary (Chapter 2).
8   Cylinder head warped. This is caused by overheating or improperly tightened head bolts. Machine shop resurfacing or head replacement is necessary (Chapter 2).
9   Valve spring broken or weak. Caused by component failure or wear; the springs must be replaced (Chapter 2).
10   Valve not seating properly. This is caused by a bent valve (from over-revving or improper valve adjustment), burned valve or seat (improper carburation) or an accumulation of carbon deposits on the seat (from carburation or lubrication problems). The valves must be cleaned and/or replaced and the seats serviced if possible (Chapter 2).

8 Stalls after starting

1   Improper choke action. Make sure the choke linkage shaft is getting a full stroke and staying in the out position (Chapter 4).
2   Ignition malfunction. See Chapter 5.
3   Carburetor malfunction. See Chapter 4.
4   Fuel contaminated. The fuel can be contaminated with either dirt or water, or can change chemically if the machine is allowed to sit for several months or more. Drain the tank and float chambers (Chapter 4).
5   Intake air leak. Check for loose carburetor-to-intake manifold connections, loose or missing vacuum gauge adapter screws or hoses, or loose carburetor tops (Chapter 4).
6   Engine idle speed incorrect. Turn idle adjusting screw until the engine idles at the specified rpm (Chapter 1).

9 Rough idle

1   Ignition malfunction. See Chapter 5.
2   Idle speed incorrect. See Chapter 1.

3 Carburetors not synchronized. Adjust carburetors with vacuum gauge or manometer set as described in Chapter 1.
4 Carburetor malfunction. See Chapter 4.
5 Fuel contaminated. The fuel can be contaminated with either dirt or water, or can change chemically if the machine is allowed to sit for several months or more. Drain the tank and float chambers (Chapter 4).
6 Intake air leak. Check for loose carburetor-to-intake manifold connections, loose or missing vacuum gauge adapter screws or hoses, or loose carburetor tops (Chapter 4).
7 Air filter clogged. Replace the air filter element (Chapter 1).

Poor running at low speeds

10 Spark weak

1 Battery voltage low. Check and recharge battery (Chapter 9).
2 Spark plugs fouled, defective or worn out. Refer to Chapter 1 for spark plug maintenance.
3 Spark plug cap or HT wiring defective. Refer to Chapters 1 and 5 for details on the ignition system.
4 Spark plug caps not making contact.
5 Incorrect spark plugs. Wrong type, heat range or cap configuration. Check and install correct plugs listed in Chapter 1.
6 Ignition control module defective. See Chapter 5.
7 Pulse generator defective. See Chapter 5.
8 Ignition HT coils defective. See Chapter 5.

11 Fuel/air mixture incorrect

1 Pilot screws out of adjustment (Chapter 4).
2 Pilot jet or air passage clogged. Remove and overhaul the carburetors (Chapter 4).
3 Air bleed holes clogged. Remove carburetor and blow out all passages (Chapter 4).
4 Air filter clogged, poorly sealed or missing (Chapter 1).
5 Air filter housing poorly sealed. Look for cracks, holes or loose clamps and replace or repair defective parts.
6 Fuel level too high or too low. Check the float height (Chapter 4).
7 Fuel tank breather hose obstructed (not California models).
8 Carburetor intake manifolds loose. Check for cracks, breaks, tears or loose clamps. Replace the rubber intake manifold joints if split or perished.

12 Compression low

1 Spark plugs loose. Remove the plugs and inspect their threads. Reinstall and tighten to the specified torque (Chapter 1).
2 Cylinder head not sufficiently tightened down. If the cylinder head is suspected of being loose, then there's a chance that the gasket and head are damaged if the problem has persisted for any length of time. The head bolts should be tightened to the proper torque in the correct sequence (Chapter 2).
3 Improper valve clearance. This means that the valve is not closing completely and compression pressure is leaking past the valve. Check and adjust the valve clearances (Chapter 1).
4 Cylinder and/or piston worn. Excessive wear will cause compression pressure to leak past the rings. This is usually accompanied by worn rings as well. A top end overhaul is necessary (Chapter 2).
5 Piston rings worn, weak, broken, or sticking. Broken or sticking piston rings usually indicate a lubrication or carburation problem that causes excess carbon deposits or seizures to form on the pistons and rings. Top-end overhaul is necessary (Chapter 2).
6 Piston ring-to-groove clearance excessive. This is caused by excessive wear of the piston ring lands. Piston replacement is necessary (Chapter 2).

7 Cylinder head gasket damaged. If the head is allowed to become loose, or if excessive carbon build-up on the piston crown and combustion chamber causes extremely high compression, the head gasket may leak. Retorquing the head is not always sufficient to restore the seal, so gasket replacement is necessary (Chapter 2).
8 Cylinder head warped. This is caused by overheating or improperly tightened head bolts. Machine shop resurfacing or head replacement is necessary (Chapter 2).
9 Valve spring broken or weak. Caused by component failure or wear; the springs must be replaced (Chapter 2).
10 Valve not seating properly. This is caused by a bent valve (from over-revving or improper valve adjustment), burned valve or seat (improper carburation) or an accumulation of carbon deposits on the seat (from carburation, lubrication problems). The valves must be cleaned and/or replaced and the seats serviced if possible (Chapter 2).

13 Poor acceleration

1 Carburetors leaking or dirty. Overhaul the carburetors (Chapter 4).
2 Timing not advancing. The pulse generator or the ignition control module may be defective. If so, they must be replaced with new ones, as they can't be repaired.
3 Carburetors not synchronized. Adjust them with a vacuum gauge set or manometer (Chapter 1).
4 Engine oil viscosity too high. Using a heavier oil than that recommended in Chapter 1 can damage the oil pump or lubrication system and cause drag on the engine.
5 Brakes dragging. Usually caused by debris which has entered the brake piston seals, or from a warped disc or bent axle. Repair as necessary (Chapter 7).

Poor running or no power at high speed

14 Firing incorrect

1 Air filter restricted. Clean or replace filter (Chapter 1).
2 Spark plugs fouled, defective or worn out. See Chapter 1 for spark plug maintenance.
3 Spark plug caps or HT wiring defective. See Chapters 1 and 5 for details of the ignition system.
4 Spark plug caps not in good contact. See Chapter 5.
5 Incorrect spark plugs. Wrong type, heat range or cap configuration. Check and install correct plugs listed in Chapter 1.
6 Ignition control module defective. See Chapter 5.
7 Ignition coils defective. See Chapter 5.

15 Fuel/air mixture incorrect

1 Main jet clogged. Dirt, water or other contaminants can clog the main jets. Clean the fuel tap filter, the in-line filter, the float chamber area, and the jets and carburetor orifices (Chapter 4).
2 Main jet wrong size. The standard jetting is for sea level atmospheric pressure and oxygen content.
3 Throttle shaft-to-carburetor body clearance excessive. Refer to Chapter 4 for inspection and part replacement procedures.
4 Air bleed holes clogged. Remove and overhaul carburetors (Chapter 4).
5 Air filter clogged, poorly sealed, or missing (Chapter 1).
6 Air filter housing poorly sealed. Look for cracks, holes or loose clamps, and replace or repair defective parts.
7 Fuel level too high or too low. Check the float height (Chapter 4).
8 Fuel tank breather hose obstructed (not California models).
9 Carburetor intake manifolds loose. Check for cracks, breaks, tears or loose clamps. Replace the rubber intake manifolds if they are split or perished (Chapter 4).

16 Compression low

1   Spark plugs loose. Remove the plugs and inspect their threads. Reinstall and tighten to the specified torque (Chapter 1).
2   Cylinder head not sufficiently tightened down. If the cylinder head is suspected of being loose, then there's a chance that the gasket and head are damaged if the problem has persisted for any length of time. The head bolts should be tightened to the proper torque in the correct sequence (Chapter 2).
3   Improper valve clearance. This means that the valve is not closing completely and compression pressure is leaking past the valve. Check and adjust the valve clearances (Chapter 1).
4   Cylinder and/or piston worn. Excessive wear will cause compression pressure to leak past the rings. This is usually accompanied by worn rings as well. A top-end overhaul is necessary (Chapter 2).
5   Piston rings worn, weak, broken, or sticking. Broken or sticking piston rings usually indicate a lubrication or carburation problem that causes excess carbon deposits or seizures to form on the pistons and rings. Top-end overhaul is necessary (Chapter 2).
6   Piston ring-to-groove clearance excessive. This is caused by excessive wear of the piston ring lands. Piston replacement is necessary (Chapter 2).
7   Cylinder head gasket damaged. If the head is allowed to become loose, or if excessive carbon build-up on the piston crown and combustion chamber causes extremely high compression, the head gasket may leak. Retorquing the head is not always sufficient to restore the seal, so gasket replacement is necessary (Chapter 2).
8   Cylinder head warped. This is caused by overheating or improperly tightened head bolts. Machine shop resurfacing or head replacement is necessary (Chapter 2).
9   Valve spring broken or weak. Caused by component failure or wear; the springs must be replaced (Chapter 2).
10   Valve not seating properly. This is caused by a bent valve (from over-revving or improper valve adjustment), burned valve or seat (improper carburation) or an accumulation of carbon deposits on the seat (from carburation or lubrication problems). The valves must be cleaned and/or replaced and the seats serviced if possible (Chapter 2).

17 Knocking or pinging

1   Carbon build-up in combustion chamber. Use of a fuel additive that will dissolve the adhesive bonding the carbon particles to the crown and chamber is the easiest way to remove the build-up. Otherwise, the cylinder head will have to be removed and decarbonized (Chapter 2).
2   Incorrect or poor quality fuel. Old or improper grades of fuel can cause detonation. This causes the piston to rattle, thus the knocking or pinging sound. Drain old fuel and always use the recommended fuel grade.
3   Spark plug heat range incorrect. Uncontrolled detonation indicates the plug heat range is too hot. The plug in effect becomes a glow plug, raising cylinder temperatures. Install the proper heat range plug (Chapter 1).
4   Improper air/fuel mixture. This will cause the cylinder to run hot, which leads to detonation. Clogged jets or an air leak can cause this imbalance. See Chapter 4.

18 Miscellaneous causes

1   Throttle valve doesn't open fully. Adjust the throttle grip freeplay (Chapter 1).
2   Clutch slipping. May be caused by loose or worn clutch components. Refer to Chapter 2 for clutch overhaul procedures.
3   Timing not advancing.
4   Engine oil viscosity too high. Using a heavier oil than the one recommended in Chapter 1 can damage the oil pump or lubrication system and cause drag on the engine.

5   Brakes dragging. Usually caused by debris which has entered the brake piston seals, or from a warped disc or bent axle. Repair as necessary.

Overheating

19 Engine overheats

1   Coolant level low. Check and add coolant (Chapter 1).
2   Leak in cooling system. Check cooling system hoses and radiator for leaks and other damage. Repair or replace parts as necessary (Chapter 3).
3   Thermostat sticking open or closed. Check and replace as described in Chapter 3.
4   Faulty radiator cap. Remove the cap and have it pressure tested.
5   Coolant passages clogged. Have the entire system drained and flushed, then refill with fresh coolant.
6   Water pump defective. Remove the pump and check the components (Chapter 3).
7   Clogged radiator fins. Clean them by blowing compressed air through the fins from the backside.
8   Cooling fan or fan switch fault (Chapter 3).

20 Firing incorrect

1   Spark plugs fouled, defective or worn out. See Chapter 1 for spark plug maintenance.
2   Incorrect spark plugs.
3   Faulty ignition HT coils (Chapter 5).

21 Fuel/air mixture incorrect

1   Main jet clogged. Dirt, water and other contaminants can clog the main jets. Clean the fuel tap filter, the fuel pump in-line filter, the float chamber area and the jets and carburetor orifices (Chapter 4).
2   Main jet wrong size. The standard jetting is for sea level atmospheric pressure and oxygen content.
3   Air filter clogged, poorly sealed or missing (Chapter 1).
4   Air filter housing poorly sealed. Look for cracks, holes or loose clamps and replace or repair.
5   Fuel level too low. Check float height (Chapter 4).
6   Fuel tank breather hose obstructed (not California models).
7   Carburetor intake manifolds loose. Check for cracks, breaks, tears or loose clamps. Replace the rubber intake manifold joints if split or perished.

22 Compression too high

1   Carbon build-up in combustion chamber. Use of a fuel additive that will dissolve the adhesive bonding the carbon particles to the piston crown and chamber is the easiest way to remove the build-up. Otherwise, the cylinder head will have to be removed and decarbonized (Chapter 2).
2   Improperly machined head surface or installation of incorrect gasket during engine assembly.

23 Engine load excessive

1   Clutch slipping. Can be caused by damaged, loose or worn clutch components. Refer to Chapter 2 for overhaul procedures.
2   Engine oil level too high. The addition of too much oil will cause pressurization of the crankcase and inefficient engine operation. Check Specifications and drain to proper level (Chapter 1).
3   Engine oil viscosity too high. Using a heavier oil than the one

recommended in Chapter 1 can damage the oil pump or lubrication system as well as cause drag on the engine.

4 Brakes dragging. Usually caused by debris which has entered the brake piston seals, or from a warped disc or bent axle. Repair as necessary.

24 Lubrication inadequate

1 Engine oil level too low. Friction caused by intermittent lack of lubrication or from oil that is overworked can cause overheating. The oil provides a definite cooling function in the engine. Check the oil level (Chapter 1).
2 Poor quality engine oil or incorrect viscosity or type. Oil is rated not only according to viscosity but also according to type. Some oils are not rated high enough for use in this engine. Check the Specifications section and change to the correct oil (Chapter 1).

25 Miscellaneous causes

Modification to exhaust system. Most aftermarket exhaust systems cause the engine to run leaner, which make them run hotter. When installing an accessory exhaust system, always rejet the carburetors.

Clutch problems

26 Clutch slipping

1 Cable freeplay insufficient. Check and adjust cable (Chapter 1).
2 Friction plates worn or warped. Overhaul the clutch assembly (Chapter 2).
3 Plain plates warped (Chapter 2).
4 Clutch springs broken or weak. Old or heat-damaged (from slipping clutch) springs should be replaced with new ones (Chapter 2).
5 Clutch release mechanism defective. Replace any defective parts (Chapter 2).
6 Clutch center or outer drum unevenly worn. This causes improper engagement of the plates. Replace the damaged or worn parts (Chapter 2).

27 Clutch not disengaging completely

1 Cable freeplay excessive. Check and adjust cable (Chapter 1).
2 Clutch plates warped or damaged. This will cause clutch drag, which in turn will cause the machine to creep. Overhaul the clutch assembly (Chapter 2).
3 Clutch spring tension uneven. Usually caused by a sagged or broken spring. Check and replace the springs as a set (Chapter 2).
4 Engine oil deteriorated. Old, thin, worn out oil will not provide proper lubrication for the plates, causing the clutch to drag. Replace the oil and filter (Chapter 1).
5 Engine oil viscosity too high. Using a heavier oil than recommended in Chapter 1 can cause the plates to stick together, putting a drag on the engine. Change to the correct weight oil (Chapter 1).
6 Clutch outer drum guide seized on mainshaft. Lack of lubrication, severe wear or damage can cause the guide to seize on the shaft. Overhaul of the clutch, and perhaps transmission, may be necessary to repair the damage (Chapter 2).
7 Clutch release mechanism defective. Overhaul the clutch cover components (Chapter 2).
8 Loose clutch center nut. Causes drum and center misalignment putting a drag on the engine. Engagement adjustment continually varies. Overhaul the clutch assembly (Chapter 2).

Gear shifting problems

28 Doesn't go into gear or lever doesn't return

1 Clutch not disengaging. See Section 27.
2 Shift fork(s) bent or seized. Often caused by dropping the machine or from lack of lubrication. Overhaul the transmission (Chapter 2).
3 Gear(s) stuck on shaft. Most often caused by a lack of lubrication or excessive wear in transmission bearings and bushings. Overhaul the transmission (Chapter 2).
4 Gearshift drum binding. Caused by lubrication failure or excessive wear. Replace the drum and bearing (Chapter 2).
5 Gearshift lever return spring weak or broken (Chapter 2).
6 Gearshift lever broken. Splines stripped out of lever or shaft, caused by allowing the lever to get loose or from dropping the machine. Replace necessary parts (Chapter 2).
7 Gearshift mechanism stopper arm broken or worn. Full engagement and rotary movement of shift drum results. Replace the arm (Chapter 2).
8 Stopper arm spring broken. Allows arm to float, causing sporadic shift operation. Replace spring (Chapter 2).

29 Jumps out of gear

1 Shift fork(s) worn. Overhaul the transmission (Chapter 2).
2 Gear groove(s) worn. Overhaul the transmission (Chapter 2).
3 Gear dogs or dog slots worn or damaged. The gears should be inspected and replaced. No attempt should be made to service the worn parts.

30 Overshifts

1 Stopper arm spring weak or broken (Chapter 2).
2 Gearshift shaft return spring post broken or distorted (Chapter 2).

Abnormal engine noise

31 Knocking or pinging

1 Carbon build-up in combustion chamber. Use of a fuel additive that will dissolve the adhesive bonding the carbon particles to the piston crown and chamber is the easiest way to remove the build-up. Otherwise, the cylinder head will have to be removed and decarbonized (Chapter 2).
2 Incorrect or poor quality fuel. Old or improper fuel can cause detonation. This causes the pistons to rattle, thus the knocking or pinging sound. Drain the old fuel and always use the recommended grade fuel (Chapter 4).
3 Spark plug heat range incorrect. Uncontrolled detonation indicates that the plug heat range is too hot. The plug in effect becomes a glow plug, raising cylinder temperatures. Install the proper heat range plug (Chapter 1).
4 Improper air/fuel mixture. This will cause the cylinders to run hot and lead to detonation. Clogged jets or an air leak can cause this imbalance. See Chapter 4.

32 Piston slap or rattling

1 Cylinder-to-piston clearance excessive. Caused by improper assembly. Inspect and overhaul top-end parts (Chapter 2).
2 Connecting rod bent. Caused by over-revving, trying to start a badly flooded engine or from ingesting a foreign object into the combustion chamber. Replace the damaged parts (Chapter 2).
3 Piston pin or piston pin bore worn or seized from wear or lack of lubrication. Replace damaged parts (Chapter 2).

4 Piston ring(s) worn, broken or sticking. Overhaul the top-end (Chapter 2).
5 Piston seizure damage. Usually from lack of lubrication or overheating. Replace the pistons and bore the cylinders, as necessary (Chapter 2).
6 Connecting rod upper or lower end clearance excessive. Caused by excessive wear or lack of lubrication. Replace worn parts.

33 Valve noise

1 Incorrect valve clearances. Adjust the clearances by referring to Chapter 1.
2 Valve spring broken or weak. Check and replace weak valve springs (Chapter 2).
3 Camshaft or cylinder head worn or damaged. Lack of lubrication at high rpm is usually the cause of damage. Insufficient oil or failure to change the oil at the recommended intervals are the chief causes. Since there are no replaceable bearings in the head, the head itself will have to be replaced if there is excessive wear or damage (Chapter 2).

34 Other noise

1 Cylinder head gasket leaking.
2 Exhaust pipe leaking at cylinder head connection. Caused by improper fit of pipe(s) or loose exhaust flange. All exhaust fasteners should be tightened evenly and carefully. Failure to do this will lead to a leak.
3 Crankshaft runout excessive. Caused by a bent crankshaft (from over-revving) or damage from an upper cylinder component failure. Can also be attributed to dropping the machine on either of the crankshaft ends.
4 Engine mounting bolts loose. Tighten all engine mount bolts (Chapter 2).
5 Crankshaft bearings worn (Chapter 2).
6 Cam chain tensioner defective. Replace according to the procedure in Chapter 2.
7 Cam chain, sprockets or guides worn (Chapter 2).

Abnormal driveline noise

35 Clutch noise

1 Clutch outer drum/friction plate clearance excessive (Chapter 2).
2 Loose or damaged clutch pressure plate and/or bolts (Chapter 2).

36 Transmission noise

1 Bearings worn. Also includes the possibility that the shafts are worn. Overhaul the transmission (Chapter 2).
2 Gears worn or chipped (Chapter 2).
3 Metal chips jammed in gear teeth. Probably pieces from a broken clutch, gear or shift mechanism that were picked up by the gears. This will cause early bearing failure (Chapter 2).
4 Engine oil level too low. Causes a howl from transmission. Also affects engine power and clutch operation (Chapter 1).

37 Final drive noise

1 Chain not adjusted properly (Chapter 1).
2 Engine sprocket or rear sprocket loose. Tighten fasteners (Chapter 6).
3 Sprocket(s) worn. Replace sprocket(s) (Chapter 6).
4 Rear sprocket warped. Replace (Chapter 6).
5 Wheel coupling damper worn. Replace damper (Chapter 6).

Abnormal frame and suspension noise

38 Front end noise

1 Low fluid level or improper viscosity oil in forks. This can sound like spurting and is usually accompanied by irregular fork action (Chapter 6).
2 Spring weak or broken. Makes a clicking or scraping sound. Fork oil, when drained, will have a lot of metal particles in it (Chapter 6).
3 Steering head bearings loose or damaged. Clicks when braking. Check and adjust or replace as necessary (Chapters 1 and 6).
4 Fork triple clamps loose. Make sure all clamp pinch bolts are tight (Chapter 6).
5 Fork tube bent. Good possibility if machine has been dropped. Replace tube with a new one (Chapter 6).
6 Front axle or axle clamp bolt loose. Tighten them to the specified torque (Chapter 6).

39 Shock absorber noise

1 Fluid level incorrect. Indicates a leak caused by defective seal. Shock will be covered with oil. Replace shock or seek advice on repair from a Honda dealer (Chapter 6).
2 Defective shock absorber with internal damage. This is in the body of the shock and can't be remedied. The shock must be replaced with a new one (Chapter 6).
3 Bent or damaged shock body. Replace the shock with a new one (Chapter 6).

40 Brake noise

1 Squeal caused by pad shim not installed or positioned correctly - rear caliper (Chapter 7).
2 Squeal caused by dust on brake pads. Usually found in combination with glazed pads. Clean using brake cleaning solvent (Chapter 7).
3 Contamination of brake pads. Oil, brake fluid or dirt causing brake to chatter or squeal. Clean or replace pads (Chapter 7).
4 Pads glazed. Caused by excessive heat from prolonged use or from contamination. Do not use sandpaper, emery cloth, carborundum cloth or any other abrasive to roughen the pad surfaces as abrasives will stay in the pad material and damage the disc. A very fine flat file can be used, but pad replacement is suggested as a cure (Chapter 7).
5 Disc warped. Can cause a chattering, clicking or intermittent squeal. Usually accompanied by a pulsating lever and uneven braking. Replace the disc (Chapter 7).
6 Loose or worn wheel bearings. Check and replace as needed (Chapter 7).

Oil pressure indicator light comes on

41 Engine lubrication system

1 Engine oil pump defective, blocked oil strainer gauze or failed relief valve. Carry out oil pressure check (Chapter 2).
2 Engine oil level low. Inspect for leak or other problem causing low oil level and add recommended oil (Chapter 1).
3 Engine oil viscosity too low. Very old, thin oil or an improper weight of oil used in the engine. Change to correct oil (Chapter 1).
4 Camshaft or journals worn. Excessive wear causing drop in oil pressure. Replace cam and/or cylinder head. Abnormal wear could be caused by oil starvation at high rpm from low oil level or improper weight or type of oil (Chapter 1).
5 Crankshaft and/or bearings worn. Same problems as paragraph 4. Check and replace crankshaft and/or bearings (Chapter 2).

42 Electrical system

1   Oil pressure switch defective. Check the switch according to the procedure in Chapter 9. Replace it if it is defective.
2   Oil pressure indicator light circuit defective. Check for pinched, shorted, disconnected or damaged wiring (Chapter 9).

Excessive exhaust smoke

43 White smoke

1   Piston oil ring worn. The ring may be broken or damaged, causing oil from the crankcase to be pulled past the piston into the combustion chamber. Replace the rings with new ones (Chapter 2).
2   Cylinders worn, cracked, or scored. Caused by overheating or oil starvation. The cylinders will have to be rebored and new pistons installed.
3   Valve oil seal damaged or worn. Replace oil seals with new ones (Chapter 2).
4   Valve guide worn. Perform a complete valve job (Chapter 2).
5   Engine oil level too high, which causes the oil to be forced past the rings. Drain oil to the proper level (Chapter 1).
6   Head gasket broken between oil return and cylinder. Causes oil to be pulled into the combustion chamber. Replace the head gasket and check the head for warpage (Chapter 2).
7   Abnormal crankcase pressurization, which forces oil past the rings. Clogged breather hose is usually the cause.

44 Black smoke

1   Air filter clogged. Clean or replace the element (Chapter 1).
2   Main jet too large or loose. Compare the jet size to the Specifications (Chapter 4).
3   Choke cable or linkage shaft stuck, causing fuel to be pulled through choke circuit (Chapter 4).
4   Fuel level too high. Check and adjust the float height(s) as necessary (Chapter 4).
5   Float needle valve held off needle seat. Clean the float chambers and fuel line and replace the needles and seats if necessary (Chapter 4).

45 Brown smoke

1   Main jet too small or clogged. Lean condition caused by wrong size main jet or by a restricted orifice. Clean float chambers and jets and compare jet size to Specifications (Chapter 4).
2   Fuel flow insufficient. Float needle valve stuck closed due to chemical reaction with old fuel. Float height incorrect. Restricted fuel line. Clean line and float chamber and adjust floats if necessary.
3   Carburetor intake manifold clamps loose (Chapter 4).
4   Air filter poorly sealed or not installed (Chapter 1).

Poor handling or stability

46 Handlebar hard to turn

1   Steering head bearing adjuster nut too tight. Check adjustment as described in Chapter 1.
2   Bearings damaged. Roughness can be felt as the bars are turned from side-to-side. Replace bearings and races (Chapter 6).
3   Races dented or worn. Denting results from wear in only one position (eg, straightahead), from a collision or hitting a pothole or from dropping the machine. Replace races and bearings (Chapter 6)
4   Steering stem lubrication inadequate. Causes are grease getting hard from age or being washed out by high pressure car washes.

Disassemble steering head and repack bearings (Chapter 6).
5   Steering stem bent. Caused by a collision, hitting a pothole or by dropping the machine. Replace damaged part. Don't try to straighten the steering stem (Chapter 6).
6   Front tire air pressure too low (Chapter 1).

47 Handlebar shakes or vibrates excessively

1   Tires worn or out of balance (Chapter 7).
2   Swingarm bearings worn. Replace worn bearings by referring to Chapter 6.
3   Rim(s) warped or damaged. Inspect wheels for runout (Chapter 7).
4   Wheel bearings worn. Worn front or rear wheel bearings can cause poor tracking. Worn front bearings will cause wobble (Chapter 7).
5   Handlebar clamp bolts loose (Chapter 6).
6   Fork triple clamp bolts loose. Tighten them to the specified torque (Chapter 6).
7   Engine mounting bolts loose. Will cause excessive vibration with increased engine rpm (Chapter 2).

48 Handlebar pulls to one side

1   Frame bent. Definitely suspect this if the machine has been dropped. May or may not be accompanied by cracking near the bend. Replace the frame (Chapter 6).
2   Wheels out of alignment. Caused by improper location of axle spacers or from bent steering stem or frame (Chapter 6).
3   Swingarm bent or twisted. Caused by age (metal fatigue) or impact damage. Replace the arm (Chapter 6).
4   Steering stem bent. Caused by impact damage or by dropping the motorcycle. Replace the steering stem (Chapter 6).
5   Fork tube bent. Disassemble the forks and replace the damaged parts (Chapter 6).
6   Fork oil level uneven. Check and add or drain as necessary (Chapter 6).

49 Poor shock absorbing qualities

1   Too hard:
  a) *Fork oil level excessive (Chapter 6).*
  b) *Fork oil viscosity too high. Use a lighter oil (see the Specifications in Chapter 6).*
  c) *Fork tube bent. Causes a harsh, sticking feeling (Chapter 6).*
  d) *Shock shaft or body bent or damaged (Chapter 6).*
  e) *Fork internal damage (Chapter 6).*
  f) *Shock internal damage.*
  g) *Tire pressure too high (Chapter 1).*
2   Too soft:
  a) *Fork or shock oil insufficient and/or leaking (Chapter 6).*
  b) *Fork oil level too low (Chapter 6).*
  c) *Fork oil viscosity too light (Chapter 6).*
  d) *Fork springs weak or broken (Chapter 6).*
  e) *Shock internal damage or leakage (Chapter 6).*

Braking problems

50 Brakes are spongy, don't hold

1   Air in brake line. Caused by inattention to master cylinder fluid level or by leakage. Locate problem and bleed brakes (Chapter 7).
2   Pad or disc worn (Chapters 1 and 7).
3   Brake fluid leak. See paragraph 1.
4   Contaminated pads. Caused by contamination with oil, grease, brake fluid, etc. Clean or replace pads. Clean disc thoroughly with brake cleaner (Chapter 7).

5   Brake fluid deteriorated. Fluid is old or contaminated. Drain system, replenish with new fluid and bleed the system (Chapter 7).
6   Master cylinder internal parts worn or damaged causing fluid to bypass (Chapter 7).
7   Master cylinder bore scratched by foreign material or broken spring. Repair or replace master cylinder (Chapter 7).
8   Disc warped. Replace disc (Chapter 7).

51   Brake lever or pedal pulsates

1   Disc warped. Replace disc (Chapter 7).
2   Axle bent. Replace axle (Chapter 7).
3   Brake caliper bolts loose (Chapter 7).
4   Brake caliper sliders damaged or sticking (rear caliper), causing caliper to bind. Lube the sliders or replace them if they are corroded or bent (Chapter 7).
5   Wheel warped or otherwise damaged (Chapter 7).
6   Wheel bearings damaged or worn (Chapter 7).

52   Brakes drag

1   Master cylinder piston seized. Caused by wear or damage to piston or cylinder bore (Chapter 7).
2   Lever balky or stuck. Check pivot and lubricate (Chapter 7).
3   Brake caliper binds. Caused by inadequate lubrication or damage to caliper sliders (Chapter 7).
4   Brake caliper piston seized in bore. Caused by wear or ingestion of dirt past deteriorated seal (Chapter 7).
5   Brake pad damaged. Pad material separated from backing plate.

Usually caused by faulty manufacturing process or from contact with chemicals. Replace pads (Chapter 7).
6   Pads improperly installed (Chapter 7).

Electrical problems

53   Battery dead or weak

1   Battery faulty. Caused by sulfated plates which are shorted through sedimentation. Also, broken battery terminal making only occasional contact (Chapter 9).
2   Battery cables making poor contact (Chapter 9).
3   Load excessive. Caused by addition of high wattage lights or other electrical accessories.
4   Ignition (main) switch defective. Switch either grounds (earths) internally or fails to shut off system. Replace the switch (Chapter 9).
5   Regulator/rectifier defective (Chapter 9).
6   Alternator stator coil open or shorted (Chapter 9).
7   Wiring faulty. Wiring grounded (earthed) or connections loose in ignition, charging or lighting circuits (Chapter 9).

54   Battery overcharged

1   Regulator/rectifier defective. Overcharging is noticed when battery gets excessively warm (Chapter 9).
2   Battery defective. Replace battery with a new one (Chapter 9).
3   Battery amperage too low, wrong type or size. Install manufacturer's specified amp-hour battery to handle charging load (Chapter 9).

Chapter 1 Tune-up and routine maintenance

Contents

Specifications

Engine

Spark plugs

Type	NGK CR9EH-9 or ND U27FER-9
Electrode gap	0.8 to 0.9 mm (0.031 to 0.035 in)

Valve clearances (COLD engine)

Intake	0.13 to 0.19 mm (0.005 to 0.007 in)
Exhaust	0.19 to 0.25 mm (0.007 to 0.010 in)
Engine idle speed	1100 ± 100 rpm
Cylinder compression pressure	171 psi (11.8 Bars)

Carburetor synchronization

Maximum vacuum difference between any two cylinders.	20 mm (0.8 in) Hg
Cylinder numbering (from left side to right side of the bike)	1-2-3-4
Firing order	1-2-4-3

Miscellaneous

Brake pad minimum thickness	See text

Freeplay adjustments

Throttle grip	2 to 6 mm (0.08 to 0.24 in)
Clutch lever	10 to 20 mm (0.4 to 0.8 in)
Drive chain	25 to 35 mm (1.0 to 1.4 in)

Minimum tire tread depth*

Front	1.5 mm (0.06 in)
Rear	2.0 mm (0.08 in)

Tire pressures (cold)

Front	36 psi (2.5 Bars)
Rear	42 psi (2.9 Bars)

*At the time of writing, UK law requires that tread depth must be at least 1 mm over 3/4 of the tread breadth all the way around the tire, with no bald patches.

Torque settings

	Nm	ft-lbs
Crankshaft cap	18	13
Engine oil pan drain plug	36	26
Oil filter (see text)	10	7
Spark plugs	12	9
Rear wheel axle nut	95	69
Steering stem adjuster nut	31	22
Steering stem nut	105	76
Upper triple clamp bolts	23	17

Recommended lubricants and fluids

Engine/transmission oil
  Type .... API grade SF or SG
  Viscosity .... SAE 10W40
  Capacity (US 1993 and 1994 models, UK CBR900RR-N and RR-P)
    Oil change only .... 3.1 liters (3.3 US qt, 5.5 Imp pts)
    With filter change .... 3.2 liters (3.4 US qt, 5.7 Imp pts)
    After engine rebuild .... 4.0 liters (4.2 US qt, 7.0 Imp pts)
  Capacity (US 1995 model, UK CBR900RR-R and RR-S)
    Oil change only .... 3.5 liters (3.7 US qt, 6.2 Imp pts)
    With filter change .... 3.6 liters (3.8 US qt, 6.3 Imp pts)
    After engine rebuild .... 4.4 liters (4.7 US qt, 7.7 Imp pts)
Coolant
  Mixture type .... 50% distilled water, 50% corrosion inhibited ethylene glycol antifreeze

  Capacity
    Radiator and engine .... 2.8 liters (3.0 US qt, 5.0 Imp pt)
    Coolant reservoir .... 0.45 liters (0.48 US qt, 0.8 Imp pt)
Brake fluid .... DOT 4

Miscellaneous

Drive chain	SAE 80 to 90W gear oil
Wheel bearings	Medium weight, lithium-based multi-purpose grease
Swingarm pivot bearings	Molybdenum disulfide grease
Suspension linkage bearings	Molybdenum disulfide grease
Shock absorber mounting bearings	Molybdenum disulfide grease
Cables and lever pivots	Chain and cable lubricant or 10W40 motor oil
Side stand pivot	Medium-weight, lithium-based multi-purpose grease
Brake pedal/shift lever pivots	Chain and cable lubricant or 10W40 motor oil
Throttle grip	Multi-purpose grease or dry film lubricant

1 Introduction to tune-up and routine maintenance

This Chapter covers in detail the checks and procedures necessary for the tune-up and routine maintenance of your motorcycle. Section 1 includes the routine maintenance schedule, which is designed to keep the machine in proper running condition and prevent possible problems. The remaining Sections contain detailed procedures for carrying out the items listed on the maintenance schedule, as well as additional maintenance information designed to increase reliability.

Since routine maintenance plays such an important role in the safe and efficient operation of your motorcycle, it is presented here as a comprehensive check list. For the rider who does all his/her own maintenance, these lists outline the procedures and checks that should be done on a routine basis.

Maintenance information is printed on decals attached to the motorcycle. If the information on the decals differs from that included here, use the information on the decal.

Deciding where to start or plug into the routine maintenance schedule depends on several factors. If you have a motorcycle whose warranty has recently expired, and if it has been maintained according to the warranty standards, you may want to pick up routine maintenance as it coincides with the next mileage or calendar interval. If you have owned the machine for some time but have never performed any maintenance on it, then you may want to start at the nearest interval and include some additional procedures to ensure that nothing important is overlooked. If you have just had a major engine overhaul, then you may want to start the maintenance routine from the beginning. If you have a used machine and have no knowledge of its history or maintenance record, you may desire to combine all the checks into one large service initially and then settle into the maintenance schedule prescribed.

The Sections which outline the inspection and maintenance procedures are written as step-by-step comprehensive guides to the performance of the work. They explain in detail each of the routine inspections and maintenance procedures on the check list. References to additional information in applicable Chapters is also included and should not be overlooked.

Before beginning any maintenance or repair, the machine should be cleaned thoroughly, especially around the oil filter, spark plugs, cylinder head cover, side covers, carburetors, etc. Cleaning will help ensure that dirt does not contaminate the engine and will allow you to detect wear and damage that could otherwise easily go unnoticed.

2 Honda CBR900RR Routine maintenance intervals

Note: *The pre-ride inspection outlined in the owner's manual covers the checks and maintenance that should be carried out on a daily basis. It's condensed and included here to remind you of its importance. Always perform the pre-ride inspection at every maintenance interval (in addition to the procedures listed). The intervals listed below are the shortest intervals recommended by the manufacturer for each particular operation during the model years covered in this manual. Your owner's manual may have different intervals for your model.*

Daily or before riding

Check the engine oil level
Check the fuel level and inspect for leaks
Check the engine coolant level and look for leaks
Check the operation of both brakes - also check the fluid levels and look for leakage
Check the tires for damage, the presence of foreign objects and correct air pressure
Check the throttle for smooth operation
Check the operation of the clutch - make sure the freeplay is correct
Check that drive chain slack is not excessive
Make sure the steering operates smoothly, without looseness or binding
Check for proper operation of the headlight, taillight, brake light, turn signals, indicator lights, speedometer and horn
Make sure the sidestand returns to its fully up position and stays there under spring tension
Make sure the engine KILL switch works properly

After the initial 600 miles (1000 km)

Note: *This check is usually performed by a Honda dealer after the first 600 miles (1,000 km) from new. Thereafter, maintenance is carried out according to the following intervals of the schedule.*
Replace the engine oil and oil filter
Check and adjust the idle speed
Check the brake fluid levels
Check and adjust the clutch
Check the tightness of all fasteners
Check the steering head bearings

Every 600 miles (1000 km)

Check, adjust and lubricate the drive chain

Every 4000 miles (6000 km) or 6 months (whichever comes sooner)

Check and adjust the idle speed
Check the brake pads

Check the brake fluid levels
Check and adjust the clutch
Check the spark plug gaps - US models
Lubricate the clutch and brake lever pivots
Lubricate the gearshift/brake lever pivots and the sidestand pivot

Every 8000 miles (12,000 km) or 12 months (whichever comes sooner)

Replace the engine oil and filter
Check the fuel system hoses
Check the battery terminals
Check and adjust the throttle and choke cables
Check the spark plug gaps - UK models
Replace the spark plugs - US models
Check/adjust the carburetor synchronization
Check condition of the PAIR system hoses - California models only
Check the cooling system hoses
Check the coolant level and inspect the cooling system for leaks or damage
Check the brake system and the brake light switch operation
Check and adjust the headlight aim
Check the sidestand switch operation
Check the suspension
Check and adjust the steering head bearings
Check the tightness of all nuts, bolts and fasteners
Check the condition of the wheels and tires
Check the drive chain slider for wear
Check the condition of the exhaust system

Every 12,000 miles (18,000 km) or 18 months (whichever comes first)

Replace the air filter element
Check the EVAP system hoses - California models only
Change the brake fluid

Every 16,000 miles (24,000 km) or two years (whichever comes sooner)

Check and adjust the valve clearances
Replace the spark plugs - UK models

Every 24,000 miles (36,000 km) or three years (whichever comes sooner)

Replace the coolant

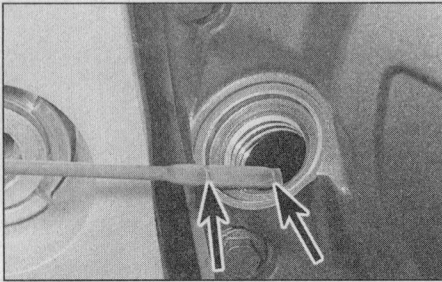

3.3 Engine oil level must lie between two marks on dipstick (arrows)

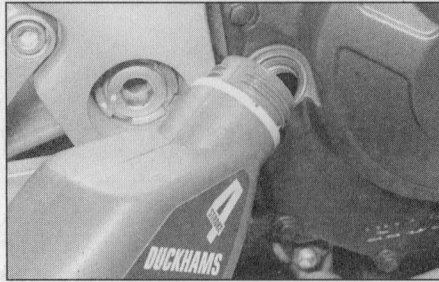

3.4 Topping up the engine oil

3.7 Fluid must lie between UPPER and LOWER lines on reservoir body (arrows)

3 Fluid levels - check

Engine oil

Refer to illustrations 3.3 and 3.4

1 Place the motorcycle on its sidestand on level ground, then start the engine and allow it to reach operating temperature. **Caution:** *Do not run the engine in an enclosed space such as a garage or shop.*

2 Stop the engine and hold the motorcycle upright for a minute or two to allow the oil level to stabilize.

3 Unscrew the filler cap/dipstick from the right crankcase cover and wipe it clean. With the motorcycle held upright, insert the dipstick until the filler cap threads are resting on the crankcase cover, then remove it and check the oil level. **Note:** *Do not screw the filler cap into position.* The oil level should be between the marks on the end of the dipstick **(see illustration)**.

4 If the level is below the lower (minimum) dipstick mark, top up with oil of the recommended grade and type, to bring the level up to the upper (maximum) mark **(see illustration)**. **Note:** *Do not overfill.*

Brake fluid

5 In order to ensure proper operation of the hydraulic disc brakes, the fluid level in the master cylinder reservoirs must be maintained.

Front brake

Refer to illustrations 3.7, 3.9a and 3.9bc

6 Hold the motorcycle upright and turn the handlebars until the top

of the master cylinder is as level as possible. If necessary, tilt the motorcycle to make it level.

7 Make sure that the fluid level, visible through the reservoir body, is between the UPPER and LOWER lines **(see illustration)**.

8 If the level is low, the fluid must be replenished. Before removing the master cylinder cap, cover the surrounding components to protect them from brake fluid spills and remove all dust and dirt from the area around the cap.

9 Remove its retaining screws and lift off the cap, diaphragm plate and diaphragm **(see illustrations)**. Using a good quality brake fluid of the recommended type, from a freshly opened container, top up the reservoir to the UPPER level line.

10 When the fluid level is correct, clean and dry the diaphragm, fold it into its compressed state and install it in the reservoir followed by the diaphragm plate. Install the reservoir cap and tighten its screws.

Rear brake

Refer to illustrations 3.11, 3.12, 3.13a and 3.13b

11 With the motorcycle held upright check the fluid level in the rear brake reservoir. Make sure that the fluid level, visible through the translucent material of the reservoir, is between the UPPER and LOWER lines on the reservoir **(see illustration)**.

12 If the level is low, the fluid must be replenished. Remove the reservoir mounting bolt and swing the reservoir out from the frame **(see illustration)**. Remove all dust and dirt from the area around the cap.

13 Unscrew the reservoir cap and remove the diaphragm plate and diaphragm **(see illustrations)**. Using a good quality brake fluid of the

3.9a Remove two screws from reservoir cap . . .

3.9b . . . and lift out plate and diaphragm

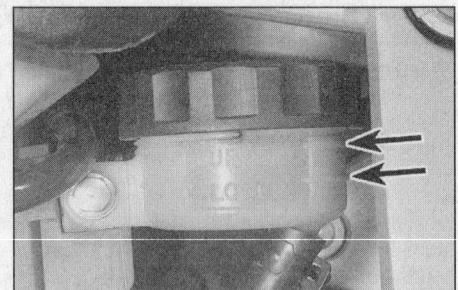

3.11 Fluid must lie between UPPER and LOWER lines on reservoir body (arrows)

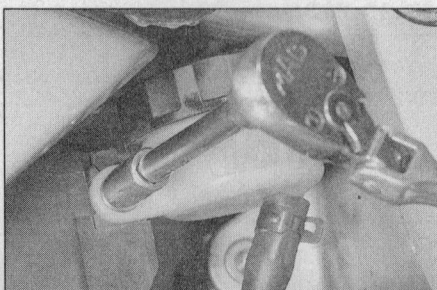

3.12 Remove bolt to free reservoir from its mounting on rear fender (mudguard)

3.13a Unscrew reservoir cap . . .

3.13b . . . and lift out plate and diaphragm

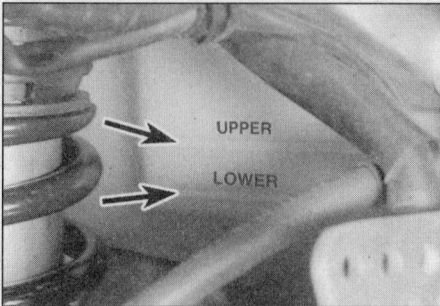

3.19 Coolant level must lie between UPPER and LOWER lines on reservoir body (arrows)

3.20 Topping up the coolant reservoir

4.1 Battery terminals must be clean, free of corrosion and the leads securely fastened

recommended type, from a freshly opened container, top up the reservoir to the UPPER level line.

14 When the fluid level is correct, clean and dry the diaphragm, fold it into its compressed state and install it in the reservoir followed by the diaphragm plate. Install the reservoir cap, tightening it securely, and bolt the reservoir back into position.

Both brakes

15 Check the operation of both brakes before taking the machine on the road; if there is evidence of air in the system, it must be bled as described in Chapter 7.

16 If the brake fluid level was low, inspect the brake system for leaks.

Coolant level

Refer to illustrations 3.19 and 3.20

17 Warm the engine up to normal operating temperature (see Step 1). **Caution:** *Do not run the engine in an enclosed space such as a garage or shop.*

18 Stop the engine. Hold the motorcycle upright and view the coolant level through the translucent body of the reservoir.

19 The coolant level should be between the UPPER and LOWER lines at the rear of the coolant reservoir; it may be necessary to clean road dirt off the reservoir to view the level due to its exposed location **(see illustration)**.

20 If the level is below the lower mark, remove the reservoir filler cap via the frame aperture, and top up the level to the upper mark using a coolant mixture of the required strength **(see illustration)**. **Note:** *Use only the specified ingredients as given in the Specifications at the start of this Chapter.* If the coolant is significantly above the upper level mark at any time, the surplus coolant should be siphoned off to prevent it from being expelled out of the breather hose when the engine is running.

21 If the coolant level falls steadily, check the system for leaks as described in Section 19. If no leaks are found and the level still continues to fall, it is recommended that the machine be taken to a Honda dealer who will pressure test the system.

4 Battery - check

Refer to illustration 4.1

1 All models covered in this manual are fitted with a sealed battery, and therefore require no maintenance. **Note:** *Do not attempt to remove the battery caps to check the electrolyte level or battery specific gravity. Removal will damage the caps, resulting in electrolyte leakage and battery damage.* All that should be done is to check that its terminals are clean and tight and that the casing is not damaged or leaking **(see illustration)**. See Chapter 9 for further details

2 If the machine is not in regular use, disconnect the battery and give it a refresher charge every month to six weeks (see Chapter 9).

5 Brake pads - wear check

Front brake

Refer to illustrations 5.1a, 5.1b and 5.1c

1 Pry the pad inspection cover from the back of each caliper **(see illustration)**. View the friction material thickness on each pad; if worn down so that the grooves are no longer visible, ie so that friction material remaining is level with the base of the grooves, both pads in each front caliper must be renewed immediately **(see illustrations)**. Clip the inspection cover back into place, so that its tabs top and bottom engage the cutouts in the caliper body.

2 If in doubt about the amount of pad material remaining or the pad condition, remove the pads for detailed inspection (see Chapter 7).

Rear brake

Refer to illustration 5.3

3 View the pads from the rear underside of the caliper. If either pad has worn down to, or beyond the groove in the friction material, both pads must be replaced as a set **(see illustration)**.

4 If in doubt about the amount of pad material remaining or the pad condition, remove the pads for detailed inspection (see Chapter 7).

5.1a Pry the pad inspection cover from the back of the front calipers . . .

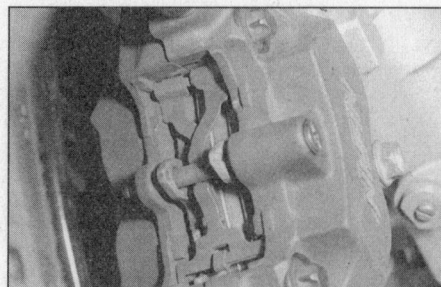

5.1b . . . to view the amount of pad material remaining

5.1c Replace the pads when or before the friction material has worn down level with the bottom of the groove(s) (arrow) - note non-original equipment pad shown

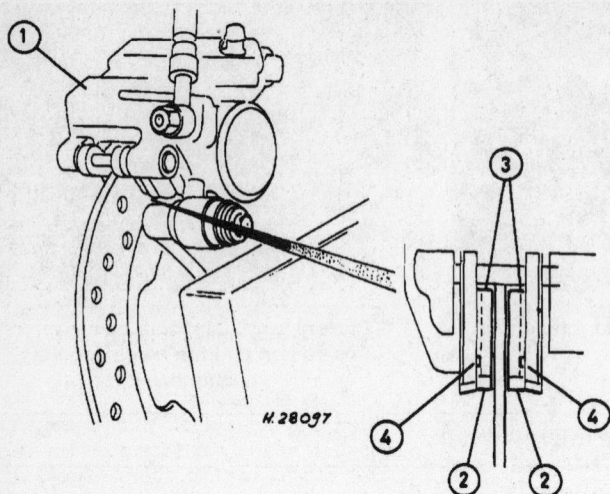

5.3 Rear brake pad wear indicator groove details

1 Caliper   *3 Wear limit*
2 Pads.   *4 Minimum thickness groove*

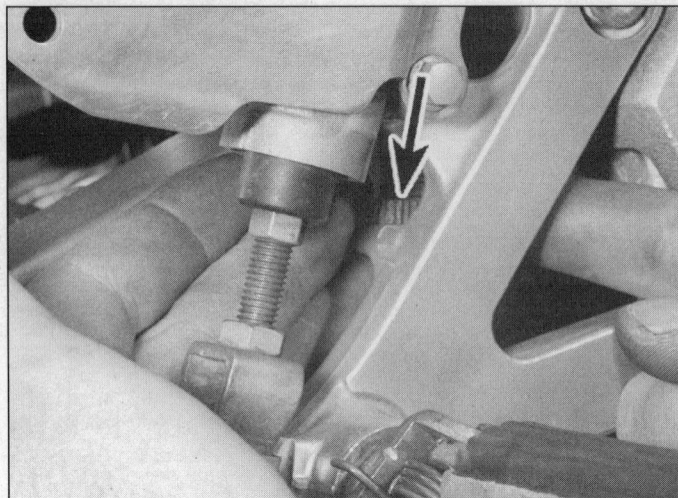

6.5 Rear brake switch knurled ring (arrow) is accessed behind rider's footpeg bracket

6   Brake system - general check

Refer to illustrations 6.5 and 6.6

1   A routine general check of the brakes will ensure that any problems are discovered and remedied before the rider's safety is jeopardized.

2   Check the brake lever and pedal for loose connections, excessive play, bends, and other damage. Replace any damaged parts with new ones (see Chapter 7).

3   Make sure all brake fasteners are tight. Check the brake pads for wear (see Section 5) and make sure the fluid level in the reservoirs is correct (see Section 3). Look for leaks at the hose connections and check for cracks in the hoses. If the lever or pedal is spongy, bleed the brakes as described in Chapter 7.

4   Make sure the brake light operates when the front brake lever is depressed. The front brake light switch is not adjustable. If it fails to operate properly, replace it with a new one (see Chapter 9).

5   Make sure the brake light is activated just before the rear brake pedal takes effect. If adjustment is necessary, hold the switch and turn the knurled adjusting ring on the switch body until the brake light is activated when required **(see illustration)**. If the switch doesn't operate the brake lights, check it as described in Chapter 9.

6   The front brake lever incorporates a span adjuster to cater for different hand sizes. Rotate the knurled wheel on the lever to adjust its position, noting that the arrow on the lever must align exactly with the notch of the knurled wheel **(see illustration)**.

7   Tires/wheels - general check

Refer to illustrations 7.2 and 7.4

1   Routine tire and wheel checks should be made with the realization that your safety depends to a great extent on their condition.

2   Check the tires carefully for cuts, tears, embedded nails or other sharp objects and excessive wear. Operation of the motorcycle with excessively worn tires is extremely hazardous, as traction and handling are directly affected. Measure the tread depth at the center of the tire and replace worn tires with new ones when the tread depth is less than specified **(see illustration)**.

3   Repair or replace punctured tires as soon as damage is noted. Do not try to patch a torn tire, as wheel balance and tire reliability may be impaired.

4   Check the tire pressures when the tires are **cold** and keep them properly inflated **(see illustration)**. Proper air pressure will increase tire life and provide maximum stability and ride comfort. Keep in mind that low tire pressures may cause the tire to slip on the rim or come off, while high tire pressures will cause abnormal tread wear and unsafe handling.

5   The cast wheels used on this machine are virtually maintenance free, but they should be kept clean and checked periodically for cracks and other damage. Never attempt to repair damaged cast wheels; they must be replaced with new ones.

6   Check the valve rubber for signs of damage or deterioration and have it replaced if necessary. Also, make sure the valve stem cap is in place and tight. If it is missing, install a new one made of metal or hard plastic.

6.6 Ensure arrow and notch align on front brake lever span adjuster

7.2 Checking the tire tread depth with a depth gauge

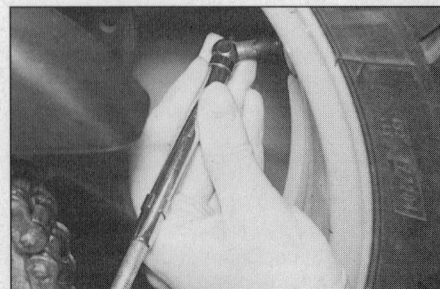

7.4 Checking the tire pressure with a pressure gauge

8.3 Throttle cable freeplay is measured in terms of twistgrip rotation at the grip flange (arrow)

8.4 Take up throttle cable freeplay at the upper adjuster . . .

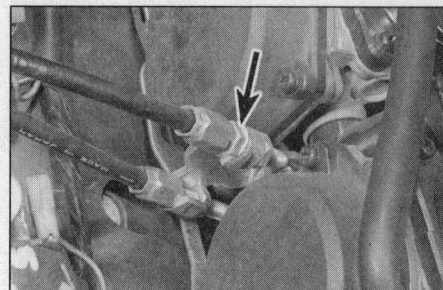

8.5 . . . or at the lower adjuster on the throttle pulley (arrow)

8 Throttle and choke operation/grip freeplay - check and adjustment

Throttle cables

Refer to illustrations 8.3, 8.4 and 8.5

1 Make sure the throttle grip rotates easily from fully closed to fully open with the front wheel turned at various angles. The grip should return automatically from fully open to fully closed when released.

2 If the throttle sticks, this is probably due to a cable fault. Remove the cables as described in Chapter 4 and lubricate them as described in Section 15. Install each cable, routing them so they take the smoothest route possible. If this fails to improve throttle operation, the cables must be replaced. Note that in very rare cases the fault could lie in the carburetors rather than the cables, necessitating the removal of the carburetors and inspection of the throttle linkage (see Chapter 4).

3 With the throttle operating smoothly, check for a small amount of freeplay at the grip **(see illustration)**. The amount of freeplay in the throttle cables, measured in terms of twistgrip rotation, should be as given in this Chapter's Specifications. If adjustment is necessary, adjust idle speed first (see Section 17).

4 Slacken the locknut on the cable upper adjuster and rotate the adjuster until the correct amount of freeplay is obtained, then tighten the locknut **(see illustration)**. If it is not possible to obtain the correct freeplay with the upper adjuster, it will also be necessary to make adjustment at the lower adjuster, situated on the carburetors.

5 To gain access to the lower adjuster remove the fuel tank and air filter housing as described in Chapter 4 **(see illustration)**. Screw the upper cable adjuster in to obtain the maximum possible freeplay, then slacken the lower adjuster locknut and set the cable freeplay using first the lower adjuster and then, if necessary, the upper adjuster. Once the

freeplay is correct tighten the locknuts securely.

6 Check that the throttle twistgrip operates smoothly and snaps shut quickly when released. **Warning:** *Turn the handlebars all the way through their travel with the engine idling. Idle speed should not change. If it does, the cables may be routed incorrectly. Correct this condition before riding the bike (see Chapter 4).*

Choke cable

Refer to illustrations 8.9a and 8.9b

7 Remove the fuel tank and air filter housing as described in Chapter 4. Operate the choke knob whilst observing the movement of the carburetor choke mechanism on the right side of the carburetor assembly. The mechanism should extend smoothly when the knob is pulled, and return home fully when the knob is returned.

8 If the choke mechanism does not operate smoothly this is probably due to a cable fault. Remove the cable as described in Chapter 4 and lubricate it as described in Section 15. Install the cable, routing it so it takes the smoothest route possible. If this fails to improve the operation of the choke, the cable must be replaced. Note that in very rare cases the fault could lie in the carburetors rather than the cable, necessitating the removal of the carburetors and inspection of the choke plungers as described in Chapter 4.

9 With the choke mechanism operating smoothly, check for a small amount of freeplay at the base of the choke knob **(see illustration)**. There is no specific setting for freeplay, but a very small amount should exist to ensure that the choke mechanism is fully off. To adjust the cable, slacken the choke cable clamping screw, situated on the carburetors, then move the lower end of the outer cable until the required amount of freeplay is obtained **(see illustration)**. Tighten the clamping screw securely.

10 Once the choke mechanism is correctly adjusted, install the air filter housing and fuel tank as described in Chapter 4.

1

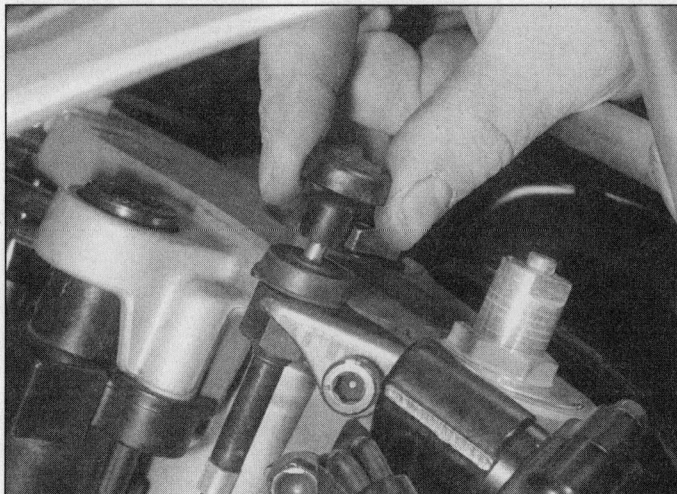

8.9a A small amount of freeplay should be felt at the choke knob

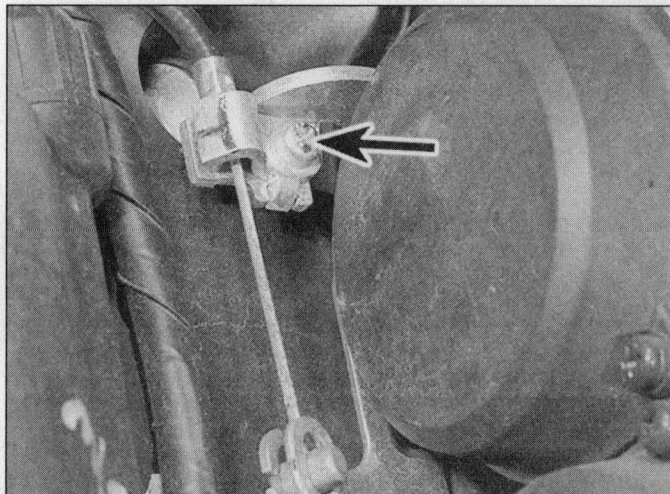

8.9b If adjustment is required, slacken the clamping screw (arrow) and reposition the outer cable

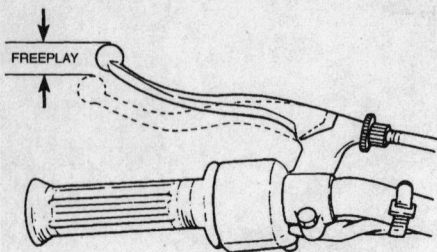

9.3a Clutch cable freeplay is measured at lever ball end

9.3b A small amount of cable adjustment can be made at the upper adjuster on the handlebar lever (arrow)

9.4 Clutch cable lower adjuster is located in bracket on clutch cover

9   Clutch - check and adjustment

Refer to illustrations 9.3a, 9.3b and 9.4

1   Check that the clutch cable operates smoothly and easily.

2   If the clutch lever operation is heavy or stiff, remove the cable as described in Chapter 2 and lubricate it as described in Section 15. Install the lubricated cable, making sure it takes the smoothest route.

3   With the cable operating smoothly, it is necessary to check that the clutch lever is correctly adjusted. Clutch cable freeplay is measured in terms of travel at the ball end of the lever, and should be as given in this Chapter's Specifications **(see illustration)**. If adjustment is required, slacken the handlebar end adjuster lockwheel on the lever mounting bracket **(see illustration)**. Reposition the adjuster as required and securely tighten the lockwheel.

4   If there is insufficient range in the handlebar adjuster it will be necessary to adjust the freeplay at the lower adjuster on the crankcase **(see illustration)**. Screw the upper adjuster fully inwards and slacken the locknut on the lower adjuster. Rotate the adjuster nut until the required freeplay is obtained at the handlebar lever, then securely tighten the lower adjuster locknut. If necessary, fine adjustments can then be made using the handlebar adjuster.

10   Drive chain and sprockets - check, adjustment and lubrication

Check

Refer to illustrations 10.3 and 10.5

1   A neglected drive chain won't last long and can quickly damage the sprockets. Routine chain adjustment and lubrication isn't difficult and will ensure maximum chain and sprocket life.

2   To check the chain, place the bike on its sidestand and shift the transmission into neutral. Make sure the ignition switch is OFF.

3   Push up on the bottom run of the chain and measure the slack midway between the two sprockets, then compare your measurements to the value listed in this Chapter's Specifications **(see illustration)**. Since the chain will rarely wear evenly, roll the bike forwards so that another section of chain can be checked; do this several times to check the entire length of chain. In some cases where lubrication has been neglected, corrosion and galling may cause the links to bind and kink, which effectively shortens the chain's length. If the chain is tight between the sprockets, rusty or kinked, it's time to replace it with a new one. If you find a tight area, mark it with felt pen or paint, and repeat the measurement after the bike has been ridden. If the chain's still tight in the same area, it may be damaged or worn. Because a tight or kinked chain can damage the transmission countershaft bearing, it's a good idea to replace it. **Caution:** *If the machine is ridden with more than 50 mm (2 inches) of slack in the drive chain, the chain will contact the frame and swingarm, causing severe damage.*

4   Check the entire length of the chain for damaged rollers, loose links and pins and replace if damage is found. **Note:** *Never install a new chain on old sprockets, and never use the old chain if you install new sprockets - replace the chain and sprockets as a set.*

5   Remove the engine sprocket cover (see Chapter 6). Check the teeth on the engine sprocket and the rear wheel sprocket for wear **(see illustration)**.

6   Inspect the drive chain slider on the swingarm for excessive wear and replace if worn down to the wear indicator arrow (see Chapter 6).

Adjustment

Refer to illustrations 10.8, 10.9a and 10.9b

7   Rotate the rear wheel until the chain is positioned with the tightest point at the center of its bottom run.

10.3 Measuring drive chain slack

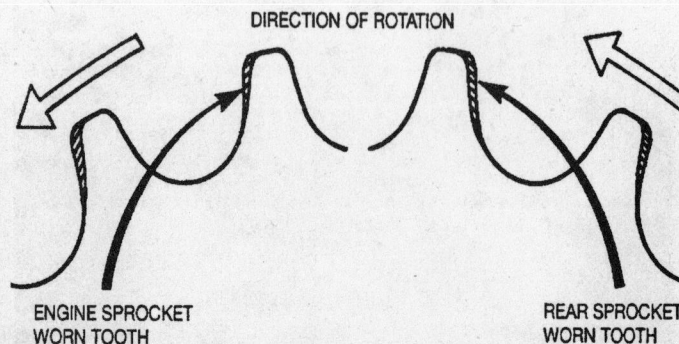

DIRECTION OF ROTATION

ENGINE SPROCKET WORN TOOTH

REAR SPROCKET WORN TOOTH

0618H

10.5 Check the sprockets in the areas indicated to see if they are worn excessively

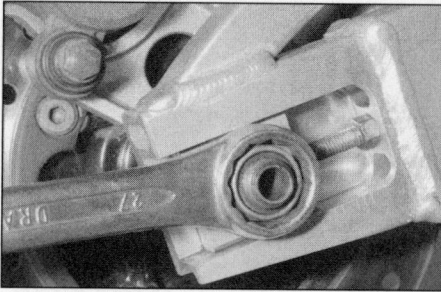

10.8 Slacken off rear axle nut

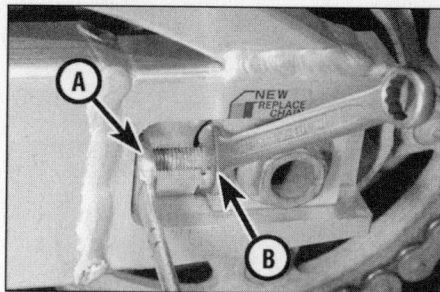

10.9a Slacken each adjuster locknut (A) and turn adjuster bolt (B) to adjust tension

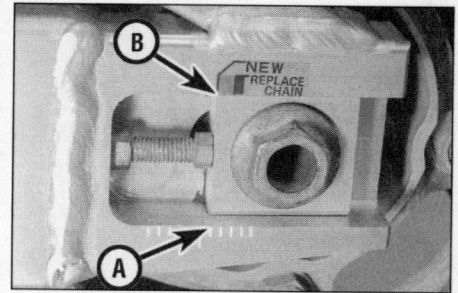

10.9b Check front edge of adjuster block in relation to alignment scale (A) and chain wear indicator decal (B)

8   Slacken the rear axle nut **(see illustration)**.

9   Back off the locknut and turn the adjusting bolts on both sides of the swingarm until the proper chain tension is obtained (get the adjuster on the chain side close, then set the adjuster on the opposite side) **(see illustration)**. Be sure to turn the adjusting bolts evenly to keep the rear wheel in alignment; the front edge of the chain adjuster block can be aligned with the graduated scale to ensure this **(see illustration)**. Having completed the adjustment, check the relationship of the front edge of the chain adjuster block with the wear decal on the swingarm ends; if it aligns with the red REPLACE CHAIN zone, the drive chain has stretched excessively and must be replaced as soon as possible.

10   Tighten the axle nut to the specified torque setting.

11   With the axle nut tightened, tighten the chain adjuster locknuts securely.

Lubrication

Note: *If the chain is extremely dirty, it should be removed and cleaned before it's lubricated (see Chapter 6).*

12   For routine lubrication, the best time to lubricate the chain is after the motorcycle has been ridden. When the chain is warm, the lubricant will penetrate the joints between the side plates better than when cold. **Note:** *Honda specifies SAE 80 to SAE 90W gear oil only; do not use chain lube, which may contain solvents that could damage the O-rings.* Apply the oil to the area where the side plates overlap - not the middle of the rollers. Apply the oil to the top of the lower chain run, so centrifugal force will work the oil into the chain when the bike is moving. After applying the lubricant, let it soak in a few minutes before wiping off any excess.

11   Engine oil/filter - change

Refer to illustrations 11.4, 11.5, 11.7 and 11.8

1   Regular oil and filter changes are the single most important maintenance procedure you can perform on a motorcycle. The oil not only lubricates the internal parts of the engine, transmission and clutch, but it also acts as a coolant, a cleaner, a sealant, and a protectant. Because of this, the oil takes a lot of abuse and should be replaced often with new oil of the recommended grade and type. Always use a good-quality brand name of oil. Trying to save money by using a cheap oil rather than a good-quality one won't pay off if the engine is damaged.

2   Before changing the oil and filter, warm up the engine so the oil will drain easily. Be careful when draining the oil, as the exhaust pipes, the engine, and the oil itself can cause severe burns.

3   Put the motorcycle on its sidestand and position a clean drain pan below the engine. Unscrew the oil filler cap to vent the crankcase and act as a reminder that there is no oil in the engine. Refer to Chapter 8 and remove the fairing lower section.

4   Next, remove the drain plug from the oil pan and allow the oil to drain into the pan **(see illustration)**. Discard the sealing washer on the drain plug; it should be replaced whenever the plug is removed.

5   When the engine oil has drained, reposition another drain pan under the oil filter on the front of the engine. Slacken the oil filter using a strap wrench or the Honda service tool (Pt. No. 07HAA-PJ70100) **(see illustration)**. **Warning:** *Take great care not to burn your hands on the exhaust system.* Unscrew the filter from the oil cooler unit and empty its contents into the drain pan. If additional maintenance is planned for this time period, check or service another component while the oil is allowed to drain completely.

6   Clean the filter thread and housing on the engine with solvent or clean shop towels. Wipe any remaining oil off the filter sealing area of the oil cooler unit.

7   Slip a new sealing washer over the drain plug **(see illustration)**. Fit the plug to the oil pan and tighten it to the specified torque setting. Avoid overtightening, as damage to the oil pan will result.

8   Apply a smear of clean oil to the sealing ring of the new filter and screw it into position on the engine **(see illustration)**. Tighten the filter firmly by hand. If access to the special Honda service tool can be gained (see Step 5), the filter can be tightened to the specified torque setting.

9   Before refilling the engine, check the old oil carefully. If the oil was drained into a clean pan, small pieces of metal or other material can be easily detected. If the oil is very metallic colored, then the engine is experiencing wear from break-in (new engine) or from insufficient lubrication. If there are flakes or chips of metal in the oil, then something is drastically wrong internally and the engine will have to be disassembled for inspection and repair.

1

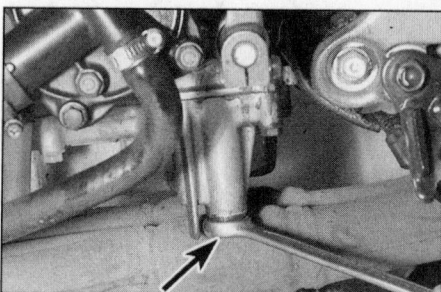

11.4 Engine oil drain plug location (arrow)

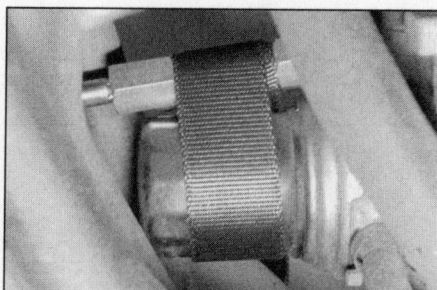

11.5 Use a strap wrench to slacken the oil filter if the Honda tool is not available

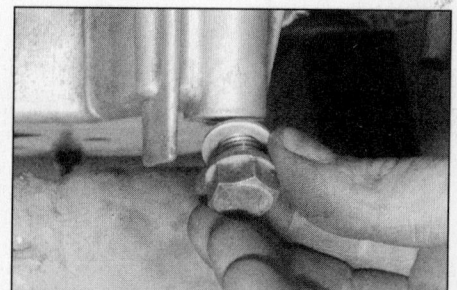

11.7 Always use a new sealing washer on the oil drain plug

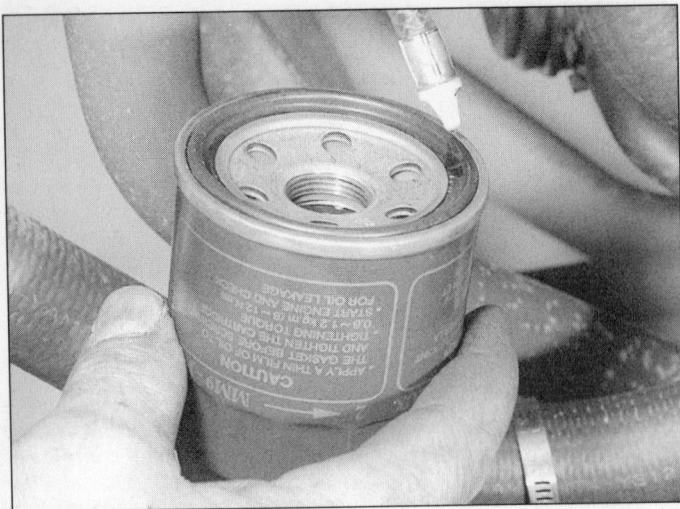

11.8 Apply oil to the filter sealing ring before installation

10 If there are pieces of fiber-like material in the oil, the clutch is experiencing excessive wear and should be checked.

11 If inspection of the oil turns up nothing unusual, refill the crankcase to the proper level with the recommended type and amount of oil and install the filler cap. Start the engine and let it run for two or three minutes. Shut it off, wait a few minutes, then check the oil level. If necessary, add more oil to bring the level up to the upper mark (see Section 3). Check around the drain plug and filter for leaks.

12 The old oil drained from the engine cannot be re-used and should be disposed of properly. Check with your local refuse disposal company, disposal facility or environmental agency to see wether they will accept the used oil for recycling. Don't pour used oil into drains or onto the ground. After the oil has cooled, it can be drained into a suitable container (capped plastic jugs, topped bottles, milk cartons, etc.) for transport to one of these disposal sites.

12 Air filter element - change

Refer to illustrations 12.2a, 12.2b and 12.3

1 Remove the fuel tank as described in Chapter 4.

2 Undo the nine retaining screws and remove the cover from the air filter housing **(see illustrations)**.

3 Lift out the air filter element **(see illustration)**.

4 Wipe out the housing with a clean rag, particularly the drain tray at the forward end into which the sediment from crankcase vapors collects.

5 Fit the new element to the housing, making sure it is the correct way up. Install the cover and securely tighten its retaining screws.

6 Install the fuel tank as described in Chapter 4.

13 Cylinder compression - check

1 Among other things, poor engine performance may be caused by leaking valves, incorrect valve clearances, a leaking head gasket, or worn pistons, rings and/or cylinder walls. A cylinder compression check will help pinpoint these conditions and can also indicate the presence of excessive carbon deposits in the cylinder heads.

2 The only tools required are a compression gauge and a spark plug wrench. Depending on the outcome of the initial test, a squirt-type oil can may also be needed.

3 Start the engine and allow it to reach normal operating temperature, then stop it.

4 Place the motorcycle on its sidestand.

5 Remove the spark plugs as described in Section 14. **Caution:** *Work carefully - don't strip the spark plug hole threads and don't burn your hands on the hot cylinder head.*

6 Disable the ignition by switching the kill switch to OFF.

7 Install the compression gauge in one of the spark plug holes and place a rag over the other three plug holes as a precaution against fire risk.

8 Hold the throttle wide open and crank the engine over a minimum of four or five revolutions (or until the gauge reading stops increasing) and observe the initial movement of the compression gauge needle as well as the final total gauge reading. Repeat the procedure for the other cylinders and compare the results to the value listed in this Chapter's Specifications.

9 If the compression in all four cylinders built up quickly and evenly to the specified amount, you can assume the engine upper end is in reasonably good mechanical condition. Worn or sticking piston rings and worn cylinders will produce very little initial movement of the gauge needle, but compression will tend to build up gradually as the engine spins over. Valve and valve seat leakage, or head gasket leakage, is indicated by low initial compression which does not tend to build up.

10 To further confirm your findings, add a small amount of engine oil to each cylinder by inserting the nozzle of a squirt-type oil can through the spark plug holes. The oil will tend to seal the piston rings if they are leaking. Repeat the test for the other cylinders.

11 If the compression increases significantly after the addition of the oil, the piston rings and/or cylinders are definitely worn. If the compression does not increase, the pressure is leaking past the valves or the head gasket. Leakage past the valves may be due to insufficient valve clearances, burned, warped or cracked valves or valve seats, or valves that are hanging up in the guides.

12 If the compression readings are considerably higher than specified, the combustion chambers are probably coated with excessive carbon deposits. It is possible (but not very likely) for carbon deposits to raise the compression enough to compensate for the effects of leakage past rings or valves. Use of a fuel additive that will dissolve the adhesive bonding the carbon particles to the crown and chamber is the easiest way to remove the build-up. Otherwise, the cylinder head will have to be removed and decarbonized (Chapter 2).

12.2a Remove its nine retaining screws . . .

12.2b . . . and lift off the air filter cover

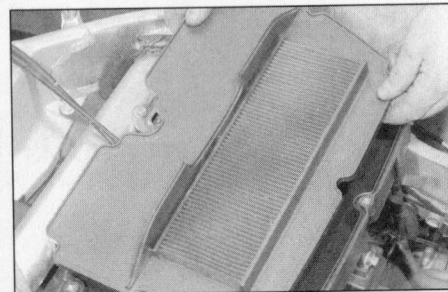

12.3 Air filter element is simply lifted out of the housing

Spark plug maintenance: Checking plug gap with feeler gauges

Altering the plug gap. Note use of correct tool

Spark plug conditions: A brown, tan or grey firing end is indicative of correct engine running conditions and the selection of the appropriate heat rating plug

White deposits have accumulated from excessive amounts of oil in the combustion chamber or through the use of low quality oil. Remove deposits or a hot spot may form

Black sooty deposits indicate an over-rich fuel/air mixture, or a malfunctioning ignition system. If no improvement is obtained, try one grade hotter plug

Wet, oily carbon deposits form an electrical leakage path along the insulator nose, resulting in a misfire. The cause may be a badly worn engine or a malfunctioning ignition system

A blistered white insulator or melted electrode indicates over-advanced ignition timing or a malfunctioning cooling system. If correction does not prove effective, try a colder grade plug

A worn spark plug not only wastes fuel but also overloads the whole ignition system because the increased gap requires higher voltage to initiate the spark. This condition can also affect air pollution

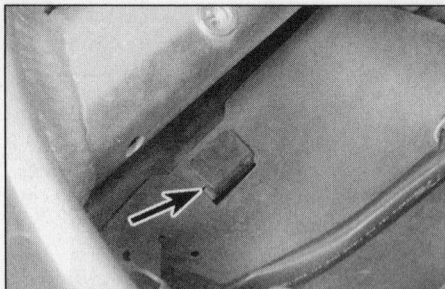

14.5a Free the rubber dust cover from the air deflector tabs (arrow) . . .

14.5b . . . remove the two retaining bolts (arrows) . . .

14.5c . . . and maneuver the air deflector out of the frame

14 Spark plugs - check and replacement

Refer to illustrations 14.5a thru 14.5c, 14.6a, 14.6b, 14.10a thru 14.10c

1   This motorcycle is equipped with spark plugs that have 10 mm threads and a 16 mm wrench hex. Make sure your spark plug socket is the correct size before attempting to remove the plugs, a suitable one is supplied in the motorcycle's toolkit.

2   Remove the fairing middle sections (see Chapter 8).

3   Remove the fuel tank and air filter housing (see Chapter 4).

4   Remove the ignition HT coils (see Chapter 5).

5   Release the rubber dust cover from the tabs of the air deflector, then remove the two bolts which retain the air deflector to the frame **(see illustrations)**. Maneuver the air deflector out from under the frame cross-member **(see illustration)**.

6   Pull the spark plug caps off the spark plugs and using a deep socket type wrench, unscrew them from the cylinder head **(see illustrations)**. Lay the plugs out in cylinder number order; if any plug shows a problem it will then be easy to identify the troublesome cylinder.

7   Inspect the electrodes for wear. Both the center and side electrodes should have square edges and the side electrode should be of uniform thickness. Look for excessive deposits and evidence of a cracked or chipped insulator around the center electrode. Compare your spark plugs to the color spark plug chart. Check the threads, the washer and the ceramic insulator body for cracks and other damage.

8   If the electrodes are not excessively worn, and if the deposits can

be easily removed with a wire brush, the plugs can be regapped and re-used (if no cracks or chips are visible in the insulator). If in doubt concerning the condition of the plugs, replace them with new ones, as the expense is minimal.

9   Cleaning spark plugs by sandblasting is permitted, provided you clean the plugs with a high flash-point solvent afterwards.

10   Before installing new plugs, make sure they are the correct type and heat range. Check the gap between the electrodes, as they are not preset. For best results, use a wire-type gauge rather than a flat (feeler) gauge to check the gap. If the gap must be adjusted, bend the side electrode only and be very careful not to chip or crack the insulator nose **(see illustrations)**. Make sure the washer is in place before installing each plug.

11   As the cylinder head is made of aluminum, which is soft and easily damaged, thread the plugs into the heads by hand. As the plugs are recessed, slip a short length of hose over the end of the plug to use as a tool to thread it into place. The hose will grip the plug well enough to turn it, but will start to slip if the plug begins to cross-thread in the hole - this will prevent damaged threads and the resultant repair costs.

12   Once the plugs are finger-tight, the job can be finished with a socket. If a torque wrench is available, tighten the spark plugs to the specified torque listed in this Chapter's Specifications. If you do not have a torque wrench, tighten the plugs finger-tight (until the washers bottom on the cylinder head) then use a wrench to tighten them a further 1/4 turn. Whichever method is used, do not over-tighten them.

13   Refit the spark plug caps and reinstall all disturbed components.

14.6a Pull the spark plug caps out of the valve cover . . .

14.6b . . . and unscrew the plugs using a long extension down through the frame, or the tool contained in the bike's toolkit

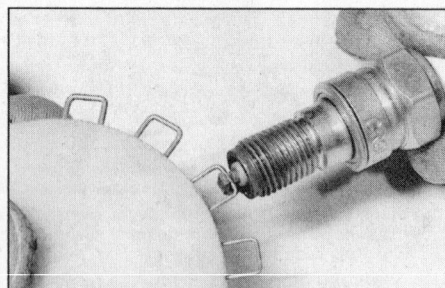

14.10a A wire type gauge is recommended to measure the spark plug electrode gap

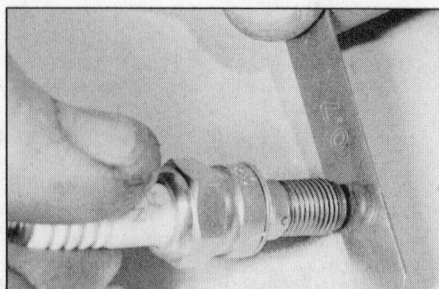

14.10b Using a feeler gauge to measure spark plug electrode gap

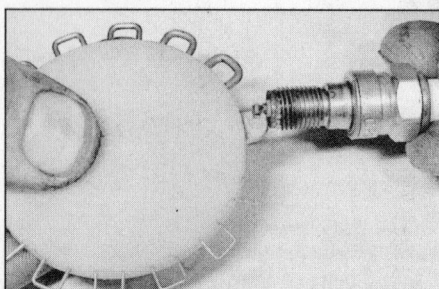

14.10c Electrode gap is adjusted by bending the side electrode

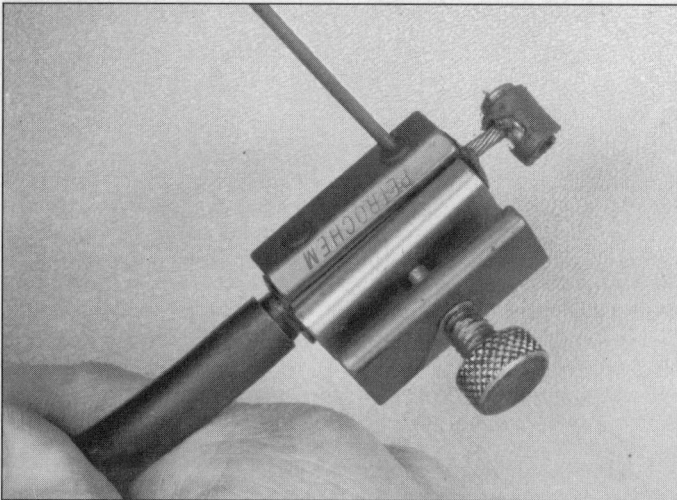

15.3 Lubricating a cable with a pressure lube adapter (make sure the tool seats around the inner cable)

15 Lubrication - general

Refer to illustration 15.3

1 Since the controls, cables and various other components of a motorcycle are exposed to the elements, they should be lubricated periodically to ensure safe and trouble-free operation.

2 The footpegs, clutch and brake lever, brake pedal, gearshift lever linkage (where applicable) and sidestand pivots should be lubricated frequently. In order for the lubricant to be applied where it will do the most good, the component should be disassembled. However, if chain and cable lubricant is being used, it can be applied to the pivot joint gaps and will usually work its way into the areas where friction occurs. If motor oil or light grease is being used, apply it sparingly as it may attract dirt (which could cause the controls to bind or wear at an accelerated rate). **Note:** *One of the best lubricants for the control lever*

pivots is a dry-film lubricant (available under several different names).

3 To lubricate the cables, disconnect the relevant cable at its upper end, then lubricate the cable with a pressure lube adapter **(see illustration)**. See Chapter 4 for the choke and throttle cable removal procedures, and Chapter 2 for clutch cable removal details.

4 The speedometer cable on US 1993 and 1994 models, and UK CBR900RR-N and RR-P models, should be removed from its housing and lubricated with motor oil or cable lubricant. Do not lubricate the upper few inches of the cable as the lubricant may travel up into the speedometer head.

16 Valve clearances - check and adjustment

Refer to illustrations 16.4a thru 16.4c, 16.6, 16.7, 16.12a and 16.12b

1 The engine must be completely cool for this maintenance procedure, so let the machine sit overnight before beginning.

2 Remove the valve cover (see Chapter 2).

3 Refer to Chapter 2 *"Cam chain tensioner - removal and installation"* and retract the tensioner plunger; there's no need to remove the tensioner.

4 Remove the fairing lower section (see Chapter 8). Unscrew the center cap to reveal the bolt head on the right end of the crankshaft. Using a socket wrench on this bolt, rotate the engine clockwise so that the line next to the T mark on the ignition rotor aligns exactly with the notch in the cover **(see illustrations)**. Check that the IN and EX marks on the camshaft sprockets are pointing outwards (in the 9 o'clock and 3 o'clock positions respectively); if they face each other, turn the crankshaft 360° and realign the T mark **(see illustration)**.

5 Draw the valve positions on a piece of paper, numbering both the intake and exhaust valves from 1 to 8, from the left end of the engine.

6 With the engine in this position, the intake valves of cylinders 1 and 3 can be checked. Insert a feeler gauge of the correct thickness (see Specifications) between each cam lobe and follower and check that it is a firm sliding fit **(see illustration)**. If it is not, use the feeler gauges to obtain the exact clearance.

7 Rotate the crankshaft clockwise half a turn (180°) so that the single scribed line on the ignition rotor is in the 12 o'clock position **(see illustration)**. The exhaust valves of cylinders 2 and 4 can now be checked. **Note:** *The clearance is different for the exhaust valves.*

16.4a Remove center cap from crankshaft right end cover and rotate crankshaft using a socket and extension bar

16.4b Position the crankshaft so the line next to the T mark is aligned with the notch (arrow) in the cover . . .

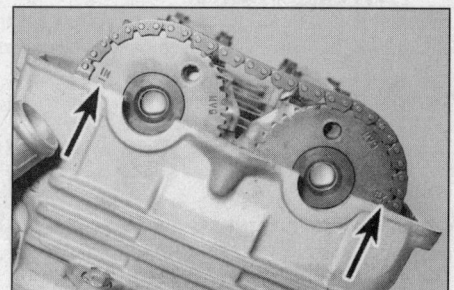

16.4c . . . and the IN and EX marks (arrows) on the camshaft sprockets are level with the cylinder head surface

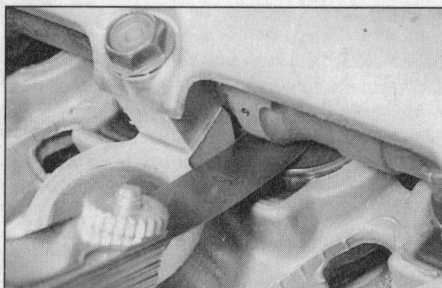

16.6 Measuring a valve clearance

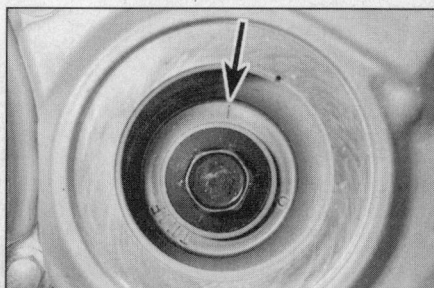

16.7 Rotate the crankshaft through 180° so that the index line (arrow) on the ignition rotor is vertical

16.12a Shim thickness is indicated by three numbers stamped on its surface; 205 here means a shim 2.050 mm thick

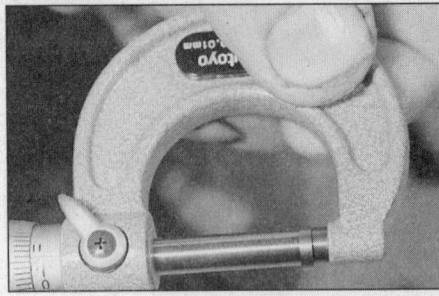

16.12b Measuring shim thickness

17.3 Engine idle speed is adjusted using idle knob (arrow)

8   Turn the crankshaft clockwise half a turn (180°) and realign the line next to the T mark with the notch. In this position check the intake valve clearances for cylinders 2 and 4.

9   Finally, turn the crankshaft a further half turn (180°) clockwise and position the single scribed line in the 12 o'clock position. Check the exhaust valve clearances on cylinders 1 and 3.

10 If any of the clearances need to be adjusted the relevant camshaft(s) must be removed as described in Chapter 2.

11 With the camshaft removed, using a magnet, lift the follower on the valve to be adjusted out of the cylinder head and remove the shim. Note that the shim is likely to stick to the inside of the follower so take great care not to lose it as the follower is removed. If more than one follower and shim is to be removed, make sure they are not interchanged.

12 The shim size should be stamped on its face, however, it is recommended that the shim is measured to check that it has not worn **(see illustrations)**. The size marking is in the form of a three figure number, eg. 180 indicating that the shim is 1.800 mm thick. Where the number does not equal a shim thickness, it should be rounded up or down, eg. 182 or 183 both indicate that the shim is 1.825 mm thick. Shims are available in 0.025 mm increments from 1.200 to 2.800 mm. The new shim thickness required can then be calculated as follows. **Note:** *Always aim to get the clearance at the mid-point of the specified range.*

13 If the valve clearance was less than specified, subtract the measured clearance from the specified clearance then deduct the result from the original shim thickness. For example:

Sample calculation - intake valve clearance too small
Clearance measured (A) - 0.08 mm
Specified clearance (B) - 0.16 mm (0.13 to 0.19 mm)
Difference (B - A) - 0.08 mm
Shim thickness fitted - 2.475 mm
Correct shim thickness required - 2.475 - 0.08 = 2.395 mm

14 If the valve clearance was greater then specified, subtract the specified clearance from the measured clearance, and add the result to the thickness of the original shim. For example:

Sample calculation - exhaust valve clearance too large
Clearance measured (A) - 0.35 mm
Specified clearance (B) - 0.22 mm (0.19 to 0.25 mm)
Difference (A - B) - 0.13 mm
Shim thickness fitted - 1.975 mm
Correct shim thickness required - 1.975 + 0.13 = 2.105 mm

15 Obtain the correct thickness shims from your Honda dealer.

16 Install the shim in position on top of the relevant valve, making sure it is correctly seated in the valve spring retainer.

17 Lubricate the followers with engine oil and insert them in their respective positions in the cylinder head, making sure each one squarely enters its bore.

18 Install the camshaft(s) as described in Chapter 2.

19 Rotate the crankshaft a few times, to settle all disturbed components, and recheck all valve clearances as described above. If necessary, repeat the adjustment procedure.

20 Release the stopper key from the cam chain tensioner. Install a new sealing washer on the tension bolt and tighten the bolt securely.

21 Install the valve cover as described in Chapter 2.

22 Apply a smear of oil to the crankshaft cap O-ring and tighten the cap to the specified torque setting.

23 Install all the remaining components in a reverse of the removal sequence.

17 Idle speed - check and adjustment

Refer to illustration 17.3

1   The idle speed should be checked and adjusted before and after the carburetors are synchronized and when it is obviously too high or too low. Before adjusting the idle speed, make sure the valve clearances and spark plug gaps are correct. Also, turn the handlebars back-and-forth and see if the idle speed changes as this is done. If it does, the throttle cables may not be adjusted correctly, or may be worn out. This is a dangerous condition that can cause loss of control of the bike. Be sure to correct this problem before proceeding.

2   The engine should be at normal operating temperature, which is usually reached after 10 to 15 minutes of stop and go riding. Place the motorcycle on its sidestand and make sure the transmission is in neutral.

3   Turn the idle speed knob, which is located in a wire guide which extends from the middle fairing section bracket on the left side, until the idle speed listed in this Chapter's Specifications is obtained **(see illustration)**.

4   Snap the throttle open and shut a few times, then recheck the idle speed. If necessary, repeat the adjustment procedure.

5   If a smooth, steady idle can't be achieved, the fuel/air mixture may be incorrect. Refer to Chapter 4 for additional carburetor information.

18 Carburetor synchronization - check and adjustment

Refer to illustrations 18.9a, 18.9b and 18.14
Warning: *Gasoline (petrol) is extremely flammable, so take extra precautions when you work on any part of the fuel system. Don't smoke or allow open flames or bare light bulbs near the work area, and don't work in a garage where a natural gas-type appliance (such as a water heater or clothes dryer) is present. If you spill any fuel on your skin, rinse it off immediately with soap and water. When you perform any kind of work on the fuel system, wear safety glasses and have a fire extinguisher suitable for a Class B type fire (flammable liquids) on hand.*
Warning: *Take great care not to burn your hand on the hot engine unit when accessing the gauge take-off points on the intake manifolds.*

1   Carburetor synchronization is simply the process of adjusting the carburetors so they pass the same amount of fuel/air mixture to each cylinder. This is done by measuring the vacuum produced in each cylinder. Carburetors that are out of synchronization will result in decreased fuel mileage, increased engine temperature, less than ideal throttle response and higher vibration levels.

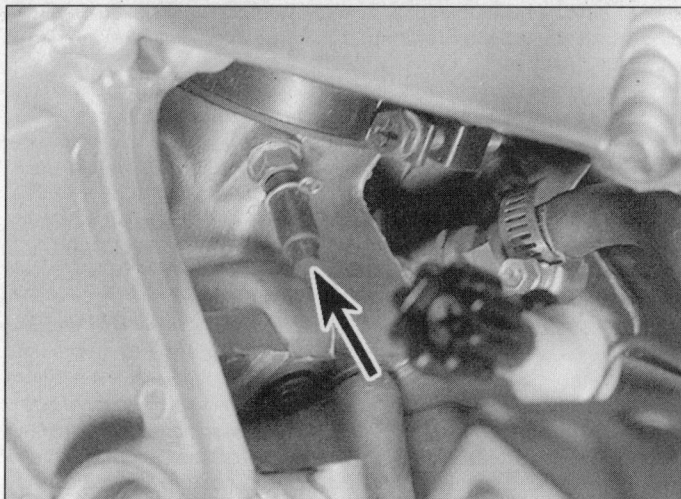

18.9a Remove rubber cap (arrow) from vacuum take-off tube on No. 1 cylinder intake manifold - all models except California

H30000

18.9b Carburetor synchronization equipment connections - all models except California

2   To properly synchronize the carburetors, you will need some sort of vacuum gauge setup, preferably with a gauge for each cylinder, or a mercury manometer, which is a calibrated tube arrangement that utilizes columns of mercury to indicate engine vacuum. Adaptors will be required to screw into the intake ports holes - see Step 9 or 10.
3   A manometer can be purchased from a motorcycle dealer or accessory shop and should have the necessary rubber hoses supplied with it for hooking into the vacuum hose fittings/adaptors.
4   A vacuum gauge setup can also be purchased from a dealer or fabricated from commonly available hardware and automotive vacuum gauges.
5   The manometer is the more reliable and accurate instrument, and for that reason is preferred over the vacuum gauge setup; however, since the mercury used in the manometer is a liquid, and extremely toxic, extra precautions must be taken during use and storage of the instrument.
6   Because of the nature of the synchronization procedure and the need for special instruments, most owners leave the task to a dealer service department or a reputable motorcycle repair shop.
7   Start the engine and let it run until it reaches normal operating temperature, then shut it off.
8   Remove the fuel tank as described in Chapter 4, and the fairing middle sections as described in Chapter 8.
9   On UK models and US models except California, remove the vacuum take-off tube rubber cap from the No. 1 cylinder intake manifold (see illustration). Grip the end of the cap only, otherwise the take-off tube may distort. Remove the vacuum take-off screws from the underside of the intake manifolds for cylinders 2, 3 and 4. Screw the gauge adaptors into the intake manifolds of cylinders 2, 3 and 4, then connect the gauge tubes to No. 1 cylinder take-off tube and the three adaptors (see illustration). Make sure there are no leaks in the setup, as false readings will result.
10   On California models, disconnect the EVAP tube from cylinder No. 1 intake manifold and the PAIR tube from cylinder No. 3 intake manifold. Remove the vacuum take-off screws from the underside of the intake manifolds of cylinders 2 and 4 and screw the gauge adaptors in their place. Connect the gauge tubes to the adaptors of cylinders 2 and 4 and the tube unions of cylinders 1 and 3. Make sure there are no leaks in the setup, as false readings will result.
11   Arrange a temporary fuel supply, either by using a small temporary tank or by using extra long fuel pipes to the now remote fuel tank on a nearby bench.
12   Start the engine and make sure the idle speed is correct. If it isn't, adjust it (see Section 17). If the gauges are fitted with damping adjustment, set this so that the needle flutter is just eliminated but so that they can still respond to small changes in pressure.
13   The vacuum readings for all of the cylinders should be the same, or

at least within the tolerance listed in this Chapter's Specifications. If the vacuum readings vary, adjust as necessary. Cylinder No. 3 is the base carburetor and the other carburetors should be balanced to it's setting.
14   The carburetors are adjusted by the three screws situated inbetween each carburetor, in the throttle linkage (see illustration). The screws are accessible from underneath the rear of the air filter housing. Note: Do not press down on the screws whilst adjusting them, otherwise a false reading will be obtained. When all the carburetors are synchronized, open and close the throttle quickly to settle the linkage, and recheck the gauge readings, readjusting if necessary.
15   When the adjustment is complete, recheck the vacuum readings and idle speed, then stop the engine. Remove the vacuum gauge or manometer. Unscrew the adaptors and fit the screws, tightening them securely. Install the rubber cap over No. 1 cylinder's take-off tube on UK and US models except California. On California models, reconnect the EVAP tube to cylinder No. 1 intake manifold and the PAIR tube to cylinder No. 3 intake manifold.
16   Detach the temporary fuel supply and install the fuel tank and fairing sections.

18.14 Carburetor synchronization screws (arrows) - cylinder numbers circled (air filter housing removed for clarity)

19 Cooling system - check

Refer to illustration 19.8

Warning: *The engine must be cool before beginning this procedure.*

1   Check the coolant level as described in Section 3.

2   Remove the left and right middle fairing sections as described in Chapter 8.

3   The entire cooling system should be checked for evidence of leakage. Examine each rubber coolant hose along its entire length, including those which link the oil cooler to the engine and water pump. Look for cracks, abrasions and other damage. Squeeze each hose at various points. They should feel firm, yet pliable, and return to their original shape when released. If they are dried out or hard, replace them with new ones.

4   Check for evidence of leaks at each cooling system joint. Tighten the hose clips carefully to prevent future leaks.

5   Check the radiator for leaks and other damage. Leaks in the radiator leave telltale scale deposits or coolant stains on the outside of the core below the leak. If leaks are noted, remove the radiator (see Chapter 3) and have it repaired at a radiator shop or replace it with a new one. **Caution:** *Do not use a liquid leak stopping compound to try to repair leaks.*

6   Check the radiator fins for mud, dirt and insects, which may impede the flow of air through the radiator. If the fins are dirty, force water or low pressure compressed air through the fins from the backside. If the fins are bent or distorted, straighten them carefully with a screwdriver.

7   Remove the right inner panel from the upper fairing (see Chapter 8).

8   Remove the pressure cap by turning it counterclockwise (anti-clockwise) until it reaches a stop **(see illustration)**. If you hear a hissing sound (indicating there is still pressure in the system), wait until it stops. Now press down on the cap and continue turning the cap until it can be removed. Check the condition of the coolant in the system. If it is rust-colored or if accumulations of scale are visible, drain, flush and refill the system with new coolant (See Section 20). Check the cap seal for cracks and other damage. If in doubt about the pressure cap's condition, have it tested by a Honda dealer or replace it with a new one. Install the cap by turning it clockwise until it reaches the first stop then push down on the cap and continue turning until it can turn further.

9   Check the antifreeze content of the coolant with an antifreeze hydrometer. Sometimes coolant looks like it's in good condition, but might be too weak to offer adequate protection. If the hydrometer indicates a weak mixture, drain, flush and refill the system as described in Section 20.

10 Start the engine and let it reach normal operating temperature, then check for leaks again. As the coolant temperature increases, the fan should come on automatically and the temperature should begin to drop. If it does not, refer to Chapter 3 and check the fan and fan circuit carefully.

11 If the coolant level is consistently low, and no evidence of leaks can be found, have the entire system pressure checked by a Honda dealer.

12 Periodically, check the drainage hole on the underside of the water pump cover (see Chapter 3). Leakage from this hole indicates failure of the pump's mechanical seal.

20 Cooling system - draining, flushing and refilling

Warning: *Allow the engine to cool completely before performing this maintenance operation. Also, don't allow antifreeze to come into contact with your skin or the painted surfaces of the motorcycle. Rinse off spills immediately with plenty of water. Antifreeze is highly toxic if ingested. Never leave antifreeze lying around in an open container or in puddles on the floor; children and pets are attracted by its sweet smell and may drink it. Check with local authorities (councils) about disposing of antifreeze. Many communities have collection centers which will see that antifreeze is disposed of safely. Antifreeze is also combustible, so don't store it near open flames.*

Draining

Refer to illustration 20.2

1   Remove the fairing lower section and right inner panel from the upper fairing (see Chapter 8).

2   Position a suitable container beneath the water pump, then remove the drain bolt and sealing washer from the pump cover **(see illustration)**.

3   Remove the pressure cap by turning it counterclockwise (anti-clockwise) until it reaches a stop. If you hear a hissing sound (indicating there is still pressure in the system), wait until it stops. Now press down on the cap and continue turning the cap until it can be removed. As the cap is removed, the flow of coolant will increase - be prepared for this.

4   Drain the coolant reservoir. Refer to Chapter 3 for reservoir removal procedure. Wash out the reservoir with fresh water.

Flushing

5   Flush the system with clean tap water by inserting a garden hose in the radiator filler neck. Allow the water to run through the system until it is clear and flows cleanly out of the drain hole. If the radiator is extremely corroded, remove it by referring to Chapter 3 and have it cleaned at a radiator shop.

6   Clean the drain hole then install the drain bolt and sealing washer.

7   Fill the cooling system with clean water mixed with a flushing

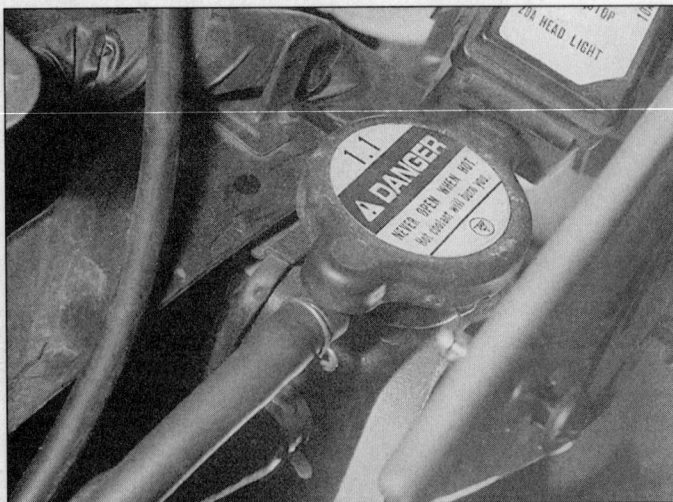

19.8 Radiator pressure cap is accessed after removing upper fairing right inner panel

20.2 Remove drain bolt in water pump cover to allow cooling system to drain

20.13 Use a new sealing washer on coolant drain bolt

20.14 A funnel and hose is recommended for filling the cooling system

23.2 Use of an auxiliary stand will enable the steering head bearings to be checked

compound. Make sure the flushing compound is compatible with aluminum components, and follow the maker's instructions carefully.

8   Start the engine and allow it reach normal operating temperature. Let it run for about ten minutes.

9   Stop the engine. Let it cool for a while, then cover the pressure cap with a heavy shop towel and turn it counterclockwise (anti-clockwise) to the first stop, releasing any pressure that may be present in the system. Once the hissing stops, push down on the cap and remove it completely.

10   Drain the system once again.

11   Fill the system with clean water and repeat the procedure in Steps 8 through 10.

Refilling

Refer to illustrations 20.13 and 20.14

13   Fit a new sealing washer to the drain bolt and tighten it securely **(see illustration)**.

14   Fill the system with the proper coolant mixture (see this Chapter's Specifications) **(see illustration). Note:** *Pour the coolant in slowly to minimise the amount of air entering the system.*

15   When the system is full (all the way up to the top of the radiator filler neck), install the pressure cap. Also top up the coolant reservoir to the UPPER level mark.

16   Start the engine and allow it to idle for 2 to 3 minutes. Flick the throttle twistgrip part open 3 or 4 times, so that the engine speed rises to approximately 4000 - 5000 rpm, then stop the engine. Any air trapped in the system should have bled back to the radiator filler neck via the small-bore air bleed hoses.

17   Let the engine cool then remove the pressure cap as described in Step 9. Check that the coolant level is still up to the radiator filler neck. If it's low, add the specified mixture until it reaches the top of the filler neck. Reinstall the cap.

18   Check the coolant level in the reservoir and top up if necessary.

19   Check the system for leaks. If all is well, install the fairing panels (Chapter 8).

20   Do not dispose of the old coolant by pouring it down the drain. Instead pour it into a heavy plastic container, cap it tightly and take it into an authorized disposal site or service station - see **Warning** at the beginning of this Section.

21   Evaporative emission control system (EVAP) and Pulse secondary air injection (PAIR) system (California models only) - check

1   These systems are installed on California models to conform to stringent state emission control standards. Both systems are explained in greater detail in Chapter 4.

2   To begin the inspection of the system, remove the fuel tank and fairing middle sections (see Chapters 4 and 8). Refer to the hose routing diagram on the air filter housing cover and trace the hoses between the system components, looking for signs of cracking, perishing or other damage. Any such hoses must be replaced.

3   Check the EVAP canister, mounted behind the carburetors on the left side of the frame, for damage or fuel leakage.

22   Exhaust system - check

1   Periodically check all of the exhaust system joints for leaks and loose fasteners. The lower fairing section and middle sections will have to be removed to do this properly (see Chapter 8). If tightening the fasteners fails to stop any leaks, replace the gaskets with new ones (a procedure which requires disassembly of the system). Refer to Chapter 4 for further information.

2   The exhaust pipe flange nuts at the cylinder heads are especially prone to loosening, which could cause damage to the head. Check them frequently and keep them tight.

23   Steering head bearings - check and adjustment

1   This motorcycle is equipped with caged ball type steering head bearings which can become dented, rough or loose during normal use of the machine. In extreme cases, worn or loose steering head bearings can cause steering wobble - a condition that is potentially dangerous.

Check

Refer to illustrations 23.2 and 23.4

2   A auxiliary stand will be needed to hold the motorcycle securely upright and allow the front wheel to be raised off the ground **(see illustration)**. Ensure that the rear of the machine is weighted or tied down so that the front wheel is off the ground

3   Point the wheel straight-ahead and slowly move the handlebars from side-to-side. Dents or roughness in the bearing races will be felt and the bars will not move smoothly.

4   Next, grasp the wheel and try to move it forward and backward **(see illustration)**. Any looseness in the steering head bearings will be

23.4 Grasp the front wheel and try to pull it back and forth; if it moves, the steering head bearings need adjustment

1

23.5 Remove the fork clamp bolt (arrow) from each side of the top triple clamp

felt as front-to-rear movement of the fork legs. If play is felt in the bearings, adjust the steering head as follows.

Adjustment

Refer to illustrations 23.5, 23.6, 23.8, 23.9, 23.10, 23.11a and 23.11b

5   Remove the front fork clamp bolts in the top triple clamp **(see illustration)**. The handlebar clamp bolts can remain in place.
6   Prise the plug out of the steering stem nut and remove the nut **(see illustration)**.
7   Gently ease the top triple clamp upwards off the fork tubes.
8   Prise the lockwasher tabs out of the slots in the locknut and adjuster nut **(see illustration)**. Unscrew the locknut using a C-spanner and discard the lockwasher; a new lockwasher must be used on reassembly.
9   Slacken the adjuster nut slightly until pressure is just released, then tighten it until all freeplay is removed, yet the steering is able to move freely **(see illustration)**. Note that Honda specify a torque setting for the adjuster nut - if this is applied, check afterwards that the steering is

still able to move freely from side to side. The object is to set the adjuster nut so that the bearings are under a very light loading, just enough to remove any freeplay. **Caution:** *Take great care not to apply excessive pressure because this will cause premature failure of the bearings.*
10 With the bearings correctly adjusted, fit a new lockwasher over the steering stem **(see illustration)**. Bend down two opposite lock washer tabs (the shorter tabs on the washer) into the grooves of the adjuster nut.
11 Hold the adjuster nut to prevent it from moving, then install the locknut and tighten it finger-tight. Tighten the locknut approximately 90° more until its slots align with the remaining lockwasher tabs **(see illustration)**. Secure the locknut in position by bending up the lock washer tabs into its slots **(see illustration)**.
12 Fit the top triple clamp to the steering stem, then install the nut and tighten it and both the fork clamp bolts to their specified torque settings (see Chapter 6). Fit the plug into the stem nut.
13 Check the bearing adjustment as described above and re-adjust if necessary.

24 Fasteners - check

1   Since vibration of the machine tends to loosen fasteners, all nuts, bolts, screws, etc. should be periodically checked for proper tightness.
2   Pay particular attention to the following:

Spark plugs
Engine oil drain plug
Gearshift lever
Footpegs and sidestand
Engine mounting bolts
Shock absorber mounting bolts
Handlebar and triple clamp bolts
Rear suspension linkage bolts
Front axle and clamp bolts
Rear axle nut
Exhaust system bolts/nuts

3   If a torque wrench is available, use it along with the torque specifications at the beginning of this, or other, Chapters.

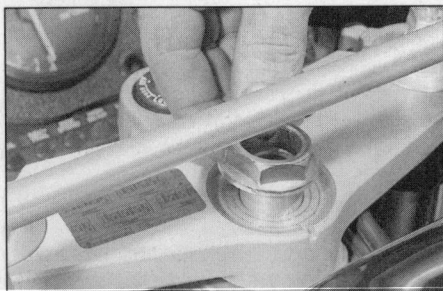

23.6 Remove the steering stem nut and lift off the top triple clamp

23.8 Bend down the lockwasher tabs to enable the locknut to be unscrewed

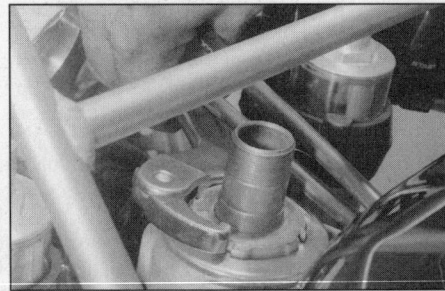

23.9 Use a C-spanner to adjust steering head bearing freeplay

23.10 Always use a new lockwasher and bend its shorter tabs down into the adjuster nut slots (arrow)

23.11a Tighten the locknut . . .

23.11b . . . and bend the remaining lockwasher tabs up into the locknut slots

25.5a Fuel tank gauze filter can be withdrawn for cleaning

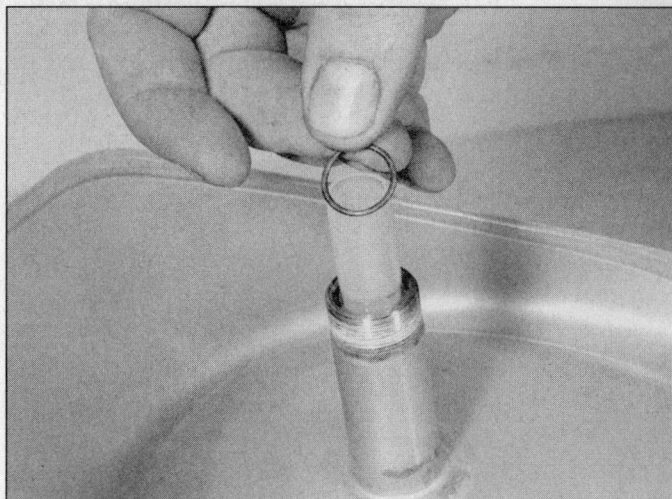

25.5b Always use a new O-ring between the tank and fuel tap to prevent leaks

---

25 Fuel system - checks and filter cleaning/replacement

Warning: *Gasoline (petrol) is extremely flammable, so take extra precautions when you work on any part of the fuel system. Don't smoke or allow open flames or bare light bulbs near the work area, and don't work in a garage where a natural gas-type appliance (such as a water heater or clothes dryer) is present. If you spill any fuel on your skin, rinse it off immediately with soap and water. When you perform any kind of work on the fuel system, wear safety glasses and have a fire extinguisher suitable for a Class B type fire (flammable liquids) on hand.*

Checks

1   Remove the fuel tank (see Chapter 4) and check the breather pipe (not California models) and drain pipe and the pipes from and to the fuel pump for signs of leakage. Replace any which are cracked or deteriorated.
2   If the carburetor gaskets are leaking, the carburetors should be disassembled and rebuilt by referring to Chapter 4.
3   If the fuel tap is leaking, tightening the retaining nut may help but if leakage persists, the tap should be disassembled and repaired or replaced with a new one.
4   Check that there is no fuel leakage from the fuel pump.

25.6 Fuel filter is fitted in pipe between tank and pump; arrow on filter body must point in direction of fuel flow

Filter cleaning/replacement

Refer to illustrations 25.5a, 25.5b and 25.6
5   The fuel filter, which is attached to the fuel tap, may become clogged and should be removed and cleaned periodically. In order to clean the filter, the fuel tank must be drained and the fuel tap removed as described in Chapter 4. Slip the filter out of the fuel tank and clean it thoroughly; if severely blocked or torn, replace the filter with a new one **(see illustration)**. Install a new O-ring on the fuel tap pipe, followed by the filter and install the fuel tap as described in Chapter 4 **(see illustration)**.
6   An in-line fuel filter is fitted in the fuel pipe from the tap to the fuel pump **(see illustration)**. Remove the fuel tank (see Chapter 4). Have a rag handy to soak up any residual fuel and disconnect the pipes from the filter. Slip the filter out of its bracket and install the new filter so that its arrow points in the direction of fuel flow (ie towards the pump). Secure the pipes to the filter with the retaining clips. Install the fuel tank (see Chapter 4), turn the tap ON and check that there are no fuel leaks.

---

26 Suspension - checks

1   The suspension components must be maintained in top operating condition to ensure rider safety. Loose, worn or damaged suspension parts decrease the vehicle's stability and control.

Front suspension

2   While standing alongside the motorcycle, apply the front brake and push on the handlebars to compress the forks several times. See if they move up-and-down smoothly without binding. If binding is felt, the forks should be disassembled and inspected (see Chapter 6).
3   Carefully inspect the area around the fork seals for any signs of fork oil leakage. If leakage is evident, the seals must be replaced as described in Chapter 6.
4   Although not specified as a maintenance item, the fork oil will degrade over a period of time and require renewal. This necessitates removal of the fork legs as described in Chapter 6.
5   Check the tightness of all suspension nuts and bolts to be sure none have worked loose.

Rear suspension

6   Inspect the rear shock for fluid leakage and tightness of its mountings. If leakage is found, the shock should be replaced.
7   Position the motorcycle on an auxiliary stand so that the rear wheel is off the ground **(see illustration 23.2)**. Grab the swingarm on each side, just ahead of the axle. Rock the swingarm from side to side - there should be no discernible movement at the rear. If there's a little

28.2a Headlight aim adjusters are situated at the top outer corner of each unit . . .

28.2b . . . and between the headlight units at the bottom (later type headlight shown)

movement or a slight clicking can be heard, make sure the swingarm pivot shaft locknut is tight. If the pivot nut is tight but movement is still noticeable, the swingarm pivot adjustment or bearings, or the suspenion linkage bearings require attention (see Chapter 6).

8　Inspect the tightness of the rear suspension nuts and bolts.

27　Sidestand - check

1　The sidestand return spring must be capable of retracting the stand fully and holding the stand retracted when the motorcycle is in use. If the spring is sagged or broken it must be replaced.

2　Lubricate the sidestand pivot regularly (see Section 15).

3　The sidestand switch prevents the motorcycle being started if the stand is extended. Check its operation by shifting the transmission into neutral, retracting the stand and starting the engine. Pull in the clutch lever and select a gear. Extend the sidestand. The engine should stop as the sidestand is extended. If the sidestand switch does not operate as follows, check its circuit as described in Chapter 9.

28　Headlight aim - check and adjustment

Refer to illustrations 28.2a and 28.2b

1　An improperly adjusted headlight may cause problems for oncoming traffic or provide poor, unsafe illumination of the road ahead. Before adjusting the headlight aim, be sure to consult with local traffic laws and regulations.

2　Adjustment is made on the back of the headlight unit, using the adjusters at the top corners and those between the individual headlights at the bottom **(see illustrations)**.

3　Due to the heat build-up in the headlight assembly on US 1995 models and UK CBR900RR-R, RR-S models (with plastic lens), vents are provided at several points in the back of the unit. Short rubber tubes fit over the vent holes and incorporate plastic mesh filters to prevent dust entering the headlight; check that these fitlers are not obstructed.

Chapter 2
Engine, clutch and transmission

Contents

Specifications

General

Capacity	893 cc
Bore	70 mm (2.76 in)
Stroke	58 mm (2.28 in)
Compression ratio	11.0 to 1

Camshafts

Intake cam lobe height
California models
Standard ... 34.940 to 35.180 mm (1.3756 to 1.3850 in)
Service limit ... 34.91 mm (1.374 in)
All other models
Standard ... 36.040 to 36.280 mm (1.4189 to 1.4283 in)
Service limit ... 36.01 mm (1.418 in)
Exhaust cam lobe height
California models
Standard ... 35.100 to 35.340 mm (1.3819 to 1.3913 in)
Service limit ... 35.07 mm (1.381 in)
All other models
Standard ... 35.800 to 36.040 mm (1.4094 to 1.4189 in)
Service limit ... 35.77 mm (1.408 in)
Camshaft bearing oil clearance
Standard ... 0.020 to 0.062 mm (0.0008 to 0.0024 in)
Service limit ... 0.1 mm (0.004 in)
Camshaft runout ... Less than 0.05 mm (0.002 in)
Camshaft follower OD
Standard ... 25.978 to 25.993 mm (1.0228 to 1.0233 in)
Service limit ... 25.97 mm (1.022 in)

Cylinder head

Maximum warpage ... 0.10 mm (0.004 in)
Cylinder head follower bore ID
Standard ... 26.010 to 26.026 mm (1.0240 to 1.0246 in)
Service limit ... 26.040 mm (1.0252 in)

Valves, guides and springs

Intake valve stem OD
    Standard.. 4.475 to 4.490 mm (0.1762 to 0.1768 in)
    Service limit... 4.465 mm (0.1758 in)
Exhaust valve stem OD
    Standard.. 4.465 to 4.480 mm (0.1758 to 0.1764 in)
    Service limit... 4.455 mm (0.1754 in)
Valve guide ID - intake and exhaust
    Standard.. 4.500 to 4.512 mm (0.1772 to 0.1776 in)
    Service limit... 4.540 mm (0.1787 in)
Valve stem-to-guide clearance
    Intake .... 0.010 to 0.037 mm (0.0004 to 0.0015 in)
    Exhaust .... 0.020 to 0.047 mm (0.0008 to 0.0019 in)
Valve seat width
    Standard.. 0.9 to 1.1 mm (0.035 to 0.043 in)
    Service limit... 1.5 mm (0.06 in)
Inner valve spring free length
    Standard.. 35.77 mm (1.408 in)
    Service limit... 34.07 mm (1.341 in)
Outer valve spring free length
    Standard.. 39.69 mm (1.563 in)
    Service limit... 37.79 mm (1.488 in)

Clutch

Friction plate thickness
    Standard.. 2.92 to 3.08 mm (0.115 to 0.121 in)
    Service limit... 2.60 mm (0.102 in)
Plain plate maximum warpage .... 0.3 mm (0.012 in)
Clutch spring free length
    Standard.. 45.5 mm (1.79 in)
    Service limit... 43.6 mm (1.72 in)
Clutch outer drum guide bush OD
    Standard.. 34.975 to 34.991 mm (1.3770 to 1.3776 in)
    Service limit... 34.965 mm (1.3766 in)
Clutch outer drum guide bush ID
    Standard.. 24.9935 to 25.0065 mm (0.98399 to 0.98451 in)
    Service limit... 24.96 mm (0.983 in)
Mainshaft OD at clutch guide bush contact point
    Standard.. 24.980 to 24.993 mm (0.9835 to 0.9840 in)
    Service limit... 24.96 mm (0.983 in)

Lubrication system

Oil pressure .... 71 psi (4.9 Bars) at 80°C (176°F)
Oil pump rotor tip-to-outer rotor clearance
    Standard.. 0.15 mm (0.006 in)
    Service limit... 0.20 mm (0.008 in)
Oil pump outer rotor-to-body clearance
    Standard.. 0.15 to 0.22 mm (0.006 to 0.009 in)
    Service limit... 0.35 mm (0.014 in)
Oil pump rotor endfloat
    Standard.. 0.02 to 0.07 mm (0.001 to 0.003 in)
    Service limit... 0.10 mm (0.004 in)

Starter clutch

Driven gear OD
    Standard.. 51.699 to 51.718 mm (2.0354 to 2.0361 in)
    Service limit... 51.684 mm (2.0348 in)

Shift drum and forks

Shift fork end thickness
    Standard.. 5.93 to 6.00 mm (0.233 to 0.236 in)
    Service limit... 5.90 mm (0.232 in)
Shift fork bore ID
    Standard.. 12.000 to 12.021 mm (0.4724 to 0.4733 in)
    Service limit... 12.030 mm (0.4736 in)
Shift fork shaft OD
    Standard.. 11.957 to 11.968 mm (0.4707 to 0.4712 in)
    Service limit... 11.95 mm (0.470 in)

Cylinder block

Cylinder bore ID	
Standard	70.000 to 70.015 mm (2.7559 to 2.7565 in)
Service limit	70.10 mm (2.760 in)
Maximum ovality (out-of-round)	0.10 mm (0.004 in)
Maximum taper	0.10 mm (0.004 in)
Cylinder-to-piston clearance	0.015 to 0.050 mm (0.0006 to 0.0020 in)
Maximum gasket face warpage	0.05 mm (0.002 in)

Pistons

Piston OD (measured 15 mm (0.6 in) up from base of skirt)	
Standard	69.965 to 69.985 mm (2.7545 to 2.7553 in)
Service limit	69.90 mm (2.752 in)
Piston pin bore OD	
Standard	17.002 to 17.008 mm (0.6694 to 0.6696 in)
Service limit	17.03 mm (0.670 in)
Piston pin OD (US 1993 and 1994 models, UK CBR900RR-N and RR-P)	
Standard	16.993 to 17.000 mm (0.6690 to 0.6693 in)
Service limit	16.98 mm (0.669 in)
Piston pin OD (US 1995 model, UK CBR900RR-R and RR-S)	
Standard	16.994 to 17.000 mm (0.6691 to 0.6693 in)
Service limit	16.98 mm (0.669 in)
Piston-to-piston pin clearance	
US 1995 model	0.002 to 0.014 mm (0.0001 to 0.0006 in)
All other models	0.002 to 0.015 mm (0.0001 to 0.0006 in)
Connecting rod to piston pin clearance	
US 1995 model	0.016 to 0.040 mm (0.0006 to 0.0016 in)
All other models	0.016 to 0.041 mm (0.0006 to 0.0016 in)

Piston rings

Top ring-to-groove clearance	
Standard	0.015 to 0.050 mm (0.0006 to 0.0020 in)
Service limit	0.10 mm (0.004 in)
Top ring end gap	
Standard	0.20 to 0.35 mm (0.008 to 0.014 in)
Service limit	0.5 mm (0.02 in)
Second ring-to-groove clearance	
Standard	0.015 to 0.045 mm (0.0006 to 0.0018 in)
Service limit	0.10 mm (0.004 in)
Second ring end gap	
Standard	0.40 to 0.55 mm (0.016 to 0.022 in)
Service limit	0.7 mm (0.03 in)
Oil control ring side rail end gap	
Standard	0.2 to 0.8 mm (0.01 to 0.03 in)
Service limit	1.0 mm (0.04 in)

Connecting rods and bearings

Connecting rod side clearance	
Standard	0.05 to 0.20 mm (0.002 to 0.008 in)
Service limit	0.3 mm (0.012 in)
Connecting rod piston pin bore ID	
Standard	17.016 to 17.034 mm (0.6699 to 0.6706 in)
Service limit	17.04 mm (0.671 in)
Connecting rod crankpin bore ID	
Size group 1	39.000 to 39.006 mm (1.5354 to 1.5357 in)
Size group 2	39.006 to 39.012 mm (1.5357 to 1.5359 in)
Size group 3	39.012 to 39.018 mm (1.5359 to 1.5361 in)
Crankshaft crankpin OD	
Size group A	35.994 to 36.000 mm (1.4171 to 1.4173 in)
Size group B	35.988 to 35.994 mm (1.4168 to 1.4171 in)
Size group C	35.982 to 35.988 mm (1.4166 to 1.4168 in)
Connecting rod bearing oil clearance	
Standard	0.030 to 0.052 mm (0.0012 to 0.0020 in)
Service limit	0.06 mm (0.002 in)

Crankshaft and main bearings

Maximum crankshaft runout	0.05 mm (0.002 in)
Crankcase main bearing bore ID	
Size group A	37.000 to 37.006 mm (1.4566 to 1.4569 in)
Size group B	37.006 to 37.012 mm (1.4569 to 1.4572 in)
Size group C	37.012 to 37.018 mm (1.4572 to 1.4574 in)

2

Crankshaft and main bearings - continued

Crankshaft journal OD
- Size group 1 .... 32.992 to 33.000 mm (1.3386 to 1.3388 in)
- Size group 2 .... 32.984 to 32.992 mm (1.3383 to 1.3386 in)
- Size group 3 .... 33.988 to 33.994 mm (1.3381 to 1.3383 in)

Main bearing oil clearance
- Standard .... 0.017 to 0.035 mm (0.0007 to 0.0014 in)
- Service limit (all models) .... 0.04 mm (0.002 in)

Transmission shafts

Ratios
- 1st .... 2.7692 to 1 (36/13T)
- 2nd .... 2.0000 to 1 (32/16T)
- 3rd .... 1.5789 to 1 (30/19T)
- 4th .... 1.4000 to 1 (28/20T)
- 5th .... 1.2500 to 1 (25/50T)
- 6th .... 1.1739 to 1 (27/23T)

Gear ID
Mainshaft 5th and 6th gears
- Standard .... 28.000 to 28.021 mm (1.1024 to 1.1032 in)
- Service limit .... 28.04 mm (1.104 in)

Countershaft 1st gear
- Standard .... 24.000 to 24.021 mm (0.9449 to 0.9457 in)
- Service limit .... 24.04 mm (0.946 in)

Countershaft 2nd, 3rd and 4th gears
- Standard .... 31.000 to 31.025 mm (1.2205 to 1.2215 in)
- Service limit .... 31.04 mm (1.222 in)

Gear bushing OD
Mainshaft 5th and 6th gears
- Standard .... 27.959 to 27.980 mm (1.1007 to 1.1016 in)
- Service limit .... 27.94 mm (1.100 in)

Countershaft 3rd and 4th gears
- Standard .... 30.955 to 30.980 mm (1.2187 to 1.2197 in)
- Service limit .... 30.93 mm (1.218 in)

Countershaft 2nd gear
- Standard .... 30.959 to 30.980 mm (1.2189 to 1.2197 in)
- Service limit .... 30.94 mm (1.218 in)

Gear bushing ID
Mainshaft 5th gear
- Standard .... 24.985 to 25.006 mm (0.9837 to 0.9845 in)
- Service limit .... 25.02 mm (0.985 in)

Countershaft 2nd gear
- Standard .... 27.985 to 28.006 mm (1.1018 to 1.1026 in)
- Service limit .... 28.02 mm (1.103 in)

Gear-to-bushing clearance
- Mainshaft 5th and 6th gear .... 0.020 to 0.062 mm (0.0008 to 0.0024 in)
- Countershaft 2nd gear .... 0.020 to 0.070 mm (0.0008 to 0.0028 in)
- Countershaft 3rd and 4th gear .... 0.020 to 0.075 mm (0.0008 to 0.0030 in)

Mainshaft OD at 5th gear bushing point
- Standard .... 24.967 to 24.980 mm (0.9830 to 0.9835 in)
- Service limit .... 24.96 mm (0.983 in)

Mainshaft OD at clutch outer guide
- Standard .... 24.980 to 24.993 mm (0.9835 to 0.9840 in)
- Service limit .... 24.95 mm (0.982 in)

Countershaft OD at 2nd gear bushing point
- Standard .... 27.967 to 27.980 mm (1.0904 to 1.1016 in)
- Service limit .... 27.96 mm (1.101 in)

Shaft-to-bushing clearance
- Mainshaft 5th gear and countershaft 2nd gear .... 0.005 to 0.039 mm (0.0002 to 0.0015 in)

Torque settings

	Nm	ft-lbs
Engine mountings		
Rear upper mounting nut	39	28
Rear upper mounting bracket bolts	27	20
Rear lower mounting nut	39	28
Rear lower mounting adjuster	11	8
Rear lower mounting adjuster locknut	54	39
Left side front mounting bolts	39	28
Right side front mounting bolt	44	32

Engine sprocket bolt	54	39
Valve cover bolts	10	7
Valve cover breather plate bolts	12	9
Cam chain tensioner bolts	12	9
Camshaft holder bolts	12	9
Camshaft sprocket bolts	20	14
Cylinder head bolts (see text)	48	35
Clutch center nut	130	94
Clutch cover bolts	12	9
Oil pump sprocket bolt	15	11
Oil cooler center bolt	64	46
Gearshift drum cam bolt	23	17
Gearshift shaft return spring anchor post	22	16
Starter clutch bolts	16	12
Alternator rotor bolt	103	74
Alternator stator bolts	12	9
Crankshaft end cover bolts	12	9
Crankshaft cap	18	13
Pulse generator rotor bolt	59	43
Oil pressure switch	12	9
Neutral switch	12	9
Crankcase bolts		
6 mm bolts	12	9
8 mm bolts	24	18
9 mm bolts	36	26
10 mm bolt	39	28
Connecting rod bearing cap nuts	34	25

1 General information

The engine/transmission unit is of water-cooled four-cylinder in-line design, fitted transversely across the frame. The sixteen valves are operated by double overhead camshafts, chain driven off the right end of the crankshaft. The engine/transmission unit is constructed in aluminum alloy with the crankcase being divided horizontally. The crankcase incorporates a wet sump, pressure fed lubrication system, and houses a chain driven dual rotor oil pump.

The alternator is situated on the left end of the crankshaft with the starter clutch being built into the rear of the alternator rotor. The water pump is mounted on the left side of the crankcase and is driven off the oil pump shaft.

The clutch is of the wet multi-plate type and is gear driven off the crankshaft. The transmission is of the six-speed constant mesh type. Final drive to the rear wheel is by chain and sprockets.

2 Operations possible with the engine in the frame

The components and assemblies listed below can be removed without having to remove the engine/transmission assembly from the frame. If however, a number of areas require attention at the same time, removal of the engine is recommended.

    Valve cover
    Cam chain tensioner
    Camshafts
    Cam chain, sprockets and guides
    Cylinder head
    Starter motor
    Starter motor clutch
    Alternator
    Clutch assembly
    Gearshift mechanism external components
    Oil pan, oil pump and relief valves
    Shift drum and forks - not advised due to poor access

3 Operations requiring engine removal

It is necessary to remove the engine/transmission assembly from the frame and separate the crankcase halves to gain access to the following components.

    Transmission shafts
    Crankshaft and bearings
    Piston/connecting rod assemblies and bearings
    Shift drum and forks (see Section 2)

4 Major engine repair - general note

1 It is not always easy to determine when or if an engine should be completely overhauled, as a number of factors must be considered.

2 High mileage is not necessarily an indication that an overhaul is needed, while low mileage, on the other hand, does not preclude the need for an overhaul. Frequency of servicing is probably the single most important consideration. An engine that has regular and frequent oil and filter changes, as well as other required maintenance, will most likely give many miles of reliable service. Conversely, a neglected engine, or one which has not been broken in properly, may require an overhaul very early in its life.

3 Exhaust smoke and excessive oil consumption are both indications that piston rings and/or valve guides are in need of attention, although make sure that the fault is not due to oil leakage. Refer to Chapter 1 and perform a cylinder compression check to determine for certain the nature and extent of the work required.

4 If the engine is making obvious knocking or rumbling noises, the connecting rod and/or main bearings are probably at fault.

5 Loss of power, rough running, excessive valve train noise and high fuel consumption rates may also point to the need for an overhaul, especially if they are all present at the same time. If a complete tune-up does not remedy the situation, major mechanical work is the only solution.

6 An engine overhaul generally involves restoring the internal parts to the specifications of a new engine. During an overhaul the piston rings are replaced and the cylinder walls are bored and/or honed. If a rebore is done, then new pistons will also be required. The main and

2

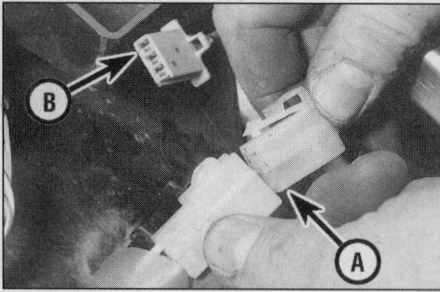

5.13a Disconnect alternator wiring (A) and side stand switch wiring (B) under seat

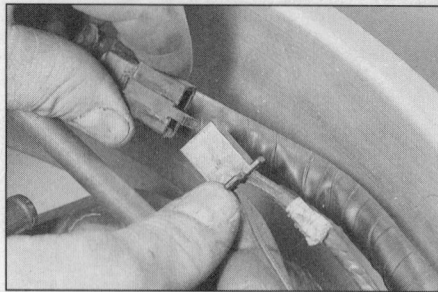

5.13b Temperature sender, neutral, and oil pressure wires share a 3-pin connector

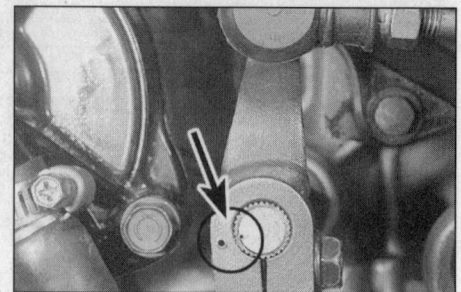

5.14 Make alignment marks (arrow) on gearshift lever and linkage crank

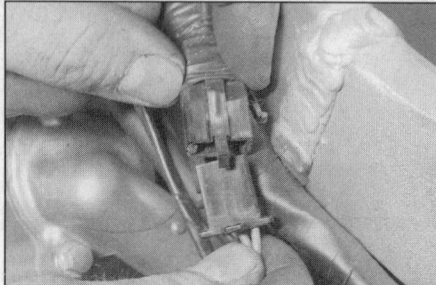

5.15a Disconnect electronic speedometer wiring at 3-pin connector inside frame . . .

5.15b . . . and free drive unit from engine sprocket cover (later models)

connecting rod bearings are usually replaced during a major overhaul. Generally the valve seats are serviced as well, since they are usually in less than perfect condition at this point. While the engine is being overhauled, other components such as the carburetors and the starter motor can also be rebuilt. The end result should be a like new engine that will give as many trouble-free miles as the original.

7   Before beginning the engine overhaul, read through the related procedures to familiarize yourself with the scope and requirements of the job. Overhauling an engine is not all that difficult, but it is time consuming. Plan on the motorcycle being tied up for a minimum of two weeks. Check on parts availability and make sure that any necessary special tools, equipment and supplies are obtained in advance.

8   Most work can be done with typical shop hand tools, although a number of precision measuring tools are required for inspecting parts to determine if they must be replaced. Often a dealer service department or motorcycle repair shop will handle the inspection of parts and offer advice concerning reconditioning and replacement. As a general rule, time is the primary cost of an overhaul so it does not pay to install worn or substandard parts.

9   As a final note, to ensure maximum life and minimum trouble from a rebuilt engine, everything must be assembled with care in a spotlessly clean environment.

5   Engine - removal and installation

Note: *Engine removal and installation should be carried out with the aid of an assistant; personal injury or damage could occur if the engine falls or is dropped. An hydraulic floor-type jack should be used to support and lower the engine to the floor if possible (they can be rented at low cost).*

Removal

Refer to illustrations 5.13a, 5.13b, 5.14, 5.15a, 5.15b, 5.17a, 5.17b, 5.17c, 5.18a through 5.18d, 5.20, 5.21, 5.22, 5.23a through 5.23e, 5.24 and 5.25

1   Since no center stand is fitted, the machine must be supported in an upright position using an auxiliary stand so it can't be accidentally knocked over while the engine is removed **(see illustration 23.2 in Chapter 1).** Work can also be made easier by raising the machine to a suitable working height on a hydraulic ramp or a suitable platform.

2   Remove the fairing lower section and both middle sections (see

Chapter 8). Additionally, to avoid damage to the inner middle panels which extend rearwards towards the frame, its is advised that they also be removed as described in Chapter 8.

3   If the engine is dirty, particularly around its mountings, wash it thoroughly before starting any major dismantling work. This will make work much easier and rule out the possibility of caked on lumps of dirt falling into some vital component.

4   Remove the fuel tank as described in Chapter 4.

5   Drain the engine oil and coolant as described in Chapter 1.

6   Remove the radiator and its mounting bracket (see Chapter 3).

7   Remove the rider's seat (see Chapter 8). Disconnect both battery cables from the battery. **Warning:** *Always disconnect the battery negative lead first and reconnect it last to prevent a battery explosion.*

8   Remove the exhaust system as described in Chapter 4.

9   Remove the air filter housing and carburetors as described in Chapter 4 and plug the intake manifold joints with clean shop towels.

10   Disconnect the spark plug caps from the plugs and pull the HT leads up through the dust cover.

11   Disconnect the clutch inner cable from the release lever as described in Section 15.

12   Peel back the rubber cover then undo the nut and disconnect the lead from the starter motor, freeing it from the wire clamp. Screw the nut back onto the starter motor terminal for safe-keeping. Remove the forward mounting bolt from the starter motor to free the earth lead; insert the bolt back into the crankcase for safe-keeping.

13   Trace the wiring back from the alternator (three yellow wires) and side stand switch (green, yellow/black and green/white wires) and disconnect the connectors where they join the harness under the seat **(see illustration).** Disconnect the three-pin wire connector on the inside of the frame right top rail; this contains the neutral, oil pressure and coolant temperature sender wires **(see illustration).** Trace the wiring up from the pulse generator (yellow and yellow/white wires) and disconnect it where it joins the main wire harness **(see illustration 4.2 , Chapter 5).** Work back along the wiring, releasing it from any relevant retaining clips so that it is free to be removed with the engine unit.

14   Remove the clamp bolt and disconnect the gearshift lever or linkage crank (as applicable) from the engine **(see illustration).**

15   On later models with an electronically-driven speedometer, disconnect the three-pin connector from inside the frame top rail, then remove the two bolts to free the drive unit from the drive sprocket cover **(see illustrations).**

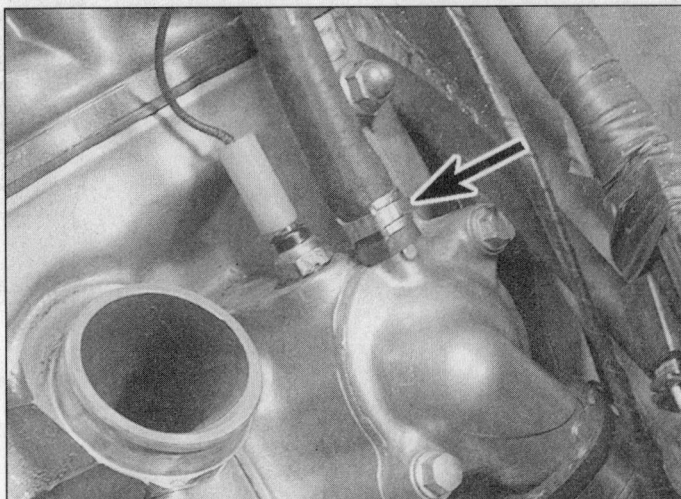

5.17a Release its clamp and pull the air bleed hose (arrow) off the thermostat housing . . .

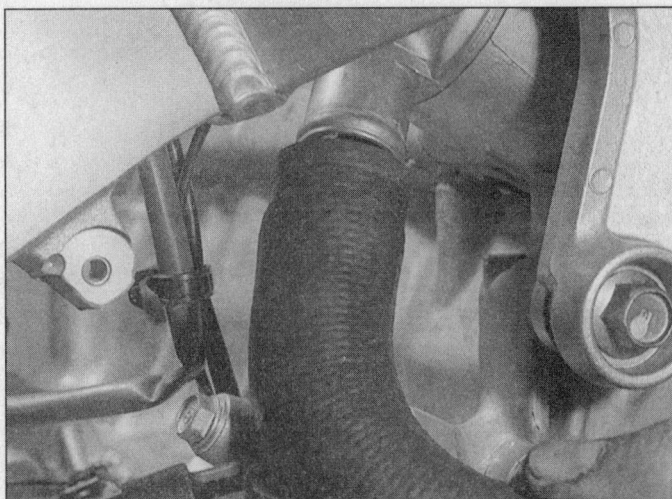

5.17b . . . radiator top hose can be pulled off the thermostat housing after slackening its clamp

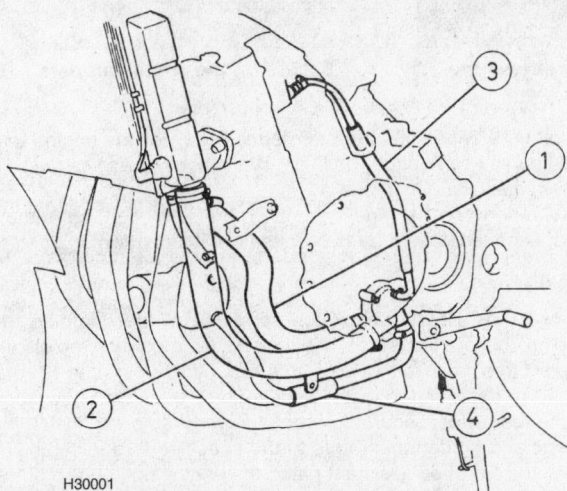

5.17c Coolant hose detail on engine left side

1 Water pump-to-coolant union hose
2 Radiator bottom hose
3 Air bleed hose
4 Oil cooler-to-water pump hose

5.18a Remove the fairing mounting brackets at the engine front right . . .

16 Remove the drive sprocket cover, chain guide plate and drive sprocket as described in Chapter 6.

17 Release their retaining clips or hose clamps and disconnect the cooling system hoses from the thermostat housing (see illustrations). The coolant hoses on the left side of the engine can be left in place, although removal will be necessary if a complete engine overhaul is planned (see illustration).

18 To prevent them being distorted as the engine is manhandled out of the frame, remove the fairing mounting brackets from the engine (front) and frame (rear) (see illustrations).

19 Position a jack and block of wood beneath the engine and raise the jack so that it is supporting the weight of the engine/transmission unit (see illustration 5.24).

20 Remove the nut from the rear lower engine mounting bolt on the

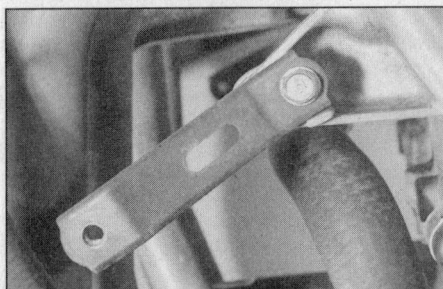

5.18b . . . and front left . . .

5.18c . . . then at the frame right . . .

5.18d . . . and frame left sides

5.20 Engine rear lower mounting bolt nut

5.21 Engine front mounting bolt - right side

5.22 Engine front mounting bolts - left side

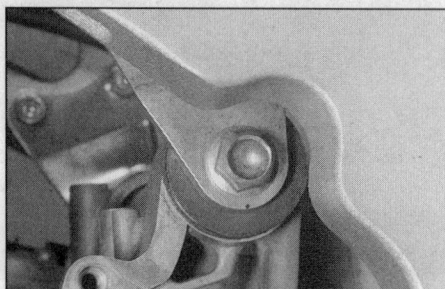

5.23a Engine rear upper mounting bolt is removed from the left side

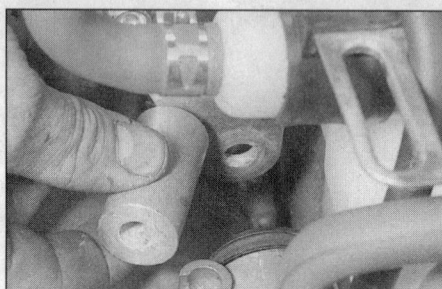

5.23b As it is withdrawn, retrieve the short collar . . .

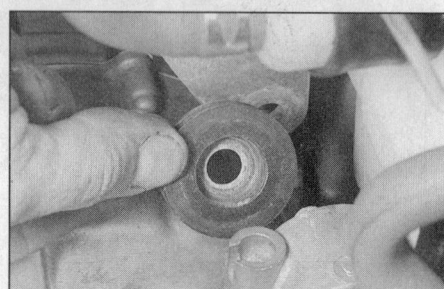

5.23c . . . the small dampers . . .

right side of the frame and withdraw the mounting bolt from the left side **(see illustration)**. Note that an adjuster and locknut are located on the right side of this mounting point. Its purpose is to take up play in the engine's lower mounting so that no stress is placed on the frame. Engine removal was found to be possible without disturbing the adjuster position, but if not, slacken off the locknut and back off the adjuster a few turns.

21 Check that the jack is supporting the weight of the engine (see Step 20) and remove the front right engine mounting bolt, retrieving its spacer from between the frame and engine **(see illustration)**.

22 Remove the two front left engine mounting bolts **(see illustration)**.

23 Remove the nut from the rear upper engine mounting bolt on the right side of the frame and withdraw the bolt from the left **(see illustration)**. Retrieve the short collar from the right side, the two small dampers positioned on each side of the engine and the long center collar between the engine lugs **(see illustrations)**. Damping rubbers are located in the engine lugs **(see illustration)**.

24 Have an assistant steady the engine as it is lowered on the jack until it is clear of the frame lugs, then move it to one side **(see illustration)**. Lift the engine unit off the jack and lower it carefully on to the work surface, taking not to break the long fins cast onto the bottom

of the oil pan. These fins are there specifically for the engine to stand on and keep it in an upright position. **Warning:** *The engine unit is heavy and may cause injury if it falls.*

25 The rear upper engine mounting brackets can be removed if desired; each is retained by two bolts **(see illustration)**.

Installation

26 Prior to installation check the condition of the damping rubbers located in the crankcase at the rear upper mounting point and the dampers fitted on each side of this mounting poin (see Step 23). If damaged or deteriorated in any way, replace them with new ones.

27 If the rear upper mounting brackets were removed, install them at this stage and tighten their bolts to the specified torque setting.

28 With the aid of an assistant place the engine unit on top of the jack and block of wood and carefully raise it into position in the frame.

29 Locate the center spacer between the damping rubbers in the engine lugs. Position a damper between the left mounting bracket and the engine, and slide the rear upper mounting bolt partway in. Position the short spacer and second damper between the right mounting bracket and engine and push the bolt fully into position. Thread the nut onto the bolt.

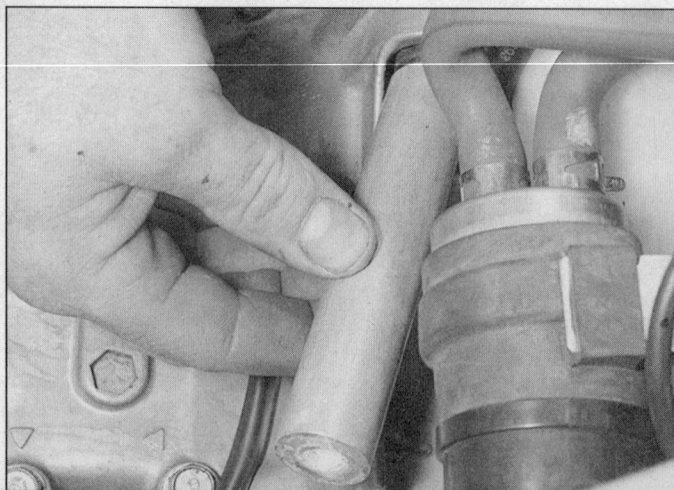

5.23d . . . and the long center collar

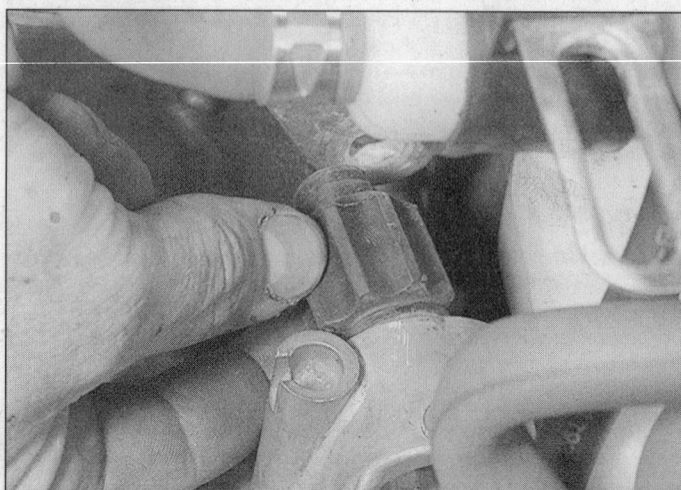

5.23e Slip the damping rubbers out of the engine mounting lugs

5.24 With the engine supported on a jack, carefully lower if out of the frame

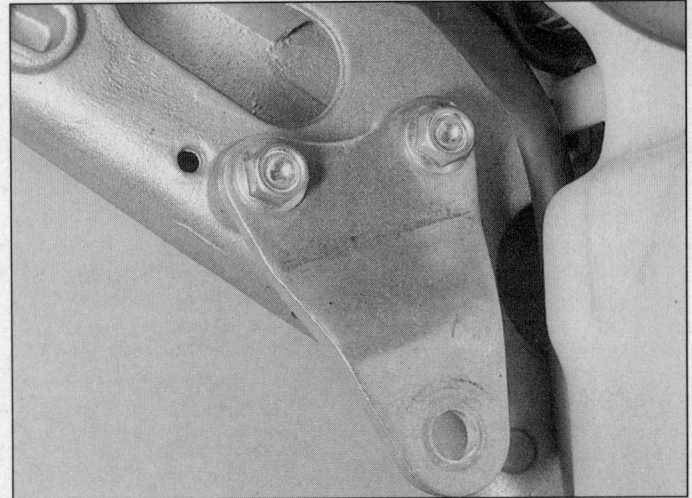

5.25 Engine rear upper mounting brackets are retained to inside of the frame by two bolts

30 Slip the spacer between the engine and frame right front mounting point and thread the right front mounting bolt into the engine unit. **Note:** *Don't mix this bolt with those on the left side - it should be 47 mm long.*

31 Install the front left mounting bolts, both 40 mm long.

32 Install the rear lower engine mounting bolt from the left side. At this point tighten the rear upper mounting bolt nut and all three front mounting bolts to their specified torque settings. If the adjuster on the rear lower mounting bolt was disturbed on removal, it must be tightened to the specified torque, then held steady while the locknut is tightened to its specified torque (Honda produce a special crowsfoot type wrench (Pt. No. 07HMA-MR70200) to enable the locknut to be secured to the correct torque). Finally, tighten the lower engine mounting bolt nut to the specified torque setting.

33 The remainder of the installation procedure is a direct reversal of the removal sequence, noting the following points.

   a) *Tighten all nuts and bolts to the specified torque settings (where given).*

   b) *Install the engine sprocket so that its marked side (ie 16T) faces outwards.*

   c) *Align the punch mark on the gearshift lever or linkage crank with that on the gearshift shaft.*

   d) *Make sure all wiring is correctly routed and retained by all the relevant clips and ties.*

   e) *Adjust the drive chain as described in Chapter 1.*

   f) *Fill the engine oil and cooling systems as described in Chapter 1.*

   g) *Prior to installing the fairing, start the engine and check for signs of coolant/oil leakage.*

6 Engine disassembly and reassembly - general information

Note: *Refer to the "Maintenance techniques, tools and working facilities" in the Introductory pages of this manual for further information.*

Disassembly

1 Before disassembling the engine, the external surfaces of the unit should be thoroughly cleaned and degreased. This will prevent contamination of the engine internals, and will also make working a lot easier and cleaner. A high flash-point solvent, such as kerosene (paraffin) can be used, or better still, a proprietary engine degreaser. Use old paintbrushes and toothbrushes to work the solvent into the various recesses of the engine casings. Take care to exclude solvent or water from the electrical components and intake and exhaust ports. **Warning:** *The use of gasoline (petrol) as a cleaning agent should be avoided because of the risk of fire.*

2 When clean and dry, arrange the unit on the workbench, leaving suitable clear area for working. Gather a selection of small containers and plastic bags so that parts can be grouped together in an easily identifiable manner. Some paper and a pen should be on hand to permit notes to be made and labels attached where necessary. A supply of clean shop towels is also required.

3 Before commencing work, read through the appropriate section so that some idea of the necessary procedure can be gained. When removing various engine components it should be noted that great force is seldom required, unless specified. In many cases, a component's reluctance to be removed is indicative of an incorrect approach or removal method. If in any doubt, re-check with the text.

4 When disassembling the engine, keep "mated" parts together (including gears, cylinders, pistons, valves, etc. that have been in contact with each other during engine operation). These "mated" parts must be reused or replaced as an assembly.

5 Engine/transmission disassembly should be done in the following general order with reference to the appropriate Sections.

   Remove the camshafts
   Remove cam chain, lower sprocket and guides
   Remove the cylinder head
   Remove the clutch
   Remove the oil pump
   Remove the external shift mechanism
   Remove the alternator
   Remove the starter motor (see Chapter 9)
   Remove the water pump (See Chapter 3)
   Remove the oil pan
   Separate the crankcase halves
   Remove the connecting rod/piston assemblies
   Remove the crankshaft
   Remove the transmission shafts/gears
   Remove the shift cam/forks

Reassembly

6 Reassembly is accomplished by reversing the general disassembly sequence.

7 Valve cover - removal and installation

Note: *The valve cover can be removed with the engine in the frame. If the engine has been removed, ignore the steps which do not apply.*

Removal

1 Remove the fairing middle sections as described in Chapter 8.

2 Remove the fuel tank and air filter housing (see Chapter 4).

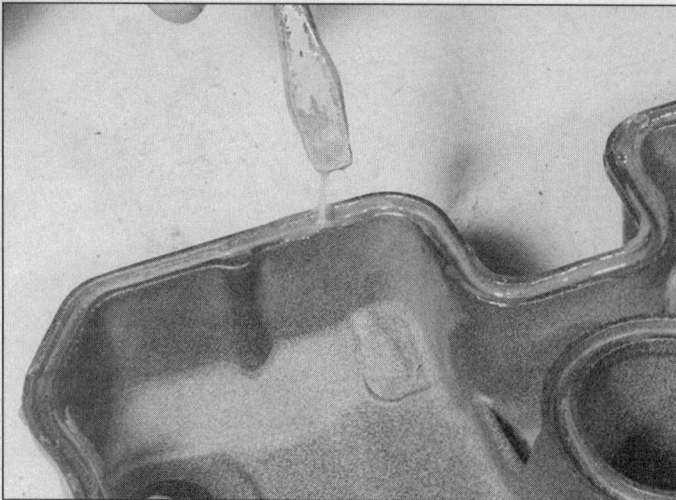

7.11a Apply adhesive to valve cover groove . . .

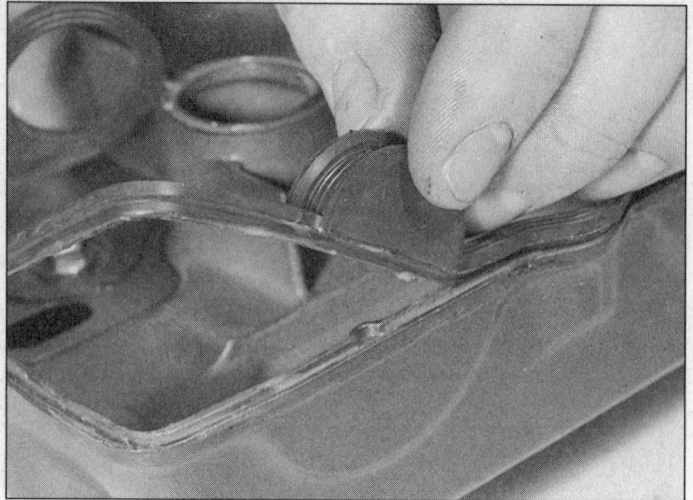

7.11b . . . when tacky press the new cover seal into place

3   Remove the ignition HT coils (see Chapter 5).

4   Release the rubber dust cover from the tabs of the air deflector, then remove the two bolts which retain the air deflector to the frame. Maneuver the air deflector out from under the frame cross-member **(see illustrations 14.5a through 14.5c in Chapter 1)**.

5   Disconnect the spark plug caps and position them clear of the cover.

6   Remove the radiator as described in Chapter 3; this will allow access for the valve cover to be withdrawn forwards out of the frame.

7   Slacken and remove the six cover bolts with their sealing washers.

8   Lift the valve cover from the head and maneuver it out of the frame.

9   Examine the rubber seal for signs of damage or deterioration and replace if it is cracked or brittle. Also check the cover bolt seals for signs of damage and replace if necessary. The crankcase breather plate can be unbolted from inside the valve cover if required.

Installation

Refer to illustrations 7.11a, 7.11b, 7.12, 7.13, 7.14a, 7.14b and 7.15

10   If the breather plate was removed, apply a drop of non-permanent

thread locking compound to the bolt threads and tighten them to the specified torque setting.

11   If a new seal is being fitted, note that it must be stuck into the cover groove using a suitable adhesive. Apply the adhesive to the cover groove, and when tacky, stick the seal in place **(see illustrations)**.

12   Apply a smear of sealant (not adhesive) to the half circle cutouts in the cylinder head. Carefully install the valve cover on the cylinder head **(see illustration)**.

13   Fit the sealing washers to the valve cover making sure the "UP" mark on each one is facing upwards **(see illustration)**.

14   Install the six cover retaining bolts and tighten them to the specified torque setting, noting that those next to the triangular markings on the cover rear outer bolt holes should be tightened first **(see illustrations)**.

15   The remainder of installation is the reverse of removal. If the breather hose was detached from the valve cover, push it on the union and secure with the spring clip **(see illustration)**.

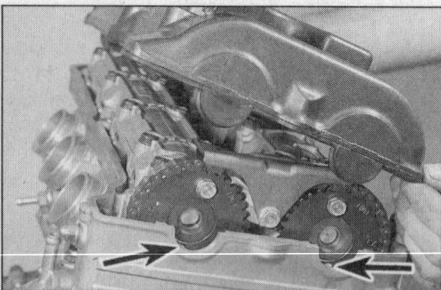

7.12 Smear sealant on the cutouts (arrows) before installing the valve cover

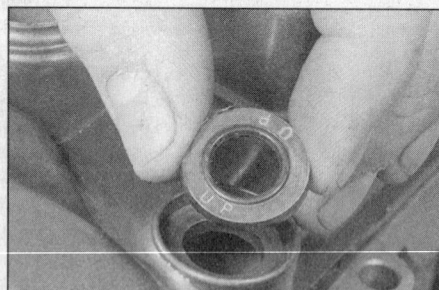

7.13 Sealing washer UP markings must face upwards

7.14a Install the valve cover bolts . . .

7.14b . . . and tighten those next to the triangular markings first (arrow)

7.15 Attach the breather hose to the valve cover - secure with the spring clip

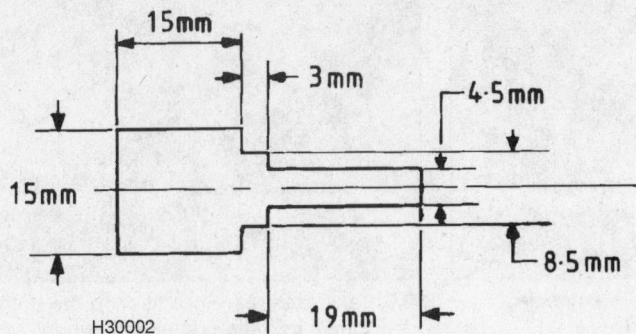

8.2a A copy of Honda's tensioner locking key can be made from a piece of 1 mm mild steel

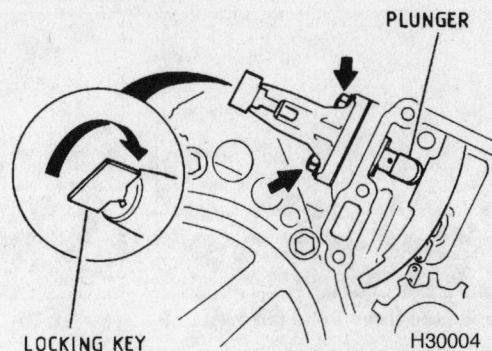

8.2b Locking key in position and tensioner plunger shown retracted. Tensioner mounting bolts arrowed

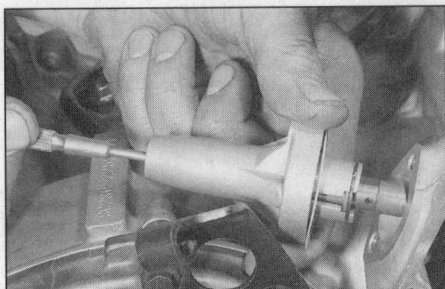

8.3 Tensioner plunger can be retracted using a flat-bladed screwdriver as shown

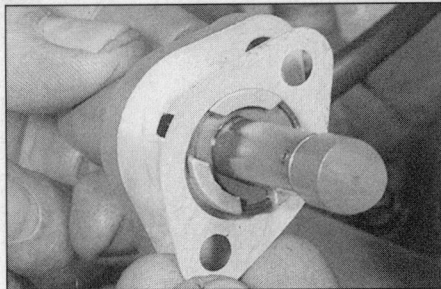

8.5 Install a new gasket at the tensioner-to-cylinder block joint

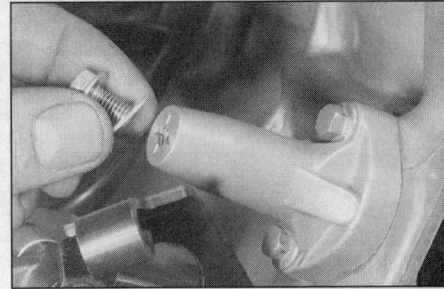

8.7 Use a new sealing washer on the tensioner end bolt

8 Cam chain tensioner - removal and installation

Removal

Refer to illustrations 8.2a, 8.2b and 8.3

1 Remove the fairing right middle section (see Chapter 8). Unbolt the fairing support bracket from the frame (see illustration 5.18c).

2 Unscrew the bolt and sealing washer from the end of the cam chain tensioner. If the Honda locking key or a home-made equivalent is available, insert it in the end of the tensioner so that it engages the slotted plunger. Turn the plunger fully clockwise, then push the key into the end of the tensioner body to hold it in position (see illustrations). Remove the two bolts to free the tensioner from the engine.

3 If the locking key is not available, use a small flat-bladed screwdriver to rotate the plunger fully clockwise and hold it in this position whilst the tensioner body bolts are removed (see illustration). The plunger will spring back out once the screwdriver is removed, but can be easily reset on installation.

4 Discard the gasket; a new one must be used on installation. Do not dismantle the tensioner.

Installation

Refer to illustrations 8.5 and 8.7

5 Ensure the tensioner and cylinder block surfaces are clean and dry and fit a new gasket to the tensioner (see illustration).

6 If the locking key described above is available, insert it in the tensioner body and rotate the plunger fully clockwise to retract it into the body, then push the key shoulders into the end of the body to lock it. Install the tensioner on the engine and tighten its bolts to the specified torque setting. Remove the key and install the tensioner end bolt with a new sealing washer.

7 If the key is not available, rotate the plunger fully clockwise with a flat-bladed screwdriver and hold it in this position whilst the tensioner bolts are installed and fully tightened to the specified torque setting. Install the tensioner end bolt with a new sealing washer (see illustration).

8 Install the fairing support bracket and fairing panel (see Chapter 8).

9 Camshafts and followers - removal, inspection and installation

Note: *This procedure can be carried out with the engine in the frame.*

Removal

Refer to illustrations 9.4, 9.5a, 9.5b, 9.6, 9.7 and 9.9

1 Remove the valve cover as described in Section 7.

2 Remove the fairing lower section as described in Chapter 8.

3 Remove the cam chain tensioner as described in Section 8.

4 Unscrew the inspection cap from the crankshaft right end cover and rotate the crankshaft bolt clockwise until the line next to the T mark on the ignition rotor aligns with the notch cut in the cover (see illustration). The valves for cylinder No. 1 should all be closed; if not, rotate the crankshaft 360° and realign the T mark with the notch.

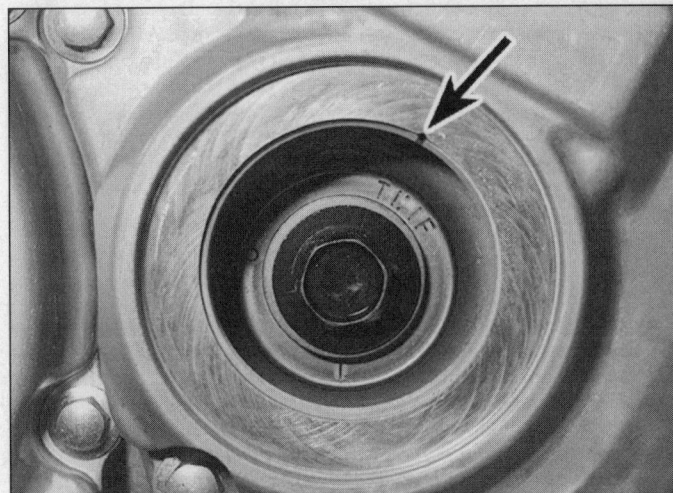

9.4 Align the line next to the T mark on the ignition rotor with the notch (arrow) in the cover

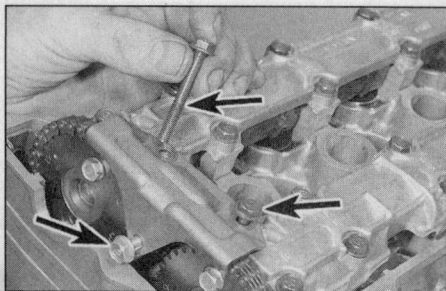

9.5a Remove the three bolts (arrows) . . .

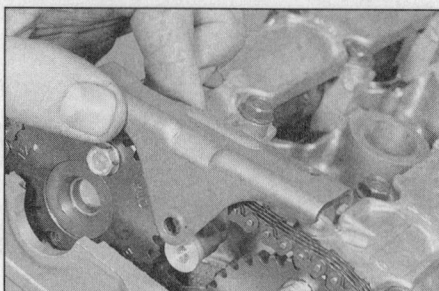

9.5b . . . to free the cam chain upper guide

9.6 Take great care not to drop the bolts down into the engine unit when unscrewing the camshaft sprocket bolts

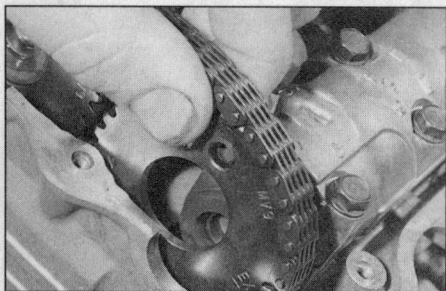

9.7 Disengage each camshaft sprocket from the chain and remove them from the engine

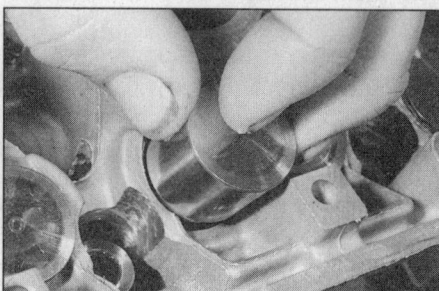

9.9 Lift each cam follower out of its bore

9.10 Check the lobes for wear - damage like this will require replacement (or repair) of the camshaft

5   Remove the three bolts which secure the cam chain upper guide and remove the guide **(see illustrations)**.

6   Unscrew the two visible camshaft sprocket retaining bolts **(see illustration)**. **Note:** *Take care not to drop the bolts down into the engine unit as they are removed. If a bolt is dropped, it must be recovered before the engine can be started - drain the engine oil and remove the right crankshaft end cover to recover the bolt.*

7   Rotate the crankshaft clockwise until the remaining two camshaft bolts are accessible, then remove them. Repositon the crankshaft as described in Step 4, then slip the sprockets off the camshaft ends and disengage them from the cam chain **(see illustration)**. **Note:** *Use a felt-tipped marker pen, label the sprockets IN and EX to aid installation - they are otherwise identical.*

8   In a **reverse** of the numbered sequence stamped on the intake camshaft holder, slacken all bolts evenly, a little at a time so that the holder is released without danger of distortion. Retrieve the two dowels if they are loose. Carry out the same procedure to remove the exhaust camshaft holder. Lift each camshaft from the cylinder head.

9   Divide a suitable container into sixteen compartments, labelling each compartment with the number of its corresponding valve in the cylinder head. Pick each follower out of the cylinder head (using a magnet if necessary) and store it in its compartment in the container **(see illustration)**. Note that the shim is likely to stick to the inside of the follower so take great care not to lose it as the follower is removed. Remove the shims and store each one with its respective follower.

Inspection

Refer to illustrations 9.10 and 9.12

Note: *Before replacing the camshafts or the cylinder head and camshaft holders because of damage, check with local machine shops specializing in motorcycle engineering work. In the case of the camshafts, it may be possible for cam lobes to be welded, reground and hardened, at a cost far lower than that of a new camshaft. If the bearing surfaces in the cylinder head are damaged, it may be possible for them to be bored out to accept bearing inserts. Due to the cost of a new cylinder head it is recommended that all options be explored before condemning it as trash!*

10 Inspect the cam bearing surfaces of the head and the holders. Look for score marks, deep scratches and evidence of spalling (a pitted

appearance). Check the lobes for heat discoloration (blue appearance), score marks, chipped areas, flat spots and spalling **(see illustration)**.

11 Camshaft runout can be checked by supporting each end of the camshaft on V-blocks, and measuring any runout using a dial gauge. If the runout exceeds the specified limit the camshaft must be replaced.

12 Measure the height of each lobe with a micrometer **(see illustration)** and compare the results to the lobe height service limit listed in this Chapter's Specifications. If damage is noted or wear is excessive, the camshaft must be replaced.

13 The camshaft bearing oil clearance should then be checked using a product known as Plastigage.

14 Clean the camshafts, the bearing surfaces in the cylinder head and the holders with a clean, lint-free cloth, then lay the camshafts in place in the cylinder head.

15 Cut strips of Plastigage and lay one piece on each bearing journal, parallel with the camshaft centerline. Make sure the camshaft holder dowels are installed and fit the holders in their proper positions. Ensuring the camshafts are not rotated at all, tighten the holder retaining bolts evenly, a little at a time, working in the tightening sequence stamped on

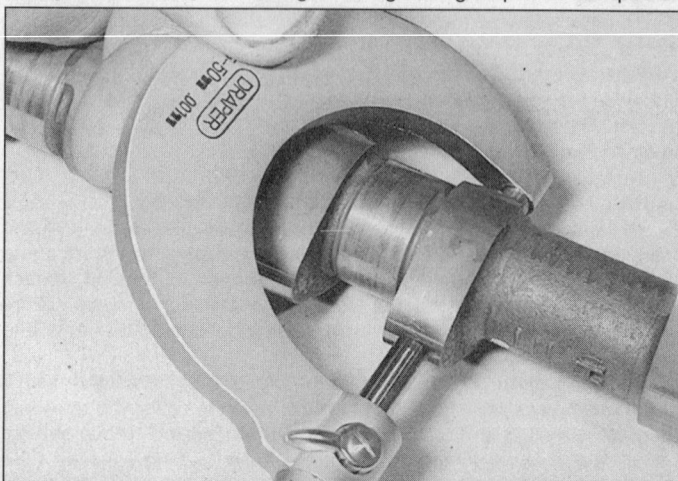

9.12 Measure the height of the camshaft lobes with a micrometer

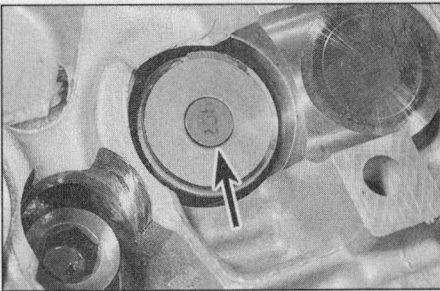
9.22 Ensure the shim (arrow) is correctly seated in the valve spring retainer

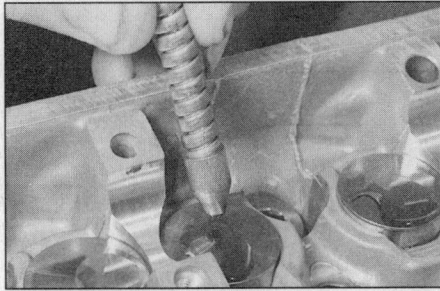
9.23 Lubricate the followers with clean engine oil prior to installing the camshafts

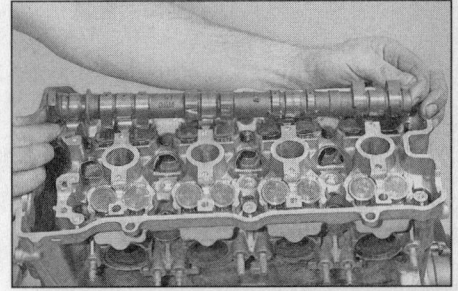
9.26a Installing the intake camshaft

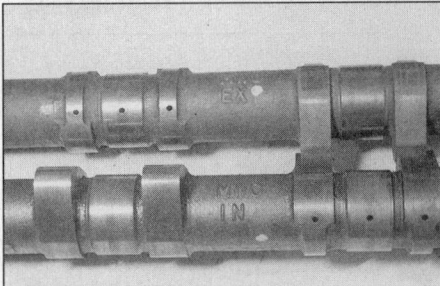
9.26b Camshafts are marked IN and EX for identification

9.28a Installing the intake camshaft holder; note the dowels (arrows)

9.28b Cam holders are also marked IN and EX for identification

the holder, until the specified torque setting is reached.

16 Unscrew the bolts in a reverse of the tightening sequence. Carefully lift off the holders, ensuring the camshafts are not rotated.

17 To determine the oil clearance, compare the crushed Plastigage (at its widest point) on each journal to the scale on the Plastigage container.

18 Compare the results to this Chapter's Specifications. If the oil clearance is greater than specified, the camshaft and/or cylinder head and camshaft holder bearing surfaces are worn.

19 Except in cases of oil starvation, the cam chain wears very little. If the cam chain has stretched excessively, which makes it difficult to maintain proper tension, replace it with a new one (see Section 10).

20 Check the sprockets for wear, cracks and other damage, replacing them if necessary. If the sprockets are worn, the cam chain is also worn, and also the sprocket on the crankshaft. If wear this severe is apparent the cam chain and all sprockets should be replaced.

21 Check the upper cam chain guide for wear or damage. If it is worn or damaged, the cam chain may be worn out. Refer to Section 10 for cam chain replacement information.

Installation

Refer to illustrations 9.22, 9.23, 9.26a, 9.26b, 9.28a, 9.28b, 9.35, 9.38, 9.42 and 9.44

22 Fit each shim to the top of its correct valve making sure it is correctly seated in the valve spring retainer **(see illustration)**. **Note:** *It is most important that the shims are returned to their original valves otherwise the valve clearances will be inaccurate.*

23 Install the followers in their respective positions in the cylinder head, making sure each one squarely enters its bore. Lubricate the followers with clean engine oil **(see illustration)**.

24 Position the crankshaft as described in Step 4.

25 Apply a smear of clean engine oil to the cylinder head camshaft bearing surfaces.

26 Lay the intake camshaft in position on the cylinder head so that its No. 1 cylinder lobes point away from the followers **(see illustration)**. **Note:** *Ensure the camshafts are fitted correctly; the intake camshaft is marked "IN" and the exhaust camshaft is marked "EX" (see illustration).*

27 Apply a smear of clean engine oil to the camshaft bearing journals and camshaft holder bearing surfaces.

28 Make sure the locating dowels are in position and install the intake camshaft holder on the head **(see illustration)**. **Note:** *Ensure the holders are fitted correctly, the intake holder is marked "IN" and the exhaust holder "EX" (see illustration).*

29 Screw in the camshaft holder bolts (noting that the longer bolts fit into the outer holes) by hand only until all bolts are contacting the holders. Do not install the No. 10 bolts in the tightening sequence yet; these will be installed after fitting the cam chain upper guide.

30 Work in the numerical sequence marked on the top of the holder and tighten its bolts by half a turn at a time to gradually draw the holder into position. Whilst tightening the bolts make sure that the holder is being pulled squarely down onto the cylinder head and is not sticking on the dowels. **Caution:** *If the bolts are carelessly tightened and the camshaft holder is not drawn squarely onto the head, it is likely to break. If this happens the complete cylinder head must be replaced; the holders are matched to the cylinder head and cannot be replaced separately.*

31 Once the camshaft holder is in contact with the head, working in the specified sequence, tighten the bolts to the specified torque setting.

32 Repeat steps 26 through 31 to install the exhaust camshaft.

34 Make sure the crankshaft is still positioned as described in Step 4.

35 Keeping the front run of the cam chain tight, locate the exhaust sprocket with the cam chain so that its EX mark is level with the upper surface of the cylinder head and pointing outwards **(see illustration)**. Locate the sprocket on the end of the exhaust camshaft.

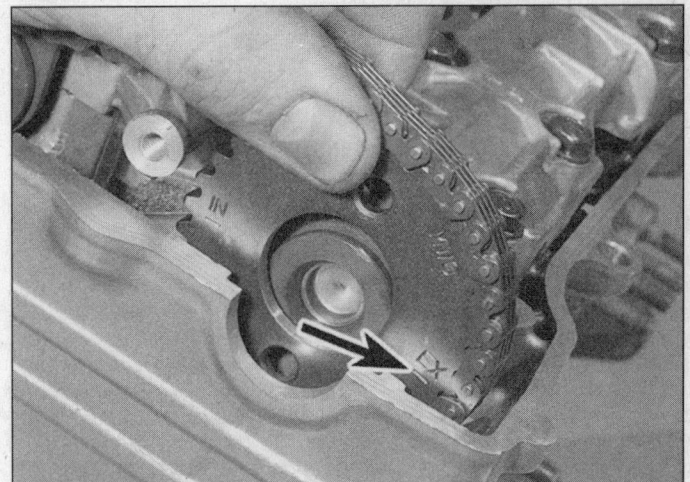
9.35 Ensure EX marking (arrow) is level with head surface and pointing outwards when installing exhaust cam sprocket

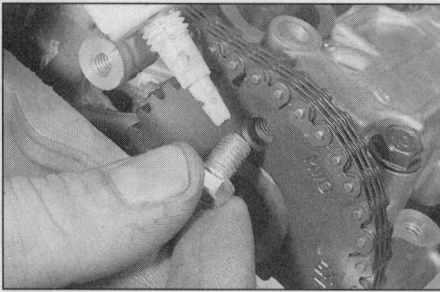

9.38 Apply non-permanent thread locking compound to the sprocket bolts

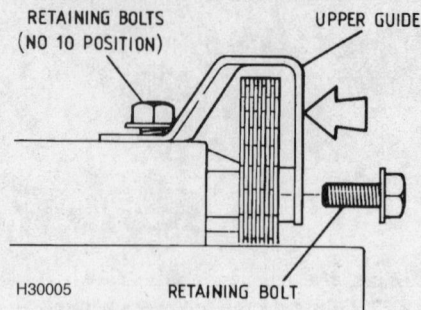

H30005

9.42 Cam chain upper guide installation

RETAINING BOLTS (NO 10 POSITION)   UPPER GUIDE   RETAINING BOLT

9.44 Sprocket timing marks (arrows) must align with head surface

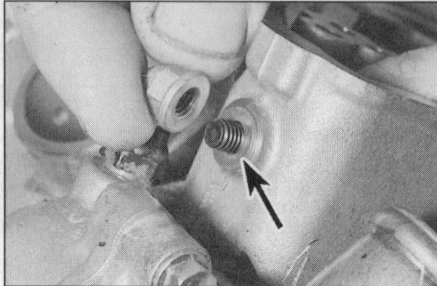

10.2a Slacken and remove the nut and sealing washer (arrow) . . .

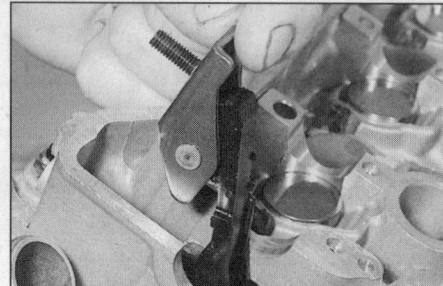

10.2b . . . and lift the tensioner blade out of the cylinder head

36 Engage the intake sprocket with the cam chain so that its IN mark is level with the upper surface of the cylinder head and pointing outwards. Locate the sprocket on the end of the intake camshaft.

37 Remove all traces of old locking compound from the camshaft sprocket bolts using a wire brush.

38 Apply a drop of non-permanent locking compound to the threads of each bolt **(see illustration)**. Fit one bolt to each visible camshaft sprocket hole, tightening them lightly only at this stage.

39 Watch the cam sprockets carefully to make sure the chain doesn't jump a tooth, and using a suitable socket, rotate the crankshaft clockwise until it is possible to insert the two remaining sprocket bolts. Apply a drop of non-permanent locking compound to the bolt threads, then screw them in and tighten to the specified torque setting.

41 Rotate the crankshaft clockwise until the other sprocket bolts are accessible then tighten them to the specified torque setting.

42 Position the cam chain upper guide on the camshaft holders. Press the side of the guide toward the cam sprockets whilst its retaining bolts are tightened **(see illustration)**. The No. 10 cam holder bolts should be tightened to the specified torque setting.

43 Install the cam chain tensioner (see Section 8).

44 Align the T mark on the crankshaft rotor with the index mark on the

cover (see Step 4) and check that both the sprocket timing marks are correctly aligned with the cylinder head upper surface **(see illustration)**. If the marks aren't lined up, disengage the sprocket(s) from the chain and adjust their position.

45 Rotate the crankshaft a few times, to settle all disturbed components, and check all valve clearances as described in Chapter 1, making adjustments as necessary.

46 Lubricate all bearing surfaces with clean engine oil and fit the valve cover as described in Section 7.

47 Install any components removed to improve access.

10 Cam chain and tensioner/guide blades - removal and installation

Note: *The cam chain and blades can be removed with the engine in the frame.*

Cam chain tensioner blade (rear)

Removal

Refer to illustrations 10.2a and 10.2b

1 Remove the intake camshaft sprocket as described in Section 9.

2 Slacken and remove the nut and sealing washer securing the guide to the cylinder head and lift the guide out of position **(see illustrations)**.

3 Check the tensioner blade for deep grooves, cracking and other obvious damage, replacing it if necessary.

Installation

Refer to illustration 10.4

4 Installation is the reverse of removal using a new sealing washer **(see illustration)**.

Cam chain guide blade (front)

Removal

Refer to illustrations 10.8, 10.9, 10.11, 10.12a, 10.12b and 10.13

5 Remove the exhaust camshaft as described in Section 9.

6 Drain the engine oil as described in Chapter 1.

7 Trace the wiring from the ignition pulse generator coil up to its two-pin connector along the top right side of the frame and disconnect it. Free the wiring from any ties.

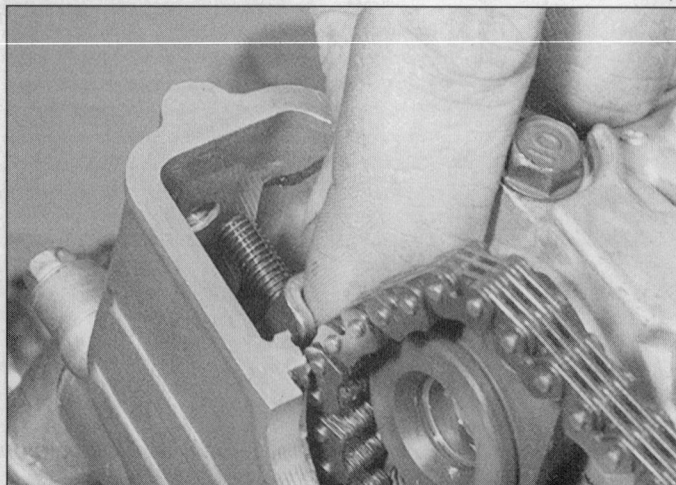

10.4 Slip the tensioner stud through the hole in the cylinder head

10.8 Crankshaft right end cover is retained by eight bolts (arrows)

10.9 Remove the pulse generator retaining bolt and washer

10.11 Remove the cam chain guide blade bolt and let the blade drop into the casing

10.12a Make a mark on the sprocket using paint or a marker pen . . .

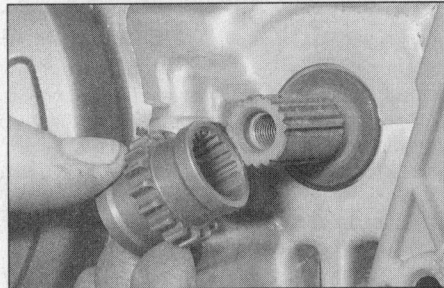

10.12b . . . then slip the sprocket off the crankshaft

10.13 Withdraw the guide blade from the crankcase

8 Remove the eight crankshaft end cover retaining and withdraw the cover squarely from the engine unit **(see illustration)**. Remove the gasket and recover the two dowels if they are loose.

9 Slacken and remove the pulse generator rotor retaining bolt and washer **(see illustration)**. To prevent the crankshaft rotating, remove the left crankshaft end cover as described in Section 21 and retain the alternator rotor using a large open-ended wrench on the rotor flats **(see illustration 21.6)**. Alternatively, if the engine is still in the frame, the engine can be locked through the transmission by selecting top gear and applying the rear brake hard.

10 Slide off the pulse generator rotor.

11 Remove the bolt from the cam chain guide blade and allow the blade to drop down to the bottom of the casing **(see illustration)**.

12 Use paint or a suitable marker pen to mark the outside edge of the crankshaft cam chain sprocket **(see illustration)**. This mark can then be used to make sure the sprocket is installed the correct way around. Disengage the chain from the sprocket and withdraw the sprocket from the crankshaft **(see illustration)**.

13 Withdraw the guide blade from the engine **(see illustration)**.

14 Check the guide blade for deep grooves, cracking and other obvious damage, replacing it if necessary.

Installation

Refer to illustrations 10.16, 10.18, 10.20 and 10.23

15 Remove all traces of gasket from the crankcase and cover mating surfaces. Make sure the oil gallery plug is in position in the crankcase.

16 Lubricate the pivot bush with engine oil and insert it into the guide, making sure the bush shoulder is on the inside of the guide blade (between the guide blade and crankcase) **(see illustration)**.

17 Insert the guide blade into the engine, but don't install its bolt at this stage.

18 Slide the sprocket on the crankshaft using the mark made on removal to ensure it is installed the correct way around. Engage the chain on the sprocket, then bring the guide blade up into position and tighten its bolt securely **(see illustration)**.

19 Align the pulse generator rotor splines with those of the crankshaft and slide on the rotor. **Note:** *Make sure the rotor timing marks are facing outwards.*

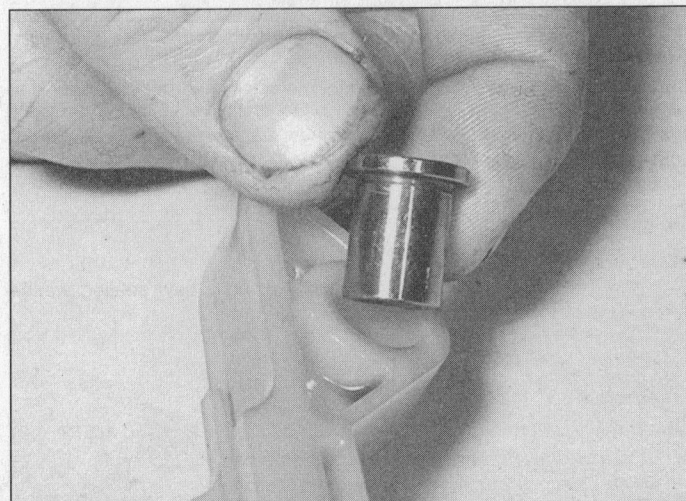

10.16 Fit the pivot bush to the guide blade so that its collar will be positioned between the guide blade and crankcase

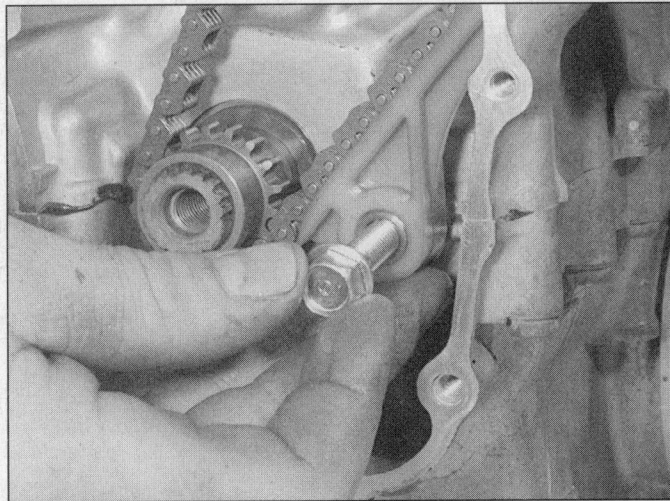

10.18 Install the guide blade bolt and tighten it securely

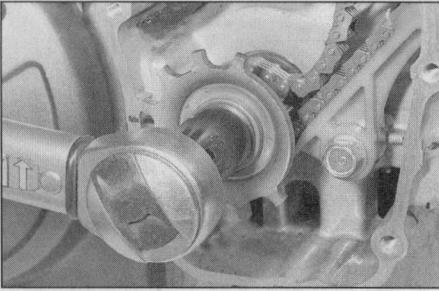

10.20 Tighten the pulse generator rotor bolt to the specified torque setting

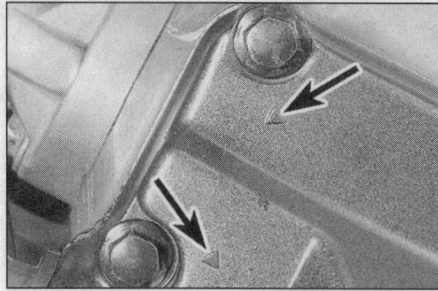

10.23 Apply thread-lock to the bolts next to the triangular markings (arrows)

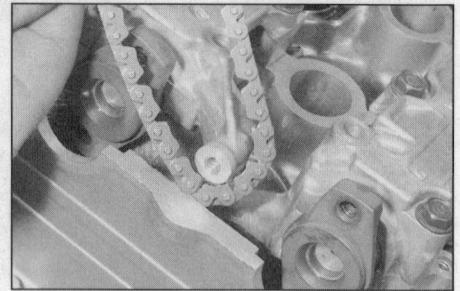

10.29 Slip cam chain over intake cam holder boss and into cam chain tunnel

20 Tighten the rotor bolt to the specified torque setting whilst preventing crankshaft rotation using the method employed on removal **(see illustration)**.

21 Install the exhaust camshaft sprocket and check the valve timing (see Section 9).

22 Apply a smear of sealant to the cover wiring grommet. Install the locating dowels and fit a new gasket to the crankcase.

23 Fit the cover to the engine noting that a drop of non-permanent thread locking compound should be applied to the two bolts which go in the holes marked with the cast triangles **(see illustration)**. Tighten the cover bolts to the specified torque setting.

24 Make sure the wiring is correctly routed up to the two-pin connector and reconnect it. Secure the wiring in position with all the necessary ties.

25 Fill the engine with the correct type and amount of oil as described in Chapter 1.

Cam chain

Removal

26 Remove the camshaft sprockets and tensioner/guide blades as described above.

27 Drop the chain down and remove it from the end of the crankshaft.

Inspection

28 Check the cam chain for binding and obvious damage, and inspect the sprocket for damage such as chipped or missing teeth. If either of these conditions are visible, or if the chain appears to be stretched, both the cam chain and sprockets (crankshaft and both camshaft sprockets) should be replaced as a set.

Installation

Refer to illustration 10.29

29 Feed the cam chain over the boss on the intake cam holder and down into the tunnel **(see illustration)**. Hang the chain over the camshafts and locate it over the crankshaft.

30 Install the tensioner/guide blades and camshaft sprockets as described above.

11 Cylinder head - removal and installation

Caution: *The engine must be completely cool before beginning this procedure or the cylinder head may become warped.*
Note: *This procedure can be performed with the engine in the frame. If the engine has already been removed, ignore the preliminary steps which don't apply.*

Removal

Refer to illustrations 11.6, 11.9 and 11.12

1 Remove the exhaust system as described in Chapter 4.

2 Remove the carburetors as described in Chapter 4.

3 On California models, remove the pulse secondary air injection system (PAIR) control valve as described in Chapter 4 and disconnect the EVAP hose and PAIR hose from intake manifold Nos. 1 and 3 respectively.

4 Remove the camshafts and followers as described in Section 9.

5 Slacken and remove the nut and sealing washer securing the cam chain tensioner blade to the cylinder head and lift out the blade **(see illustrations 10.2a and 10.2b)**.

6 Release its clip and disconnect the air bleed hose from its union on the rear of the cylinder head **(see illustration)**.

7 Pull the wire connector off the coolant temperature sender unit and disconnect the air bleed hose and radiator top hose from the thermostat housing **(see illustrations 5.17a and 5.17b)**.

8 Remove the upper of the two front left engine mounting bolts **(see illustration 5.22)**.

9 Unscrew the two external cylinder head bolts from the right end of the head **(see illustration)**.

10 Working from the outside to the inside in a criss-cross pattern, slacken the ten cylinder head bolts by half a turn at a time. Once all pressure is released from the bolts, fully unscrew them and remove along with their washers.

11 Tap around the joint faces of the cylinder head with a soft-faced mallet to free the head. Don't attempt to free the head by inserting a screwdriver between the head and cylinder block - you'll damage the sealing surfaces.

12 Lift the head off the block, and remove it from the engine **(see illustration)**. Pass a piece of wire or a screwdriver through the cam chain to prevent it falling down into its tunnel.

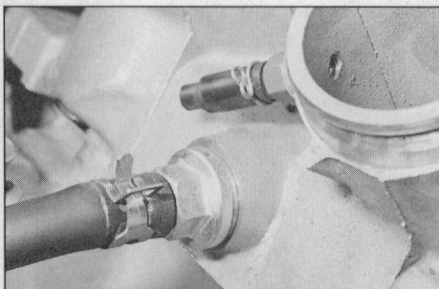

11.6 Disconnect the air bleed hose from its union on the rear of the cylinder head

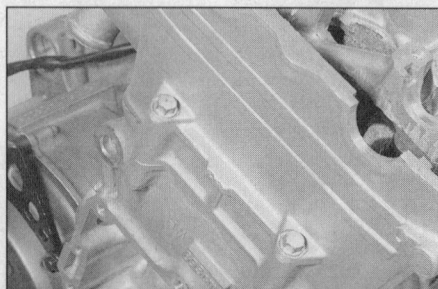

11.9 Two external cylinder head bolts are located on the right side

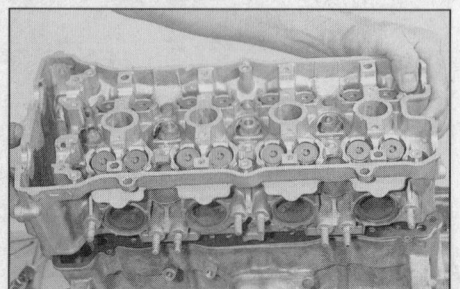

11.12 Lifting the cylinder head off the block

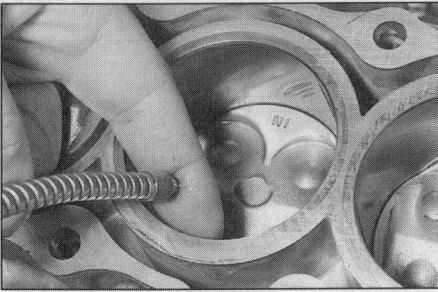

11.15 Prior to installing the head, lubricate the bores with clean engine oil

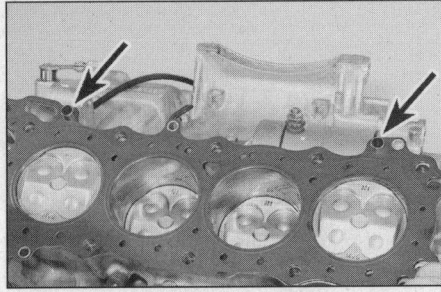

11.16 Ensure the locating dowels (arrows) are in position and fit the new head gasket

11.17 Pass a suitable tool through the cam chain to prevent it falling back into the engine

13 Remove the old head gasket and discard it. If loose, remove the cylinder head locating dowels from the cylinder block and store them with the head for safe-keeping.

14 Check the cylinder head gasket and the mating surfaces on the cylinder head and block for signs of leakage, which could indicate warpage. Check the flatness of the head as described in Section 13.

Installation

Refer to illustrations 11.15, 11.16, 11.17, 11.18 and 11.20

15 Ensure both cylinder head and block mating surfaces are clean and fit the locating dowels to the block (if removed). Apply a smear of engine oil to the surface of each cylinder bore **(see illustration)**.

16 Fit the new head gasket over the locating dowels **(see illustration)**.

17 Carefully lower the cylinder head onto the block whilst feeding the cam chain up through the head. Pass a screwdriver or piece of wire through the chain to prevent it falling back into the engine **(see illustration)**.

18 Apply a drop of clean engine oil to the threads and undersides of the cylinder head bolts **(see illustration)**.

19 Fit the cylinder head bolts and washers. Tighten them all by hand.

20 Working from the inside to the outside in a criss-cross pattern, tighten the cylinder head bolts to approximately half the specified torque setting given in the Specifications. Then go around in the same sequence and tighten the bolts to the full specified torque setting **(see illustration)**.

21 Install the two bolts to the right end of the cylinder head and tighten them securely.

22 Install the engine mounting bolt to the engine left side and tighten it to the specified torque setting.

23 Connect the coolant hose and air bleed hoses to their unions and secure with the clamp/clips.

24 Connect the wiring connector to the coolant temperature sender.

25 Lower the cam chain tensioner blade into position, fit a new sealing washer and securely tighten its retaining nut.

26 Install the camshafts and followers as described in Section 9.

27 Fit the exhaust system, carburetors and, on California models, the PAIR and EVAP components.

28 Fill the cooling system as described in Chapter 1.

12 Valves/valve seats/valve guides - servicing

1   Because of the complex nature of this job and the special tools and equipment required, servicing of the valves, the valve seats and the valve guides (commonly known as a valve job) is best left to a professional.

2   The home mechanic can, however, remove and disassemble the head, do the initial cleaning and inspection, then reassemble and deliver the head to a dealer service department or properly equipped motorcycle repair shop for the actual valve servicing. Refer to Section 13 for those procedures.

3   The dealer service department will remove the valves and springs, recondition or replace the valves and valve seats, replace the valve guides, check and replace the valve springs, spring retainers and keepers (collets) (as necessary), replace the valve seals with new ones and reassemble the valve components.

4   After the valve job has been performed, the head will be in like-new condition. When the head is returned, be sure to clean it again very thoroughly before installation on the engine to remove any metal particles or abrasive grit that may still be present from the valve service operations. Use compressed air, if available, to blow out all the holes and passages.

2

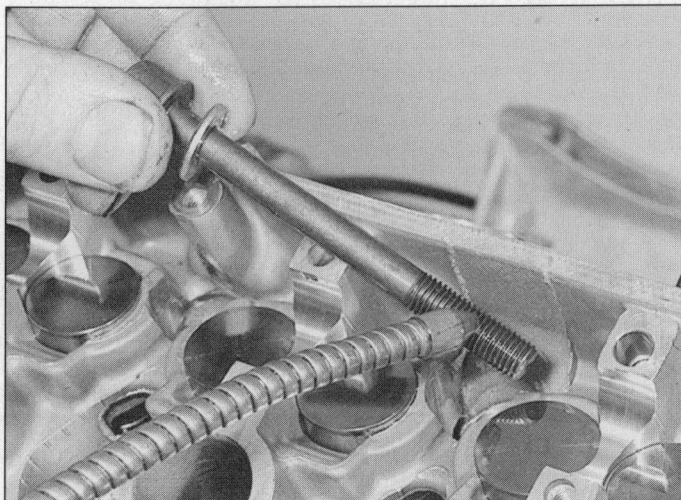

11.18 Apply a smear of oil to the threads and underside of the heads of the cylinder head bolts

11.20 Tighten the head bolts to the specified torque as described in the text

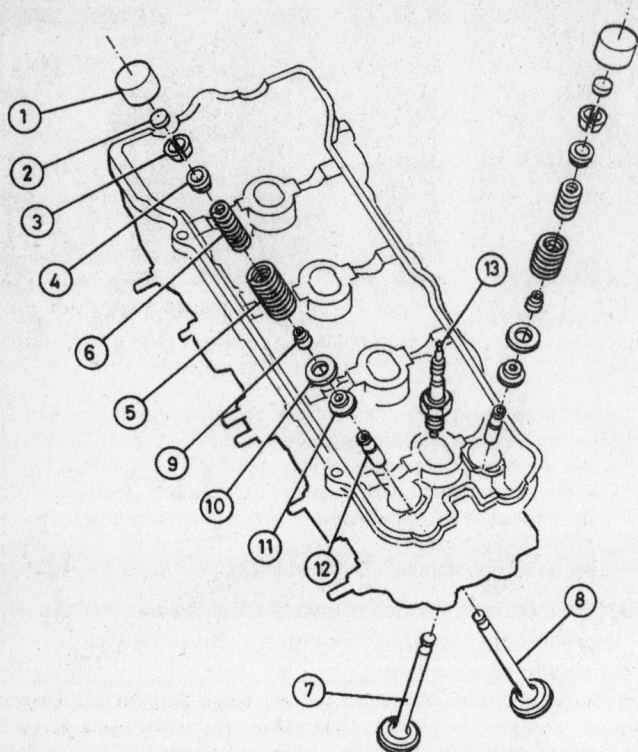

13.3 Cylinder head and valve components

1 Follower	8 Exhaust valve
2 Shim	9 Valve guide seal
3 Keepers (collets)	10 Outer spring seat
4 Spring retainer	11 Inner spring seat
5 Outer valve spring	12 Valve guide
6 Inner valve spring	13 Spark plug
7 Intake valve	

13 Cylinder head and valves - disassembly, inspection and reassembly

1   As mentioned in the previous Section, valve servicing and valve guide replacement should be left to a dealer service department or motorcycle repair shop. However, disassembly, cleaning and inspection of the valves and related components can be done (if the necessary special tools are available) by the home mechanic. This way no expense is incurred if the inspection reveals that service work is not required at this time.

2   To properly disassemble the valve components without the risk of damaging them, a valve spring compressor is absolutely necessary. This special tool can usually be rented, but if it's not available, have a dealer service department or motorcycle repair shop handle the entire process of disassembly, inspection, service or repair (if required) and reassembly of the valves.

Disassembly

Refer to illustrations 13.3, 13.7a, 13.7b, 13.8a, 13.8b and 13.11

3   Remove the followers and their shims if not already done (see Section 9). Store the parts in such a way that they can be returned to their original locations without getting mixed up **(see illustration)**.

4   Before the valves are removed, scrape away any traces of gasket material from the head gasket sealing surface. Work slowly and do not nick or gouge the soft aluminum of the head. Gasket removing solvents, which work very well, are available at most motorcycle shops and auto parts stores.

5   Carefully scrape all carbon deposits out of the combustion chamber area. A hand held wire brush or a piece of fine emery cloth can be used once the majority of deposits have been scraped away. Do not use a wire brush mounted in a drill motor, or one with extremely stiff bristles, as the head material is soft and may be eroded away or scratched by the wire brush.

6   Before proceeding, arrange to label and store the valves along with their related components so they can be kept separate and reinstalled in the same valve guides they are removed from (labelled plastic bags work well for this).

7   To prevent damage to the follower bore from the valve spring compressor, you are advised to protect the bore surface. A plastic container from a photographic 35 mm film works well since it fits between the valve spring and follower bore. Cut the container away as shown and slip it into place **(see illustrations)**.

8   Compress the valve spring on the first valve with a spring compressor, then remove the keepers (collets) and the retainer from the valve assembly. **Note:** *Take great care not to mark the cylinder head follower bore with the spring compressor.* Do not compress the springs any more than is absolutely necessary. Carefully release the valve spring compressor and remove the springs and the valve from the head **(see illustration)**. If the valve binds in the guide (won't pull through), push it back into the head and deburr the area around the keeper (collet) groove with a very fine file or whetstone **(see illustration)**.

9   Repeat the procedure for the remaining valves. Remember to keep the parts for each valve together so they can be reinstalled in the same location.

10   Once the valves have been removed and labelled, pull off the valve stem seals with pliers and discard them (the old seals should never be reused), then remove the spring seats.

11   Next, clean the cylinder head with solvent and dry it thoroughly. Compressed air will speed the drying process and ensure that all holes and recessed areas are clean. Check that the drain hole under the exhaust ports is clear; its purpose is to drain off any water which might collect in the spark plug channels **(see illustration)**.

12   Clean all of the valve springs, keepers (collets), retainers and spring seats with solvent and dry them thoroughly. Do the parts from one valve at a time so that no mixing of parts between valves occurs.

13   Scrape off any deposits that may have formed on the valve, then use a motorized wire brush to remove deposits from the valve heads and stems. Again, make sure the valves do not get mixed up.

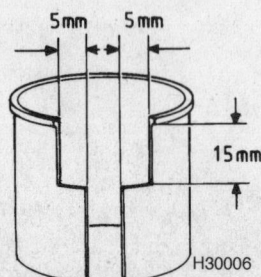

13.7a Follower bore protector can be made from an old 35 mm film container . . .

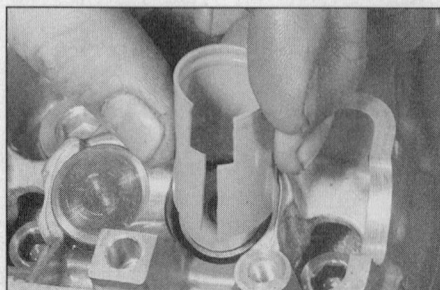

13.7b . . . and inserted into the follower bore as shown

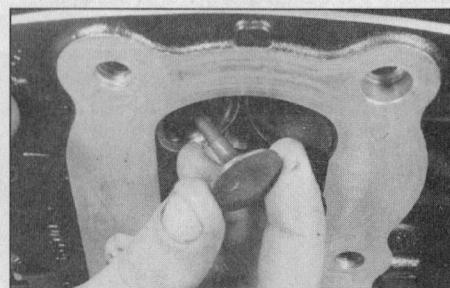

13.8a Take the valve out of the combustion chamber, but don't force it if it's stuck . . .

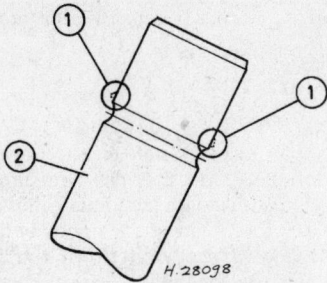

13.8b ... check the area around the keeper (collet) groove for burrs

1 Burrs (remove)   2 Valve stem

13.11 Check that each drain hole is clear (arrow)

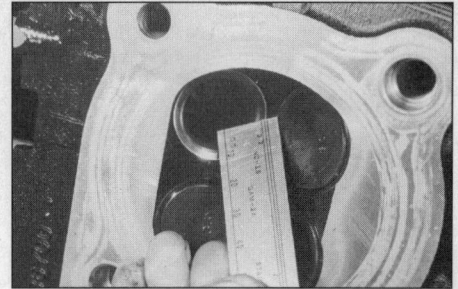

13.16 Measuring valve seat width

13.17a Insert a small hole gauge into the valve guide and expand it so there's a slight drag when it's pulled out

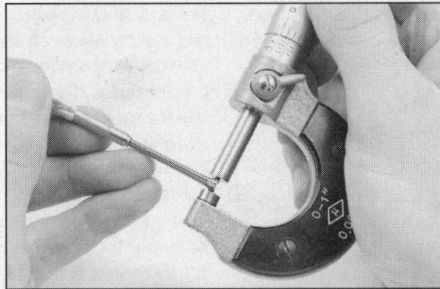

13.17b Measure the small hole gauge with a micrometer

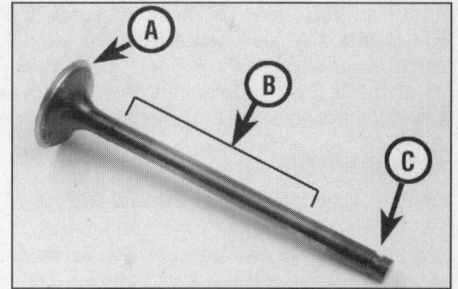

13.18 Check the valve face (A), stem (B) and keeper/collet groove (C) for signs of damage or wear

Inspection

Refer to illustrations 13.16, 13.17a, 13.17b, 13.18, 13.19, 13.20a and 13.20b

14 Inspect the head very carefully for cracks and other damage. If cracks are found, a new head will be required. Check the cam bearing surfaces for wear and evidence of seizure. Check the camshafts and followers for wear as well (see Section 9).

15 Using a precision straightedge and a feeler gauge, check the head gasket mating surface for warpage. Lay the straightedge lengthways, across the head and diagonally (corner-to-corner), intersecting the head stud holes, and try to slip a feeler gauge under it, on either side of each combustion chamber. The gauge should be the same thickness as the cylinder head warp limit listed in this Chapter's Specifications. If the feeler gauge can be inserted between the head and the straightedge, the head is warped and must either be machined or, if warpage is excessive, replaced with a new one.

16 Examine the valve seats in each of the combustion chambers. If they are pitted, cracked or burned, the head will require valve service that's beyond the scope of the home mechanic. Measure the valve seat width and compare it to this Chapter's Specifications **(see illustration)**. If it exceeds the service limit, or if it varies around its circumference, valve service work is required.

17 Clean the valve guides to remove any carbon build-up, then measure the inside diameters of the guides (at both ends and the center of the guide) with a small hole gauge and micrometer **(see illustrations)**. Record the measurements for future reference. These measurements, along with the valve stem diameter measurements, will enable you to compute the valve stem-to-guide clearance. This clearance, when compared to the Specifications, will be one factor that will determine the extent of the valve service work required. The guides are measured at the ends and at the center to determine if they are worn in a bell-mouth pattern (more wear at the ends). If they are, guide replacement is an absolute must.

18 Carefully inspect each valve face for cracks, pits and burned spots. Check the valve stem and the keeper (collet) groove area for cracks **(see illustration)**. Rotate the valve and check for any obvious indication that it is bent. Check the end of the stem for pitting and excessive wear. The presence of any of the above conditions indicates the need for valve servicing.

19 Measure the valve stem diameter **(see illustration)**. By subtracting the stem diameter from the valve guide diameter, the valve stem-to-guide clearance is obtained. If the stem-to-guide clearance is greater than listed in this Chapter's Specifications, the guides and valves will have to be replaced with new ones.

20 Check the end of each valve spring for wear and pitting. Measure the free length and compare it to this Chapter's Specifications **(see illustration)**. Any springs that are shorter than specified have sagged and should not be reused. Stand the spring on a flat surface and check it for squareness **(see illustration)**.

2

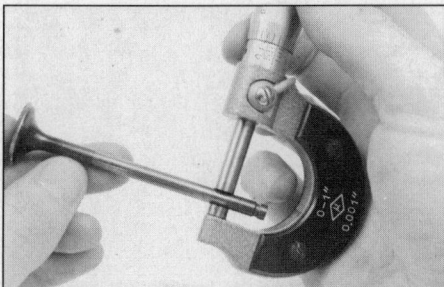

13.19 Measure the valve stem diameter with a micrometer

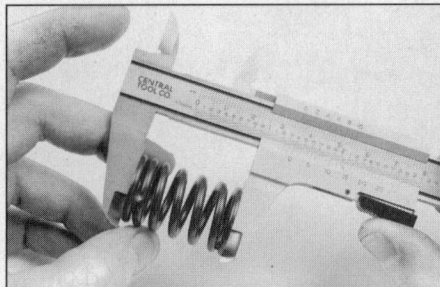

13.20a Measure the free length of the valve springs

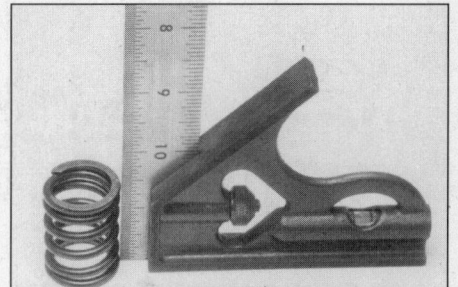

13.20b Check the valve springs for squareness

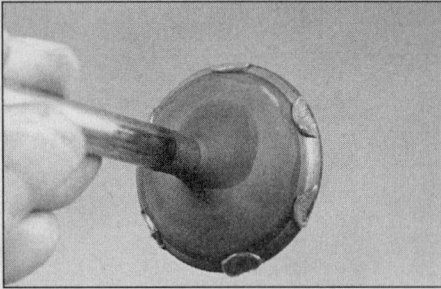

13.24 Apply lapping compound very sparingly, in small dabs, to the valve face only

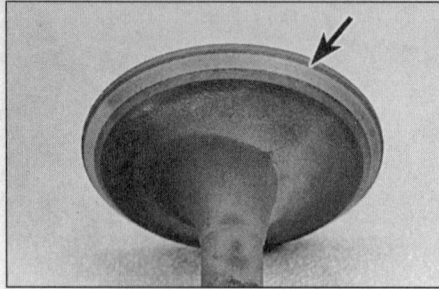

13.25a After lapping, the valve face should exhibit a uniform, unbroken contact pattern (arrow) . . .

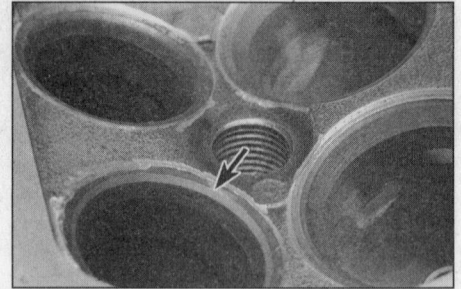

13.25b . . . and the seat should be the specified width (arrow) with a smooth, unbroken appearance

21 Check the spring retainers and keepers (collets) for obvious wear and cracks. Any questionable parts should not be reused, as extensive damage will occur in the event of failure during engine operation.

22 If the inspection indicates that no service work is required, the valve components can be reinstalled in the head.

Reassembly

Refer to illustrations 13.24, 13.25a, 13.25b, 13.29a and 13.29b

23 Before installing the valves in the head, they should be lapped to ensure a positive seal between the valves and seats. This procedure requires coarse and fine valve lapping compound (available at auto parts stores) and a valve lapping tool. If a lapping tool is not available, a piece of rubber or plastic hose can be slipped over the valve stem (after the valve has been installed in the guide) and used to turn the valve.

24 Apply a small amount of coarse lapping compound to the valve face, then slip the valve into the guide **(see illustration)**. **Note:** *Make sure the valve is installed in the correct guide and be careful not to get any lapping compound on the valve stem.*

25 Attach the lapping tool (or hose) to the valve and rotate the tool between the palms of your hands. Use a back-and-forth motion rather than a circular motion. Lift the valve off the seat and turn it at regular intervals to distribute the lapping compound properly. Continue the lapping procedure until the valve face and seat contact area is of uniform width and unbroken around the entire circumference of the valve face and seat **(see illustrations)**.

26 Carefully remove the valve from the guide and wipe off all traces of lapping compound. Use solvent to clean the valve and wipe the seat area thoroughly with a solvent soaked cloth.

27 Repeat the procedure with fine valve lapping compound, then repeat the entire procedure for the remaining valves.

28 Lay the spring seats in place in the cylinder head, then install new valve stem seals on each of the guides. Use an appropriate size deep socket to push the seals into place until they are properly seated. Don't twist or cock them, or they will not seal properly against the valve stems. Also, don't remove them again or they will be damaged.

29 Coat the valve stems with clean engine oil, then install one of them into its guide. Next, install the springs and retainer, compress the springs and install the keepers (collets). **Note:** *Install the springs with the tightly wound coils at the bottom, next to the spring seat* **(see illustration)**. When compressing the springs with the valve spring compressor, depress them only as far as is absolutely necessary to slip the keepers (collets) into place. **Note:** *Use the protector described in Step 7 to avoid damaging the cylinder head follower bore with the spring compressor.* Apply a small amount of grease to the keepers (collets) to help hold them in place as the pressure is released from the springs **(see illustrations)**. Make certain that the keepers (collets) are securely locked in their retaining grooves.

30 Support the cylinder head on blocks so the valves can't contact the workbench top, then very gently tap each of the valve stems with a soft-faced hammer. This will help seat the keepers (collets) in their grooves.

31 Once all of the valves have been installed in the head, check for proper valve sealing by pouring a small amount of solvent into each of the valve ports. If the solvent leaks past the valve(s) into the combustion chamber area, disassemble the valve(s) and repeat the lapping procedure, then reinstall the valve(s) and repeat the check. Repeat the procedure until a satisfactory seal is obtained.

14 Clutch - removal, inspection and installation

Note: *This procedure can be performed with the engine in the frame. If the engine has already been removed, ignore the preliminary steps which don't apply.*

Removal

Refer to illustrations 14.4, 14.5, 14.7, 14.8 and 14.10

1 Remove the fairing right middle section and lower section as described in Chapter 8.

2 Drain the engine oil as described in Chapter 1.

3 Disconnect the clutch cable from the release lever (see Section 15).

4 Working in a criss-cross pattern, evenly slacken the ten clutch cover retaining bolts **(see illustration)**. Three of the cover bolts are longer than the others, so take note of their location as a guide to installation.

13.29a Install the valve springs with their tightly wound coils downward (against the cylinder head)

13.29b A small dab of grease will help hold the keepers (collets) in place on the valve while the valve spring is released

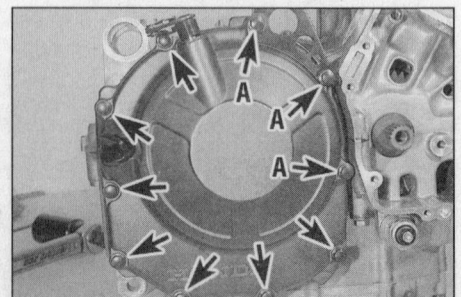

14.4 Remove the ten clutch cover retaining bolts (arrows) noting that the longer bolts are fitted in holes A

14.5 Withdraw the clutch cover

14.7 Slacken clutch spring bolts evenly, in a criss-cross pattern

5 Lift the cover away from the engine, being prepared to catch any residual oil which may be released as the cover is removed **(see illustration)**.

6 Remove the gasket and discard it. Note the two locating dowels fitted to the crankcase and remove these for safe-keeping if they are loose.

7 Working in a criss-cross pattern, gradually slacken the clutch spring retaining bolts until spring pressure is released **(see illustration)**. Unscrew the bolts, then remove the clutch pressure plate complete with release bearing and pushrod.

8 Lift out the clutch plates, followed by the anti-judder spring and its seat **(see illustration)**. Unless the plates are to be replaced, keep them in order to aid installation.

9 Relieve the staking on the clutch nut using a hammer and suitable pointed-nose chisel, taking care not to damage the mainshaft end.

10 The clutch center must be prevented from rotating while the center

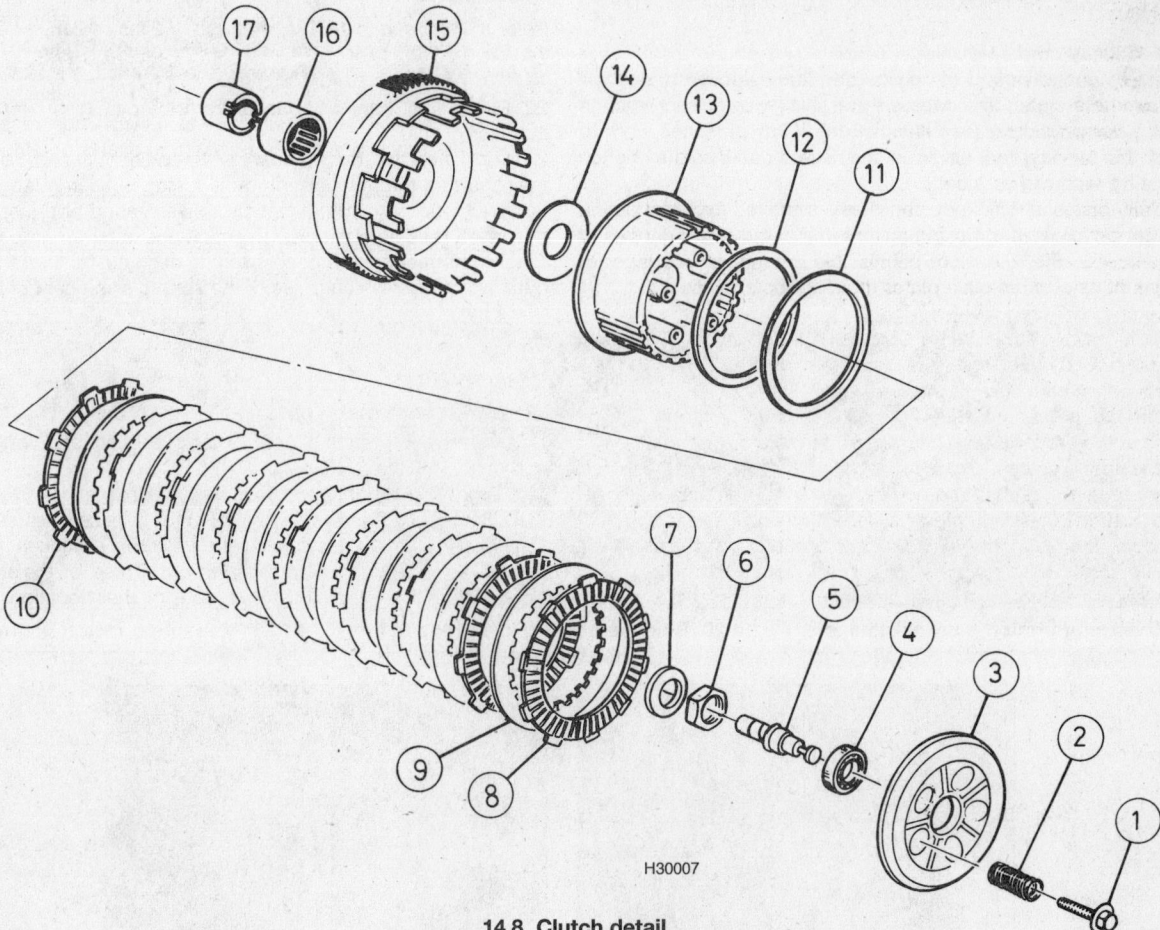

H30007

14.8 Clutch detail

1 Bolts
2 Springs
3 Pressure plate
4 Release bearing
5 Pushrod
6 Clutch nut
7 Lock washer
8 Friction plates
9 Plain plates
10 Inner friction plate
11 Anti-judder spring
12 Spring seat
13 Clutch center
14 Thrust washer
15 Outer drum
16 Needle roller bearing
17 Guide bush

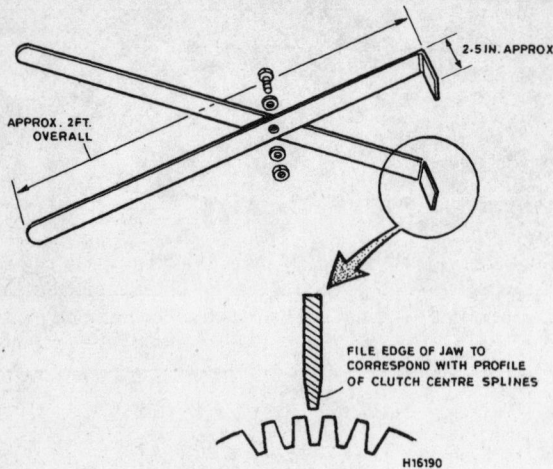

APPROX. 2FT.
OVERALL

2·5 IN. APPROX

FILE EDGE OF JAW TO
CORRESPOND WITH PROFILE
OF CLUTCH CENTRE SPLINES

H16190

14.10 Fabricated clutch holding tool

nut is slackened. The Honda service tool (Pt. No. 07724-0050001) or a home-made copy can be used to clamp the center splines and outer drum slots (see illustration). Alternatively, lock the clutch by selecting top gear and applying the rear brake hard whilst the nut is slackened.

11 Remove the nut and discard it; obtain a new one for installation.

12 Remove the washer; its outer face should be marked "OUT".

13 Withdraw the clutch center followed by the large thrust washer, outer drum and its guide bush.

Inspection

Refer to illustrations 14.14, 14.20 and 14.21

14 After an extended period of service the clutch friction plates will wear and promote clutch slip. Measure the thickness of each friction plate using a vernier caliper (see illustration). If any plate has worn to or beyond the service limit given in the Specifications, the friction plates must be replaced as a set.

15 The plain plates should not show any signs of excess heating (blueing). Check for warpage using a flat surface and feeler gauges. If any plate exceeds the maximum permissible amount of warpage, or shows signs of blueing, all plain plates must be replaced as a set.

16 Inspect the clutch assembly for burrs and indentations on the edges of the protruding tangs of the friction plates and/or slots in the edge of the outer drum with which they engage. Similarly check for wear between the inner tongues of the plain plates and the slots in the clutch center. Wear of this nature will cause clutch drag and slow disengagement during gear changes, since the plates will snag when the pressure plate is lifted. With care a small amount of wear can be corrected by dressing with a fine file, but if this is excessive the worn components can be replaced.

17 Also inspect the anti-judder spring and seat for signs of wear or distortion and replace if necessary.

18 Inspect the mainshaft, guide bush and outer drum bearing surfaces for signs of wear and damage. If access to the necessary measuring equipment can be gained, the condition of the above components can be judged by direct measurement. If any component shows signs of wear or damage, or has worn beyond its service limit given in the Specifications, it must be replaced.

19 Inspect the outer drum needle roller bearing for signs of wear or damage. If replacement is necessary, the task should be entrusted to a Honda dealer. The bearing is a press fit in the drum and a hydraulic press and suitable spacers will be required to remove the original bearing and install the new one.

20 Check the clutch release bearing for wear. The inner race of the bearing must spin freely without any notchiness. Push the bearing out of the pressure plate if replacement is required (see illustration).

21 Measure the free length of each clutch spring (see illustration). If any one has settled to less than the service limit, the clutch springs must be replaced as a set.

Installation

Refer to illustrations 14.23a, 14.23b, 14.24a, 14.24b, 14.25a, 14.25b, 14.26a, 14.26b, 14.27a, 14.27b, 14.28, 14.29a, 14.29b, 14.30a, 14.30b, 14.30c and 14.31

22 Remove all traces of gasket from the crankcase and clutch cover surfaces.

23 Lubricate the guide bush with engine oil and slide it on the mainshaft so that its slotted end is facing outwards (see illustration). Lubricate the outer drum needle roller bearing and install it over the mainshaft (see illustration). To enable the outer drum to mesh with the crankshaft gear it will be necessary to align the teeth of the main clutch outer drum gear which those of the sprung slim gear behind it; this can

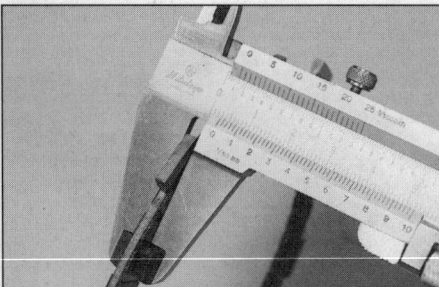

14.14 Measuring friction plate thickness

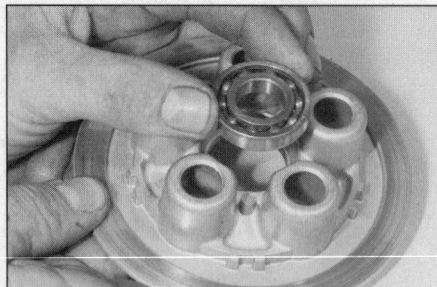

14.20 Release bearing is a press fit in pressure plate

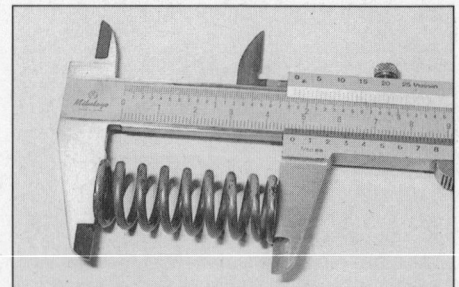

14.21 Measuring clutch spring free length

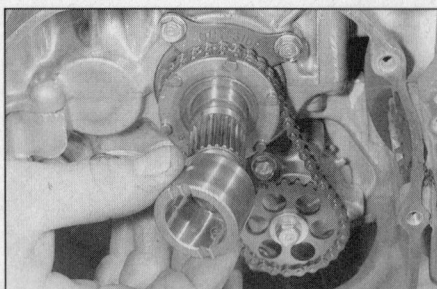

14.23a Install the guide bush making sure its slotted end is facing outwards

14.23b Install the outer drum over the mainshaft

14.24a Install the large thrust washer over the mainshaft . . .

14.24b . . . followed by the clutch center

14.25a The lockwasher must be installed so that its OUT marking faces outwards

14.25b Always fit a new clutch nut

14.26a Using the holding tool, tighten clutch nut to specified torque setting . . .

14.26b . . . and stake its shoulder into the shaft groove

be done with a flat-bladed screwdriver as the outer drum is pushed fully on the mainshaft. Note that the slots on the back of the outer drum must engage the dogs on the oil pump drive gear - check that the dogs are engaged by rotating the outer drum; the oil pump sprocket should turn as it is rotated.

24 Install the large thrust washer on the mainshaft, followed by the clutch center **(see illustrations)**.

25 Install the lock washer with its "OUT" marking facing outwards, followed by a **new** clutch nut **(see illustrations)**.

26 Using the method employed on removal (see Step 10), hold the clutch steady and tighten the clutch nut to the specified torque setting **(see illustration)**. Rotate the mainshaft so that its slot is uppermost,

then stake the shoulder of the clutch nut into the slot to lock it in position **(see illustration)**.

27 Fit the spring seat to the clutch center followed by the anti-judder spring **(see illustration)**. **Note:** *The anti-judder spring must be fitted with its convex side facing the spring seat* **(see illustration)**.

28 Install the inner friction plate in the outer drum **(see illustration)**. This plate can be identified by its larger inside diameter which allows it to locate over the anti-judder spring and seat.

29 Install a plain plate followed by one of the ordinary friction plates then alternately install all the remaining plain and friction plates, noting that the tangs of the last friction plate slot into the shallow holes in the outer drum **(see illustrations)**. **Note:** *If new clutch plates are being*

2

14.27a Install the spring seat and anti-judder spring

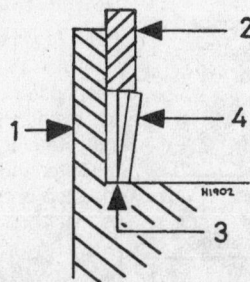

14.27b Correct fitting of anti-judder spring

1 Clutch center   3 Spring seat
2 Inner friction plate   4 Anti-judder spring

14.28 Install the inner friction plate first (it has a narrower friction surface and larger inside diameter than the others)

14.29a Fit the plain and friction plates alternately . . .

14.29b . . . tangs of last friction plate locate in shallow slots of outer drum (arrows)

14.30a Install the pushrod in the pressure plate . . .

14.30b . . . and insert the pressure plate over the clutch center posts

14.30c Install the springs, washers and bolts

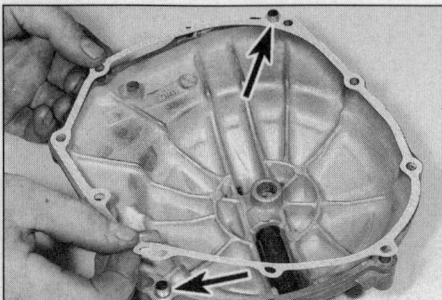

14.31 Install a new gasket over the cover dowels (arrows)

15.2a Back off the cable lower adjuster to allow disconnection of the cable nipple . . .

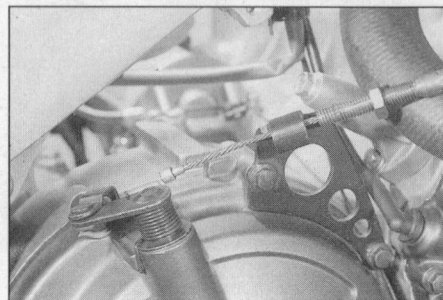

15.2b . . . then withdraw the cable and its adjuster from the bracket

fitted, *apply a coating of oil to their surfaces to prevent seizure.*
30 Install the pressure plate, complete with release bearing and pushrod **(see illustrations)**. Secure the pressure plate with the springs and headed bolts, tightening them evenly in a diagonal sequence **(see illustration)**.
31 Install the two dowels in the cover and fit a new gasket over them **(see illustration)**. Offer up the clutch cover and seat it fully onto the crankcase. Install the cover bolts, making sure the longer bolts are fitted in the correct holes **(see illustration 14.4)** and that the cable bracket is secured. Tighten the bolts evenly in a diagonal sequence to the specified torque setting.
32 Reconnect the clutch cable end to the release lever as described in Section 15 and carry out adjustment as described in Chapter 1.
33 Refill the engine with the specified quantity and type of oil (see Chapter 1).
34 Install the fairing sections (see Chapter 8).

15 Clutch cable and release lever - removal and installation

Clutch cable

Removal

Refer to illustrations 15.2a and 15.2b
1   Remove the right middle fairing section (see Chapter 8).
2   Back off the cable lower adjuster to allow the cable end to be disconnected from the release lever **(see illustration)**. Free the adjuster from its bracket on the clutch cover **(see illustration)**.
3   Align the slots of the knurled adjuster and lockring at the lever end of the cable and detach the cable from the underside of the clutch lever.
4   Take note of its routing as the old cable is removed from the bike.

Installation

5   Install the new cable making sure it is correctly routed and retained by all necessary clips.
6   Adjust the cable as described in Chapter 1 then install the fairing section as described in Chapter 8.

Clutch release lever

Removal

7   Remove the clutch cover as described in Steps 1 through 6 of Section 14.
8   Disconnect the return spring arm from the cover stop and pull the release lever out of the cover.
9   Inspect the two needle roller bearings and oil seal set in the clutch cover; these components can remain in position unless they require replacement.

Installation

Refer to illustration 15.11
10 Prior to installation of the release lever, apply engine oil to the bearings and a smear of grease to the oil seal lip.
11 Install the return spring and washer on the release lever and slide it into the clutch cover. Hook the short arm of the return spring around the lever and tension the longer end of the spring so that it rests against the stop on the cover **(see illustration)**.

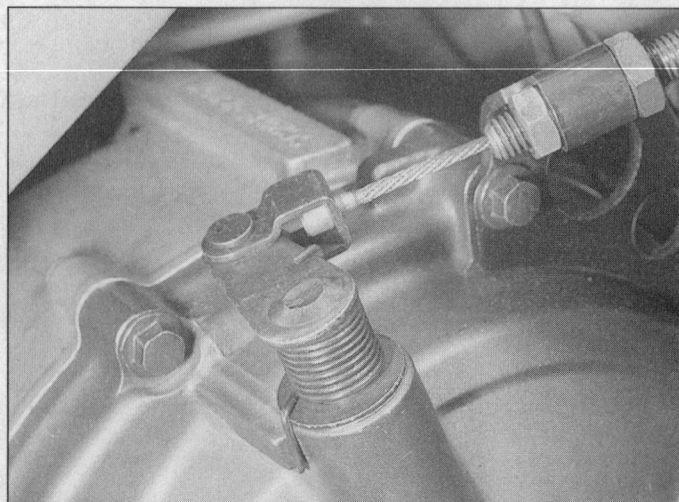

15.11 Correct installation of clutch release lever spring

16.4 Slacken the oil pan bolts evenly in a criss-cross pattern

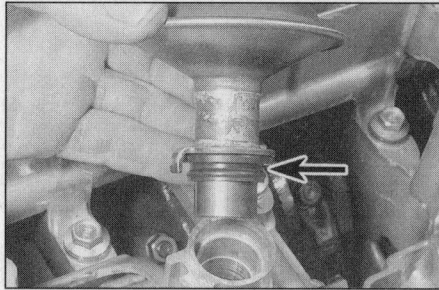

16.10a Oil pump strainer seal is installed with flange facing the strainer (arrow)

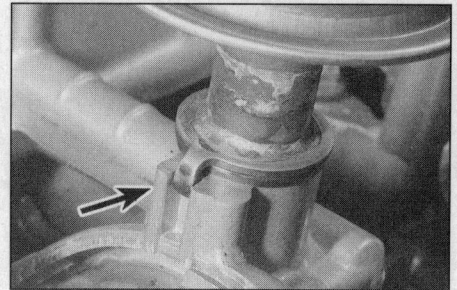

16.10b Tab on strainer must engage slot in crankcase (arrow)

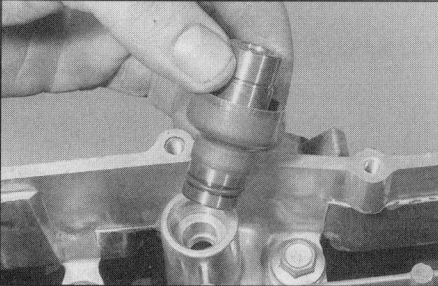

16.11 Oil pressure relief valve is a press fit in crankcase

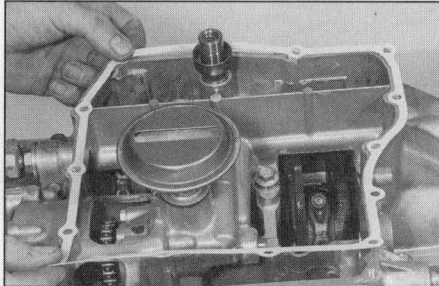

16.12a Always use a new gasket . . .

16.12b . . . when installing the oil pan

16 Oil pan - removal and installation

Note: *The oil pan can be removed with the engine in the frame. If work is being done with the engine removed ignore the preliminary steps.*

Removal

Refer to illustration 16.4

1 Remove the fairing lower section as described in Chapter 8.
2 Remove the complete exhaust system as described in Chapter 4.
3 Drain the engine oil as described in Chapter 1.
4 Working in a criss-cross pattern, gradually loosen the oil pan retaining bolts **(see illustration)**.
5 Remove the eighteen retaining bolts and lower the oil pan away from the crankcase. If the engine is in the frame, note that as the oil pan is removed, the oil pump pick-up strainer and oil pressure relief valve may fall out of the bottom of the crankcase.
6 Recover the gasket and discard it.
7 If they were not released when the oil pan was removed, remove the oil pump pick-up strainer and the oil pressure relief valve from the base of the crankcase. Remove the seal from the pick-up and the O-ring from the pressure relief valve and discard them.
8 Clean the pick-up strainer mesh in solvent. Check it for clogging or splitting and replace if necessary.

Installation

Refer to illustrations 16.10a, 16.10b, 16.11, 16.12a and 16.12b

9 Remove all traces of gasket from the oil pan and crankcase mating surfaces.
10 Fit a new seal to the oil pump pick-up strainer making sure its flange is facing the strainer **(see illustration)**. Fit the strainer to the base of the crankcase aligning its tab with the slot in the crankcase **(see illustration)**.
11 Fit a new O-ring to the oil pressure relief valve groove. Apply a smear of oil to the O-ring and ease the relief valve into position in the base of the crankcase **(see illustration)**.
12 Position a new gasket on the oil pan and fit the pan to the engine **(see illustrations)**. If the engine is in the frame, use a smear of grease on the gasket to hold it in place as the oil pan is installed.
13 Make sure the gasket is correctly positioned then insert the retaining bolts and tighten them by hand.

14 Working in a criss-cross pattern, tighten all the oil pan bolts.
15 Install the exhaust system as described in Chapter 4.
16 Fill the engine with the correct type and quantity of oil as described in Chapter 1. Start the engine and check for leaks.
17 If all is well, fit the fairing lower section as described in Chapter 8.

17 Oil pump - pressure check, removal, inspection and installation

2

Note: *The oil pump can be removed with the engine in the frame.*

Pressure check

Refer to illustration 17.3

1 To check the oil pressure, a suitable gauge and adapter piece (which screws into the oil pressure switch thread) will be needed.
2 Warm the engine up to normal operating temperature then stop it.
3 Peel back the rubber dust cover and disconnect the wire from the oil pressure switch **(see illustration)**.

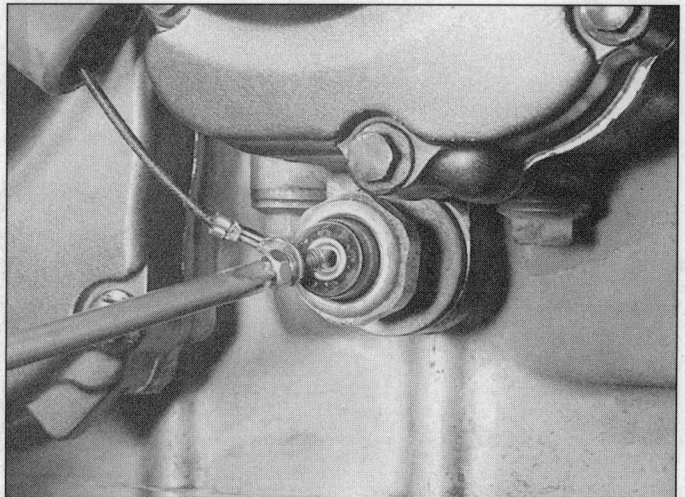

17.3 Oil pressure wire is a screw fit in end of switch

17.11 Insert a screwdriver or similar item through the oil pump sprocket hole to prevent rotation as the bolt is slackened

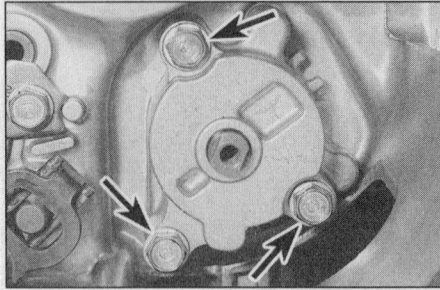

17.13a Undo the three bolts (arrows) . . .

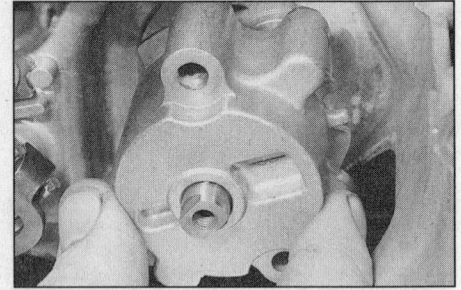

17.13b . . . and remove the oil pump

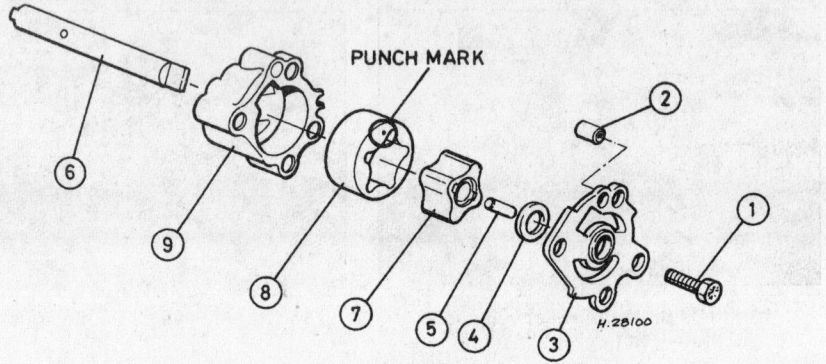

17.14 Exploded view of the oil pump

1 Cover screw	6 Driveshaft
2 Dowel	7 Inner rotor
3 Cover	8 Outer rotor
4 Thrust washer	9 Body
5 Drive pin	

4   Unscrew the switch and swiftly screw the adapter into the crankcase threads. Connect the gauge to the adapter.

5   Start the engine and increase the engine speed to 6000 rpm whilst watching the gauge reading. The oil pressure should be similar to that given in the Specifications at the start of this Chapter.

6   If the pressure is significantly lower than the standard, either the relief valve is stuck open, the oil pump is faulty, the oil pump pick-up strainer is blocked or there is other engine damage. Begin diagnosis by checking the oil pump pick-up strainer and relief valve (see Sections 16 and 18), then the oil pump. If those items check out okay, chances are the bearing oil clearances are excessive and the engine needs to be overhauled.

7   If the pressure is too high, the relief valve is stuck closed. To check it, see Section 18.

8   Stop the engine and unscrew the gauge and adapter from the crankcase.

9   Having applied a smear of sealant to its threads, install the oil pressure switch and tighten it securely. Reconnect its wire terminal and fit the dust cover over the switch.

Removal

Refer to illustrations 17.11, 17.13a and 17.13b

10   Remove the clutch as described in Section 14.

11   Slacken and remove the oil pump sprocket retaining bolt whilst preventing the sprocket turning by inserting a screwdriver through one

of the sprocket holes **(see illustration)**.

12   Slide the drive sprocket off the mainshaft and remove the drive sprocket, chain and pump sprocket as an assembly. Remove the guide bush from the center of the drive sprocket.

13   Undo the three bolts and remove the oil pump from the crankcase **(see illustrations)**.

Inspection

Refer to illustrations 17.14, 17.15, 17.16, 17.21, 17.22, 17.23, 17.26, 17.27, 17.28a, 17.28b and 17.29

14   Wash the oil pump in solvent, then dry it off **(see illustration)**.

15   Remove the pump cover screw **(see illustration)**.

16   Lift off the cover and recover the locating dowel **(see illustration)**.

17   Remove the thrust washer from the pump driveshaft.

18   Slide out the driveshaft, noting which way around it is fitted, and remove the drive pin.

19   Remove the inner and outer rotors from the pump body.

20   Check the pump body and rotors for scoring and wear. If any damage or uneven or excessive wear is evident, replace the pump (individual parts aren't available). If you are rebuilding the engine, it's a good idea to install a new oil pump.

21   Install the rotors in the pump body and measure the clearance between the outer rotor and body with a feeler gauge **(see illustration)**. Compare it to the value listed in this Chapter's Specifications. If it's excessive, replace the pump.

17.15 Undo the screw . . .

17.16 . . . and slide off the pump cover

17.21 Measuring outer rotor-to-body clearance

17.22 Measuring rotor tip clearance

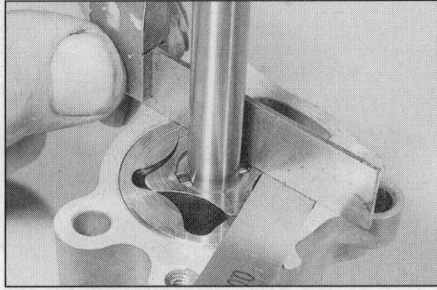

17.23 Measuring rotor endfloat

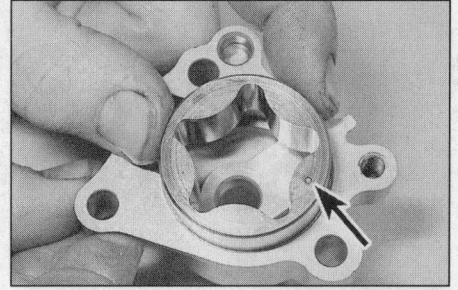

17.26 Install outer rotor, ensuring punch mark (arrow) is facing away from the body

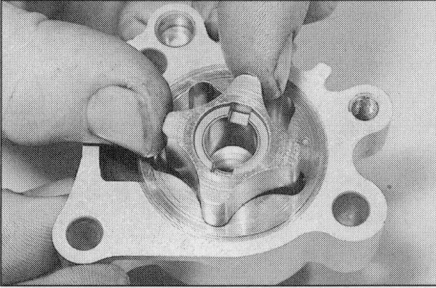

17.27 Install the inner rotor with its slot facing away from the body

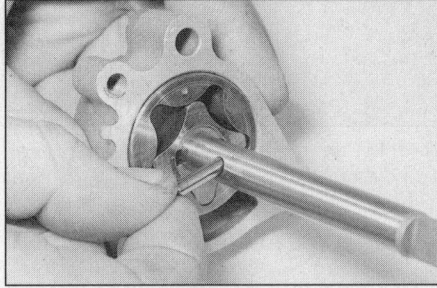

17.28a Insert the driveshaft and drive pin . . .

17.28b . . . making sure pin is correctly located in the inner rotor slot (arrow)

22 Measure the clearance between the inner rotor tip and the outer rotor tip **(see illustration)**. Again, replace the pump if the clearance is excessive.

23 Lay a straightedge across the rotors and pump body and measure the rotor endfloat (gap between the rotors and pump body) with a feeler gauge **(see illustration)**. If it's outside the limits listed in this Chapter's Specifications, replace the pump.

24 Inspect the pump drive chain and sprockets for wear and damage and replace if necessary. If replacement is necessary note that both sprockets and the chain should be replaced as a set.

25 If the pump is good make sure all components are clean and lubricate them with clean engine oil. Reassemble the pump as follows.

26 Fit the outer rotor making sure its punch mark is facing away from the pump body **(see illustration)**.

27 Install the inner rotor making sure its drive pin slot is facing away from the pump body **(see illustration)**.

28 Insert the driveshaft, making sure it is fitted the correct way around and insert the drive pin **(see illustration)**. Slide the driveshaft into

position so that the drive pin is engaged in its slot in the inner rotor **(see illustration)**.

29 Fit the thrust washer to the driveshaft **(see illustration)**.

30 Fit the locating dowel then install the pump cover and securely tighten the retaining screw.

Installation

Refer to illustrations 17.35a, 17.35b and 17.36

31 Lubricate the pump rotors with clean engine oil and check that the pump driveshaft rotates freely.

32 Install the pump in the crankcase making sure its driveshaft end is correctly aligned with the slot in the end of the water pump shaft.

33 Fit the pump bolts and tighten them securely.

34 Engage both the drive sprocket and pump sprocket with the chain noting that the "OUT" mark on the pump sprocket must be positioned so that it will be facing away from the pump when the chain is installed.

35 Fit the guide bush on the mainshaft **(see illustration)**. Install the chain and sprocket assembly over the mainshaft guide bush and pump

2

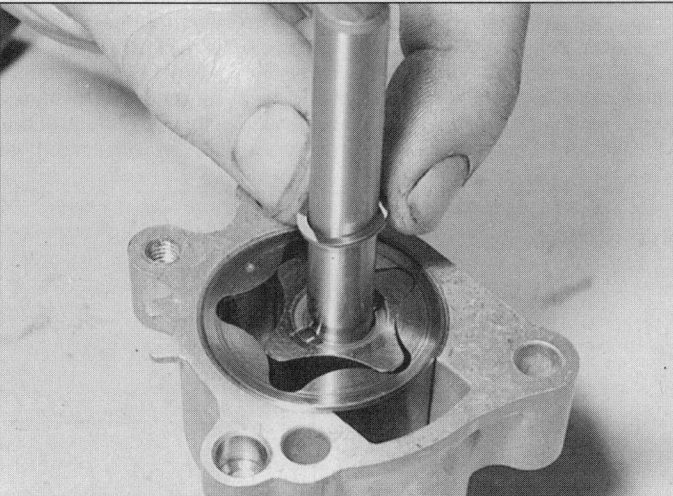

17.29 Do not omit the thrust washer from between the rotor and cover

17.35a Slide the guide bush on the mainshaft . . .

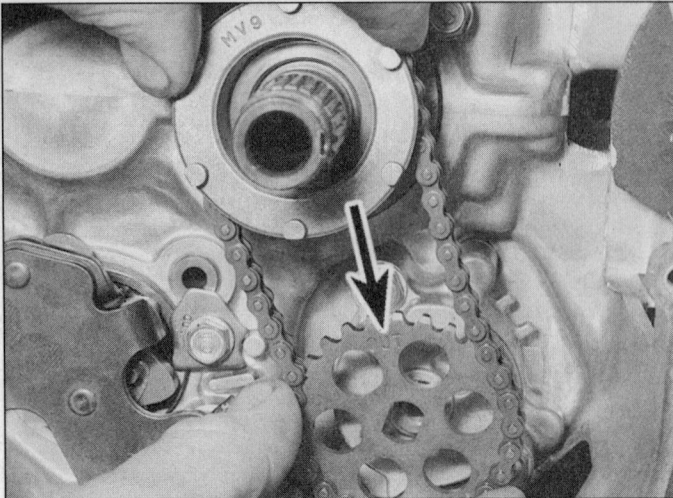

17.35b . . . and install the oil pump chain and sprocket assembly. Pump sprocket OUT mark (arrow) must face outwards

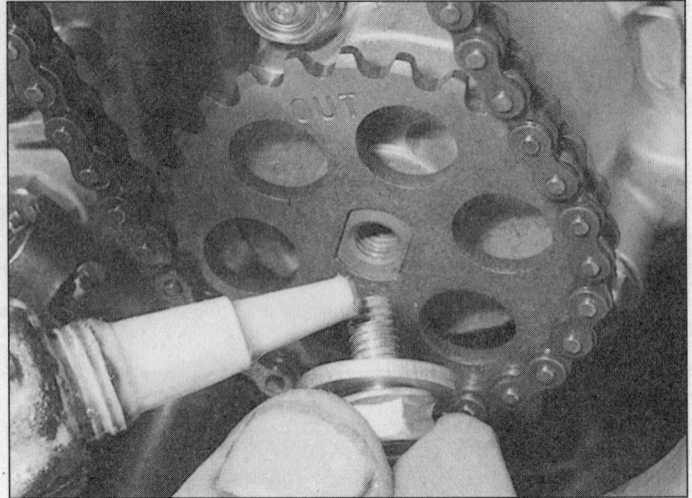

17.36 Apply thread-lock to the threads of the oil pump sprocket retaining bolt prior to installation

driveshaft and check the sprocket is fitted the correct way around **(see illustration)**.

36 Clean the sprocket bolt and apply a few drops of a suitable locking compound to its threads **(see illustration)**.

37 Fit the sprocket bolt and tighten it to the specified torque. Prevent the sprocket from rotating by inserting a screwdriver through one of the sprocket holes **(see illustration 17.11)**.

38 Install the clutch as described in Section 14.

18 Oil pressure relief valve - removal, inspection and installation

Note: *The pressure relief valve can be removed with the engine in the frame.*

Removal

1 Remove the oil pan and pressure relief valve as described in Section 16.

Inspection

2 Push the plunger into the relief valve body and check for free movement. If the valve is sticking it must be replaced (individual parts are not available).

Installation

3 Install the pressure relief valve and oil pan using the procedure in Section 16.

19 Oil cooler - removal and installation

Note: *The oil cooler can be removed with the engine in the frame. If work is being done with the engine removed ignore the preliminary steps.*

Removal

Refer to illustrations 19.4 and 19.5

1 Remove the fairing lower section and middle sections (Chapter 8).

2 Drain the engine oil as described in Chapter 1 and remove the oil filter. **Note:** *If the filter is damaged on removal it must be replaced.*

3 Drain the coolant as described in Chapter 1.

4 Slacken their clamps and disconnect the two coolant hoses from the oil cooler body **(see illustration)**.

5 Undo the bolt and withdraw the oil cooler from the front of the crankcase **(see illustration)**. Recover the large O-ring.

Installation

Refer to illustrations 19.6, 19.7a and 19.7b

6 Install a new O-ring in the oil cooler groove and smear it with engine oil **(see illustration)**.

7 Apply a few drops of non-permanent thread locking agent to the threads of the oil cooler bolt, position the oil cooler on the crankcase so that its slot engages the lug on the crankcase, and tighten the oil cooler bolt to the specified torque setting **(see illustrations)**.

8 Reconnect the coolant hoses and secure them with the clamps.

9 Apply a smear of engine oil to the oil filter O-ring and screw it onto the oil cooler bolt. If the Honda tool is available, it can be secured to a torque setting (see Chapter 1).

10 Refill the engine with oil and the cooling system with coolant as

19.4 Disconnect the water pump hose from the side of the oil cooler . . .

19.5 . . . and the coolant union hose (A) from the top of the oil cooler; oil cooler retaining bolt (B)

19.6 Fit a new O-ring in the groove on the oil cooler body

19.7a Apply thread-lock to the oil cooler bolt threads and align the slot with the crankcase lug (arrow)

19.7b Tighten the oil cooler bolt to the specified torque setting

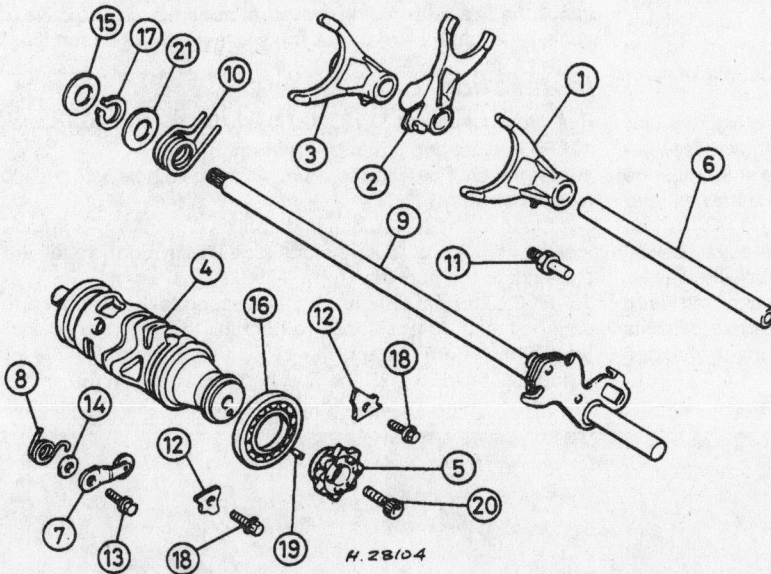

20.3 Gearshift mechanism components

1 Right shift fork
2 Center shift fork
3 Left shift fork
4 Shift drum
5 Shift drum cam
6 Shift fork shaft
7 Stopper arm
8 Spring
9 Gearshift shaft
10 Return spring
11 Spring anchor post
12 Retaining plate
13 Bolt
14 Washer
15 Thrust washer
16 Bearing
17 Snap-ring
18 Bolt
19 Locating pin
20 Bolt
21 Washer (early models only)

described in Chapter 1. Start the engine and check that there are no leaks from the oil cooler.
11 Install the fairing sections as described in Chapter 8.

20 Gearshift mechanism - removal, inspection and installation

Note: *The gearshift mechanism components can be removed with the engine in the frame. If work is being carried out with the engine removed ignore the preliminary steps.*

Removal
Refer to illustrations 20.3, 20.5a and 20.5b
1 Remove the clutch as described in Section 14.
2 Undo the clamp bolt and disconnect the gearshift lever or linkage crank (as applicable) from the engine **(see illustration 5.14)**.
3 Withdraw the gearshift shaft from the right side of the crankcase and recover the thrust washer from the shaft **(see illustration)**.
4 Unscrew the stopper arm bolt and remove the stopper arm, washer and spring, noting their correct fitted positions.
5 Unscrew the shift drum cam bolt; remove the cam **(see illustration)**.

20.5a Unscrew the bolt, remove the shift cam . . .

20.5b . . . and withdraw the locating pin

20.8a Slide the return spring along the gearshift shaft . . .

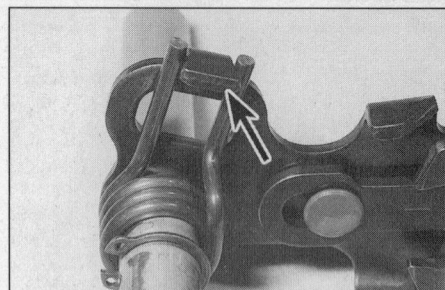

20.8b . . . and engage it with the tab (arrow). Secure it in position with the snap-ring

Remove the locating pin from the shift drum and store it with the cam for safe-keeping **(see illustration)**.

Inspection

Refer to illustrations 20.8a, 20.8b, 20.11a and 20.11b

6   Inspect the gearshift shaft return spring anchor post. If it's worn or damaged, replace it. If it's loose, unscrew it, apply a few drops of thread locking compound to the threads, reinstall the post and tighten it to the specified torque setting.

7   Check the gearshift shaft for straightness and damage to the splines. If the shaft is bent, you can attempt to straighten it, but if the splines are damaged it will have to be replaced.

8   Inspect the gearshift shaft return spring, and selector pawl and spring for damage. The return spring can be replaced individually but if the selector pawl or spring are damaged the complete shaft must be replaced. To replace the return spring, remove the snap-ring and washer (US 1993 and 1994 models, UK CBR900RR-N and RR-P), then slide off the spring. Fit the new spring, making sure it is the correct way around and thus correctly engaged with the shaft, install the washer (where fitted) and secure it in position with the snap-ring **(see illustrations)**. Note that the snap-ring should be fitted with its chamfered edge facing the return spring and must be correctly located in the shaft groove.

9   Check the condition of the stopper lever and spring. Replace the stopper lever if it's worn where it contacts the shift cam. Replace the spring if it's distorted.

10   Inspect the pins on the end of the shift cam. If they're worn or damaged, replace the cam.

11   Check the condition of the oil seal in the crankcase. If it has been leaking, pry it out **(see illustration)**. It's a good idea to replace it in any case, since gaining access to it requires a fair amount of work. Install a new seal with its spring-side facing inwards. It should be possible to install the seal with thumb pressure, but if necessary, drive it in with a socket the same diameter as the seal's outer edge **(see illustration)**.

Installation

Refer to illustrations 20.13, 20.14, 20.15a, 20.15b, 20.16 and 20.18

12   Fit the locating pin to the shift drum.

13   Fit the shift cam to the drum, aligning its hole with the locating pin **(see illustration)**.

14   Clean the shift cam bolt and apply a drop of non-permanent thread locking compound to its threads **(see illustration)**. Install the bolt and tighten it to the specified torque setting.

15   Fit the stopper arm, washer and spring to the stopper arm bolt; the spring should be positioned to hold the stopper arm against the cam **(see illustration)**. Screw the stopper arm bolt into the crankcase. Using a screwdriver, lift the arm and position it on the shift cam. With the arm correctly positioned, tighten the bolt **(see illustration)**.

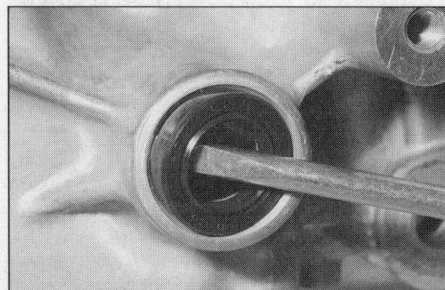

20.11a Prise out the gearshift shaft oil seal with a screwdriver . . .

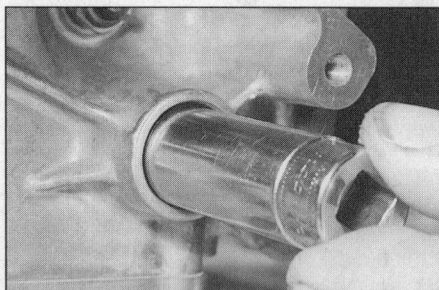

20.11b . . . and press the new seal into position with a suitable socket

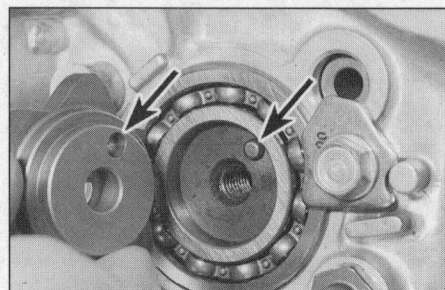

20.13 Install the shift cam making sure its hole is correctly engaged with the locating pin (arrow)

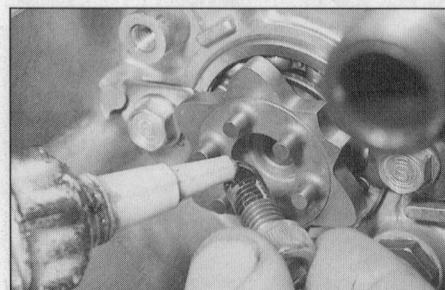

20.14 Apply thread-lock to the shift cam bolt threads before installing it

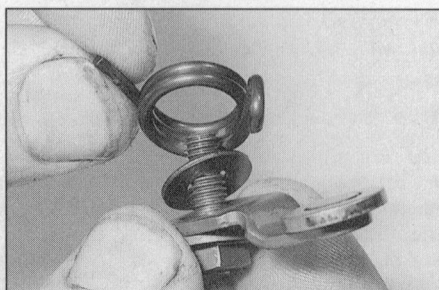

20.15a Fit the stopper arm, washer and spring to the bolt . . .

20.15b . . . screw the assembly into place, ensuring the stopper arm is engaged with the cam's neutral detent as shown

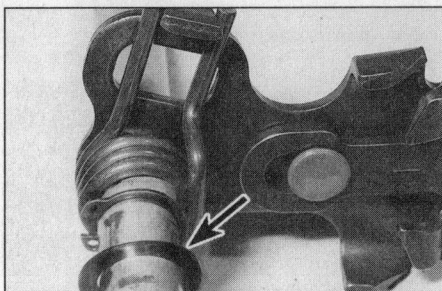

20.16 Slide the thrust washer (arrow) on the gearshift shaft . . .

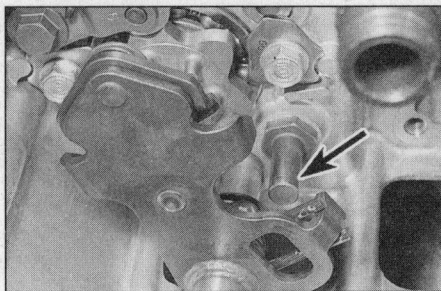

20.18 . . . install the shaft ensuring the spring engages with the anchor post (arrow) - parts shown in neutral position

21.4 Crankshaft left end cover is retained by nine bolts

21.6 Hold the rotor with a large open-ended wrench and loosen the bolt

21.7 Use the special Honda puller to release the alternator rotor

16 Slide the thrust washer on the gearshift shaft **(see illustration)**.
17 Apply a smear of grease to the lip of the gearshift shaft oil seal and wrap the splines of the gearshift shaft with electrical tape, so the splines don't damage the seal as the shaft is installed.
18 Slide the gearshift shaft into the crankcase **(see illustration)**. Ensure the return spring is correctly located over its anchor post and the selector pawls are aligned with the pins on the shift cam. Remove the tape.
19 Align the punch mark on the gearshift lever or linkage crank (as applicable) with the punch mark on the shaft and locate the gearshift lever/linkage crank on the shaft splines **(see illustration 5.14)**. Fit the clamp bolt and tighten it securely.
20 Check the operation of the external shift mechanism then install the clutch as described in Section 14.

---

21 Alternator - removal and installation

Note: *To remove the alternator rotor the special Honda rotor puller, Part Number 07733-0020001, or a pattern equivalent will be required. Do not attempt to remove the rotor using any other method. The alternator rotor can be removed with the engine in the frame. If work is being carried out with the engine removed, ignore the preliminary steps.*

Removal
Refer to illustrations 21.4, 21.6, 21.7 and 21.8

1 Remove the fairing lower section, left middle section and rider's seat as described in Chapter 8.
2 Drain the engine oil as described in Chapter 1.
3 Trace the wiring back from the left crankshaft end cover to the three-pin connector under the seat which contains the yellow alternator wires **(see illustration 5.13a)**. Disconnect the alternator wiring connector and free the wiring from the clips on the inside of the frame and fuel pump bracket.
4 Unscrew the nine cover bolts and withdraw the cover squarely from the engine unit **(see illustration)**. **Note:** *Due to the magnetic pull of the rotor, the cover may prove difficult to remove. Do not pry the cover away with a screwdriver as the mating surfaces will be damaged.*
5 Remove the cover locating dowel from the crankcase and discard the gasket.

6 Slacken and remove the rotor retaining bolt and washer whilst holding the rotor to prevent it turning. The rotor can be retained using a large open-ended wrench on the rotor flats **(see illustration)**. Alternatively, if the engine is still in the frame, the engine can be locked through the transmission by selecting top gear and applying the rear brake hard.
7 Screw the rotor puller tool into the center of the rotor and tighten it securely **(see illustration)**. Sharply tap on the end of puller tool to release the rotor's grip on the tapered crankshaft end. Remove the rotor.
8 Recover the Woodruff key from the crankshaft and store it safely inside the alternator rotor **(see illustration)**.
9 Refer to Chapter 9 for details of alternator stator coil removal.

Installation
Refer to illustrations 21.13, 21.14 and 21.15

10 Degrease the rotor and crankshaft tapers and remove any metal particles of swarf from the rotor magnet. Remove all traces of gasket from the cover and crankcase mating surfaces.
11 Install the Woodruff key in the crankshaft taper.

21.8 Pick the Woodruff key out of the crankshaft

21.13 Oil the threads and underside of the rotor bolt head prior to installation

12 Align the slot in the rotor taper with the Woodruff key and gently push the rotor on the crankshaft making sure the starter gear teeth are correctly meshed. Gently tap the rotor center with a soft-faced hammer to seat it on the crankshaft taper.

13 Oil the rotor bolt threads and underside if its flanged head. Install the washer on the bolt and tighten it to the specified torque setting whilst holding the rotor using the method employed on removal **(see illustration)**.

14 Apply a smear of sealant to the crankshaft left end cover wiring grommet **(see illustration)**.

15 Install the locating dowel in the end cover and fit a new gasket over it **(see illustration)**.

16 Install the end cover on the crankcase and tighten its bolts evenly in a criss-cross pattern to the specified torque setting.

17 Make sure the wiring is correctly routed up to the three-pin connector and secure it with the ties.

18 Fill the engine with the correct type and amount of oil as described in Chapter 1.

19 Install the seat and fairing sections as described in Chapter 8.

22 Starter motor clutch - removal, inspection and installation

Note: *The starter motor clutch can be removed with the engine in the frame.*

Removal

Refer to illustrations 22.2 and 22.3

1 Remove the alternator rotor as described in Section 21.

2 Remove the starter driven gear from the rear of the rotor **(see illustration)**.

3 Slide the idle gear shaft out of the crankcase and remove the idler gear **(see illustration)**.

Inspection

Refer to illustrations 22.5a, 22.5b, 22.7, 22.8 and 22.9

4 Inspect the starter idler gear and driven gear teeth and replace them as a pair if any teeth are chipped or missing. Check the idler shaft and gear bearing surfaces for signs of wear or damage, and replace if necessary.

5 Inspect the starter clutch rollers and driven gear contact surfaces for signs of wear and scoring **(see illustrations)**. The starter clutch rollers should be unmarked with no signs of wear such as pitting or flat spots. The degree of wear on the driven gear can be assessed by measuring the outside diameter of its boss and comparing it to the service limit given in the Specifications. Replace any damaged parts as necessary.

6 Inspect the crankshaft needle roller bearing and driven gear contact surfaces for wear or scoring. Note that at the time of writing; the bearing was not available separately from the crankshaft. Therefore, if the bearing is worn the complete crankshaft assembly must be replaced. Refer to your Honda dealer for the latest information on bearing availability. If the bearing is not available separately, seek the advice of a local machine shop specializing in motorcycle engineering work, to see if they can replace the bearing before replacing the crankshaft assembly. If a replacement bearing can be obtained, the bearing can be drawn off the crankshaft using a suitable puller. Install the new bearing, using a suitable tubular drift which contacts only the bearing inner race.

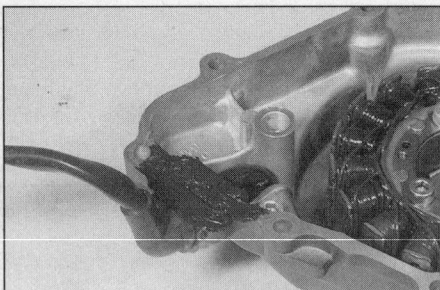

21.14 Smear sealant over the wiring grommet face

21.15 Install a new gasket over the dowel (arrow)

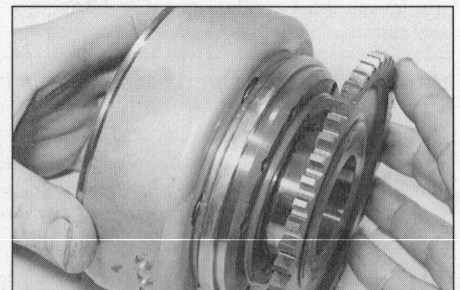

22.2 Remove the starter driven gear from the rear of the alternator rotor

22.3 Withdraw its shaft and remove the starter idler gear from the engine

22.5a Inspect the starter clutch rollers . . .

22.5b . . . and the driven gear surface for signs of wear or damage

22.7 Starter clutch is retained by six bolts (arrows)

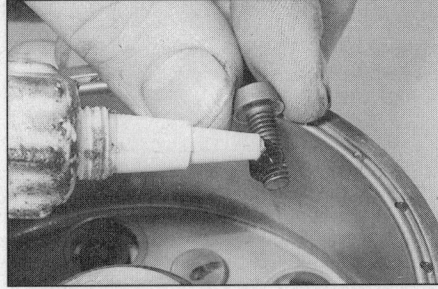

22.8 Apply thread-lock to the starter clutch bolts prior to installation . . .

22.9 . . . and tighten them to the specified torque setting

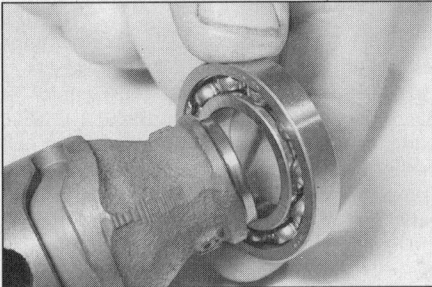

23.11 Fit the bearing to the shift drum . . .

23.12 . . . and insert the drum and bearing into the crankcase

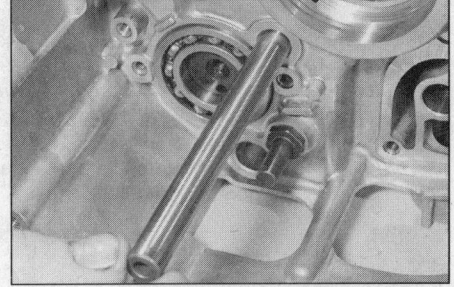

23.13 Locate the shift fork shaft in the crankcase

7   To replace the starter clutch roller assembly, clamp the rotor in a vise equipped with soft jaws then undo the six bolts and remove the clutch from the back off the rotor **(see illustration)**.

8   Clean the starter clutch bolts and apply a drop of non-permanent thread locking compound to their threads **(see illustration)**.

9   Oil the starter clutch rollers, install on the rear of the rotor, then install the bolts and tighten them to the specified torque **(see illustration)**.

Installation

10   Insert the idler gear shaft into the crankcase.

11   Apply a little clean engine oil to the shaft and slide on the idler gear.

12   Fit the driven gear to the rear of the alternator rotor, turning it counterclockwise (anti-clockwise) to help it engage with the starter clutch rollers.

13   Fit the alternator rotor as described in Section 21.

23   Shift drum and forks - removal, inspection and installation

Note: *The shift drum and forks can be removed with the engine in the frame via the oil pan aperture. However, if a complete overhaul is being carried out the shift drum and forks may be left in position until after the crankcase halves have been separated.*

Removal

1   Remove the oil pan as described in Section 16.

2   Remove the gearshift mechanism components (see Section 20).

3   Unscrew the bolts securing the shift drum bearing retaining plates to the crankcase and remove both plates, noting which way around they are fitted.

4   Withdraw the shift fork shaft slowly and remove the shift forks from the crankcase as they are released from the end of the shaft. When all three forks are removed, fully withdraw the shaft from the crankcase.

5   Remove the bearing and withdraw the shift drum from the crankcase.

Inspection

6   The shift forks and shaft should be closely inspected to ensure that they are not badly damaged or worn.

7   Measure the width of both fork ends and the internal diameter of the shaft bore. If either fork end or the shaft bore has worn beyond its service limit the shift fork(s) must be replaced.

8   The shift fork shaft can be checked for trueness by rolling it along a flat surface. A bent shaft will cause difficulty in selecting gears and make the gearshift action heavy. Measure the diameter of the shaft at the points where it is in contact with the shift forks. If the shaft is bent or has worn beyond its service limit at any point it must be replaced.

9   Inspect the shift drum grooves and selector fork guide pins for signs of wear or damage. If either component shows signs of wear or damage the shift fork(s) and drum must be replaced.

10   Check that the shift drum bearing rotates freely, with no freeplay between its inner and outer race. Replace the bearing if necessary.

Installation

Refer to illustrations 23.11, 23.12, 23.13, 23.14, 23.15a thru 23.15c, and 23.17

11   Fit the bearing to the shift drum **(see illustration)**.

12   Lubricate the bearing and shift drum grooves with clean engine oil and insert the shift drum and bearing into position in the crankcase **(see illustration)**.

13   Apply a smear of engine oil to the shift fork shaft and slide the shaft partially into the crankcase **(see illustration)**.

14   The shift forks can be identified by the letter cast on each one; L denotes the left fork, C the center, and R the right **(see illustration)**. **Note:** *All shift forks must be installed in the crankcase so that the letter*

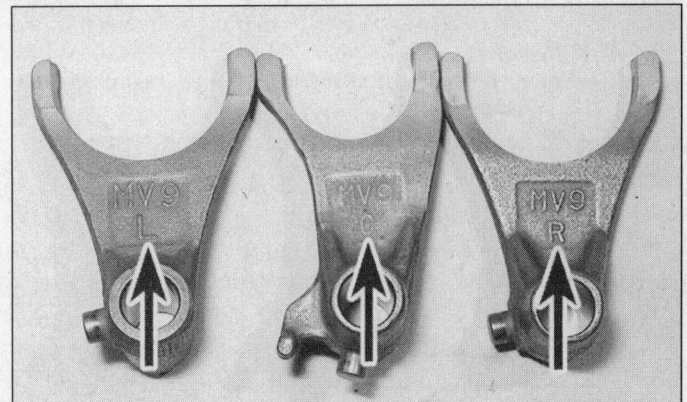

23.14 Each shift fork is marked with a letter to identify its correct fitted position (arrows)

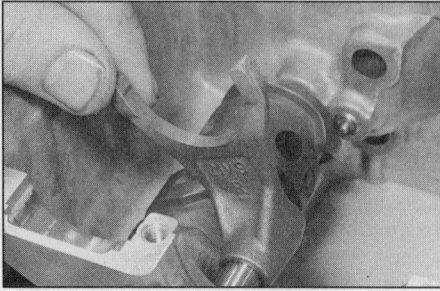

23.15a Making sure all shift forks are fitted with their markings facing the right, install the right fork . . .

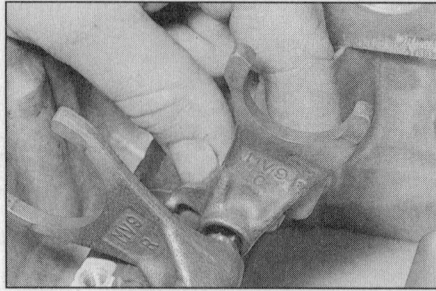

23.15b . . . followed by the center fork . . .

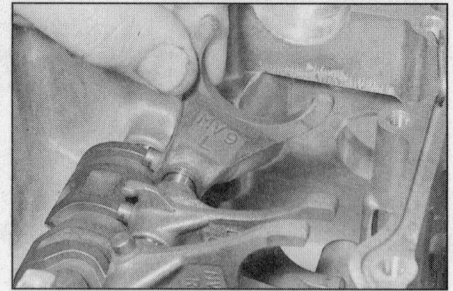

23.15c . . . and finally the left fork

on each one faces towards the right side of the casing (clutch).

15 Locate the right fork with its grooves in the gear and shift drum and slide in the shift fork shaft until it engages with the fork. Repeat the process for the center and left fork and push the shaft fully home, ensuring each fork is positioned as in Step 14 **(see illustrations)**.

16 Clean the shift drum bearing plate bolts and apply a drop of non-permanent locking compound to their threads.

17 Fit the bearing retaining plates making sure the "OUT" marks are facing outwards (away from the crankcase) **(see illustration)**. Clean the threads of the retaining plate bolts and apply a drop of non-permanent locking compound to them. Install the bolts and tighten them securely.

18 Install the gearshift mechanism components (see Section 20).

19 Install the oil pan as described in Section 16.

24 Crankcase - separation and reassembly

Separation

Refer to illustrations 24.5, 24.6 and 24.9

1 To examine and repair or replace the crankshaft, pistons and connecting rods, bearings and transmission components, the crankcase must be split into two parts.

2 To enable the crankcases to be split the engine must be removed from the frame (see Section 5) and the following components first removed with reference to the relevant Sections.

a) Camshafts and followers
b) Cam chain and blades
c) Cylinder head
d) Clutch
e) Oil pan
f) Oil pump
g) Gearshift mechanism components
h) Alternator rotor

Note: *If the crankcase halves are being separated just to examine the transmission shafts or crankshaft then there is no need to remove the cylinder head.*

3 If a complete engine overhaul is planned, remove the starter motor (Chapter 9), the starter clutch idler gear (Section 22), the oil galley plug **(see illustration 24.27)**, the water pump and coolant union (Chapter 3), the oil cooler (Section 19) and the neutral switch (Chapter 9).

4 With all the relevant components removed proceed as follows.

5 With the crankcase the right way up, slacken and remove the four 6 mm and two 8 mm bolts from the top of the crankcase **(see illustration)**. **Note:** *As each bolt is removed, store it in its relative position in a cardboard template of the crankcase halves. This will ensure all bolts are installed in the correct location on reassembly. Discard the sealing washers fitted to five of the bolts; new ones must be fitted on reassembly.*

6 Turn the crankcase upside down. Working around the crankcase, unscrew the fourteen 6 mm lower crankcase bolts **(see illustration)**. **Note:** *As each bolt is removed, store it in its relative position in a cardboard template of the crankcase halves. This will ensure all bolts are installed in the correct location on reassembly. Unscrew the single 10 mm bolt from the left rear corner of the crankcase.*

7 Working in a criss-cross pattern, starting from the outside and working inwards, gradually slacken the ten 9 mm lower crankcase bolts **(see illustration 24.6)**. Once all the bolts are loose, unscrew and remove them along with their sealing washers.

8 Carefully lift the lower crankcase half, leaving the crankshaft and transmission shafts in the upper half of the crankcase. As the upper half is lifted away take care not to dislodge or lose any main bearing inserts. **Note:** *If it won't come easily away, make sure all fasteners have been removed. Don't pry against the crankcase mating surfaces or they will leak; initial separation can be achieved by tapping gently with a soft-faced mallet.*

9 Remove the three locating dowels from the upper crankcase half **(see illustration)**.

10 Remove the two oil jets from the upper crankcase half noting which way around they are fitted. **Note:** *The jets differ in profile - take note of their exact location.*

Reassembly

Refer to illustrations 24.15, 24.17, 24.21 and 24.27

11 Install the crankshaft and transmission shafts in the upper

23.17 Fit bearing retaining plates (OUT mark facing outwards) ensuring each is engaged with the crankcase peg (arrow)

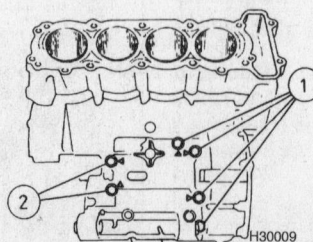

24.5 Crankcase upper half bolts
1 6 mm bolts   2 8 mm bolts

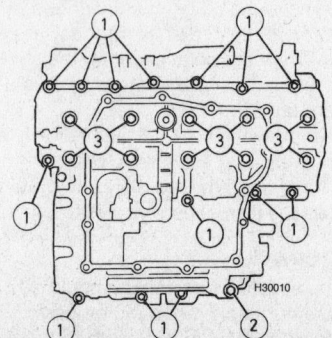

24.6 Crankcase lower half bolts
1 6 mm bolts   2 10 mm bolt   3 9 mm bolts

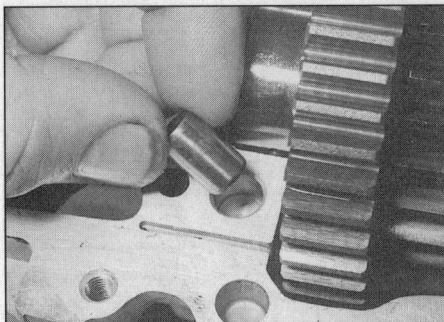

24.9 **Remove the three dowels if they are loose**

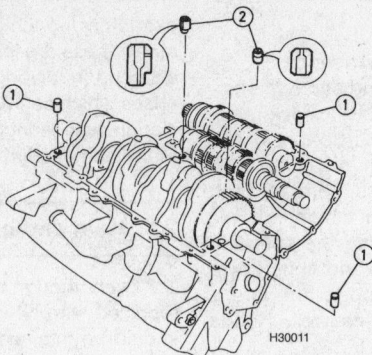

24.15 **Crankcase dowels (1) and oil jets (2)**

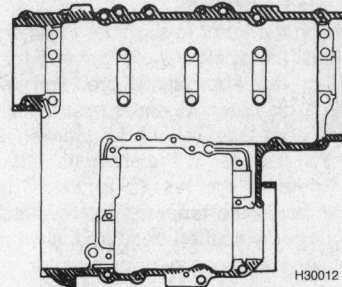

24.17 **Apply sealant to the shaded areas of the upper crankcase half**

crankcase as described in Sections 29 and 30.

12 Remove all traces of sealant from the crankcase mating surfaces, being careful not to let any fall into the case as this is done.

13 Check that all components are installed and that they can rotate smoothly and easily.

14 Lubricate the transmission shafts and crankshaft with clean engine oil then use a rag soaked in high flash-point solvent to wipe over the gasket surfaces of both halves to remove all traces of oil.

15 Make sure the oil jet holes are clear. Fit both jets to the upper crankcase half **(see illustration)**.

16 Install the three locating dowels in the upper crankcase half **(see illustration 24.15)**.

17 Apply a small amount of suitable sealant to the mating surface of the upper crankcase half **(see illustration)**. **Caution:** *Don't apply too much sealant, as it will ooze out when the case halves are assembled and may obstruct oil passages and prevent the bearings from seating.*

18 Check the position of the shift cam, shift forks and transmission shafts - make sure they're in the neutral position (ie the mainshaft and countershaft rotate independently of each other).

19 Make sure that the main bearing inserts are in position and carefully lower the lower crankcase half onto the upper half. The shift forks must engage with their respective slots in the transmission gears as the halves are joined.

20 Check that the lower crankcase half is correctly seated and that all shafts are free to rotate. **Note:** *If the casings are not correctly seated, remove the lower crankcase half and investigate the problem. Do not attempt to pull them together using the crankcase bolts as the casing will crack and be ruined.*

21 Clean the threads of the ten 9 mm lower crankcase bolts thoroughly, then apply engine oil to their threads and undersides of their heads. Install new sealing washers on the bolts and insert them in their original locations **(see illustration)**.

22 Starting from the center and working outwards in a criss-cross pattern, tighten all the bolts to approximately half the specified torque. Go around a second time in the same sequence and tighten them to the full torque setting given in the Specifications at the start of this Chapter.

23 Fit the single 10 mm crankcase bolt and tighten it to the specified torque setting.

24 Fit the fourteen 6 mm bolts in their original locations and tighten them to the specified torque setting.

25 Turn the crankcase over so that it is upright and install the two 8 mm upper crankcase bolts with new sealing washers; tighten them to the specified torque. Install new sealing washers on the three 6 mm bolt which locate in the holes marked with the triangles **(see illustration 24.5)**, then install all four 6 mm bolts and tighten them to the specified torque setting.

26 With all crankcase fasteners tightened, check that the crankshaft and transmission shafts rotate smoothly and easily. If there are any signs of undue stiffness or of any other problem, the fault must be rectified before proceeding further.

27 Fit a new O-ring to the oil gallery plug. Apply a smear of oil to the O-ring to ease installation and insert the plug into position in the crankcase **(see illustration)**.

28 Install all other removed assemblies in the reverse of the sequence given in Steps 3 and 2.

25 Crankcase - inspection and servicing

1 After the crankcases have been separated and the crankshaft and transmission components have been removed, the crankcases should be cleaned thoroughly with new solvent and dried with compressed air.

24.21 **Lubricate the threads and underside of the heads of the 9 mm lower crankcase bolts prior to installation**

24.27 **Fit a new O-ring (arrow) to the oil galley plug and install it in the right end of the crankcase**

Cylinder bores

Note: *Don't attempt to separate the liners from the cylinder block.*

2   Check the cylinder walls carefully for scratches and score marks.

3   Using the appropriate precision measuring tools, check each cylinder's diameter. Measure near the top, center and bottom of the cylinder bore, parallel to the crankshaft axis. Next, measure each cylinder's diameter at the same three locations across the crankshaft axis. Compare the results to this Chapter's Specifications. If the cylinder bores are tapered, out-of-round, worn beyond the specified limits, or badly scuffed or scored, have them rebored and honed by a dealer service department or a motorcycle repair shop. If a rebore is done, oversize pistons and rings will be required as well. Honda produce two sizes of oversize pistons (see Section 27).

4   As an alternative, if the precision measuring tools are not available, a dealer service department or motorcycle repair shop will make the measurements and offer advice concerning servicing of the cylinders.

5   If they are in reasonably good condition and not worn to the outside of the limits, and if the piston-to-cylinder clearances can be maintained properly (see Section 27), then the cylinders do not have to be rebored; honing is all that is necessary.

6   To perform the honing operation you will need the proper size flexible hone with fine stones, or a "bottle brush" type hone, plenty of light oil or honing oil, some shop towels and an electric drill motor. Hold the upper crankcase half in a vise (cushioned with soft jaws or wood blocks) when performing the honing operation. Mount the hone in the drill motor, compress the stones and slip the hone into the top of the cylinder. Lubricate the cylinder thoroughly, turn on the drill and move the hone up and down in the cylinder at a pace which will produce a fine crosshatch pattern on the cylinder wall with the crosshatch lines intersecting at approximately a 60° angle. Be sure to use plenty of lubricant and do not take off any more material than is absolutely necessary to produce the desired effect. Do not withdraw the hone from the cylinder while it is running. Instead, shut off the drill and continue moving the hone up and down in the cylinder until it comes to a complete stop, then compress the stones and withdraw the hone. Wipe the oil out of the cylinder and repeat the procedure on the other cylinders. Remember, do not remove too much material from the cylinder wall. If you do not have the tools, or do not desire to perform the honing operation, a dealer service department or motorcycle repair shop will generally do it for a reasonable fee.

7   Next, the cylinders must be thoroughly washed with warm soapy water to remove all traces of the abrasive grit produced during the honing operation. Be sure to run a brush through the bolt holes and flush them with running water. After rinsing, dry the cylinders thoroughly and apply a coat of light, rust-preventative oil to all machined surfaces.

Crankcase castings

8   Remove any oil passage plugs that haven't already been removed. All oil passages should be blown out with compressed air.

9   All traces of old gasket sealant should be removed from the mating surfaces. Minor damage to the surfaces can be cleaned up with a fine sharpening stone or grindstone. **Caution:** *Be very careful not to nick or gouge the crankcase mating surfaces or leaks will result. Check both crankcase halves very carefully for cracks and other damage.*

10 Small cracks or holes in aluminum castings may be repaired with an epoxy resin adhesive as a temporary measure. Permanent repairs can only be effected by argon-arc welding, and only a specialist in this process is in a position to advise on the economy or practical aspect of such a repair. If any damage is found that can't be repaired, replace the crankcase halves as a set.

11 Damaged threads can be economically reclaimed by using a diamond section wire insert, of the Helicoil type, which is easily fitted after drilling and re-tapping the affected thread. Most motorcycle dealers and small engineering firms offer a service of this kind.

12 Sheared studs or screws can usually be removed with screw

extractors, which consist of a tapered, left thread screws of very hard steel. These are inserted into a pre-drilled hole in the stud, and usually succeed in dislodging the most stubborn stud or screw. If a problem arises which seems beyond your scope, it is worth consulting a professional engineering firm before condemning an otherwise sound casing. Many of these firms advertise regularly in the motorcycle press.

26 Main and connecting rod bearings - general note

1   Even though main and connecting rod bearings are generally replaced with new ones during the engine overhaul, the old bearings should be retained for close examination as they may reveal valuable information about the condition of the engine.

2   Bearing failure occurs mainly because of lack of lubrication, the presence of dirt or other foreign particles, overloading the engine and/or corrosion. Regardless of the cause of bearing failure, it must be corrected before the engine is reassembled to prevent it from happening again.

3   When examining the bearings, remove the main bearings from the case halves and the rod bearings from the connecting rods and caps and lay them out on a clean surface in the same general position as their location on the crankshaft journals. This will enable you to match any noted bearing problems with the corresponding crankshaft journal.

4   Dirt and other foreign particles get into the engine in a variety of ways. It may be left in the engine during assembly or it may pass through filters or breathers. It may get into the oil and from there into the bearings. Metal chips from machining operations and normal engine wear are often present. Abrasives are sometimes left in engine components after reconditioning operations such as cylinder honing, especially when parts are not thoroughly cleaned using the proper cleaning methods. Whatever the source, these foreign objects often end up imbedded in the soft bearing material and are easily recognized. Large particles will not imbed in the bearing and will score or gouge the bearing and journal. The best prevention for this cause of bearing failure is to clean all parts thoroughly and keep everything spotlessly clean during engine reassembly. Frequent and regular oil and filter changes are also recommended.

5   Lack of lubrication or lubrication breakdown has a number of interrelated causes. Excessive heat (which thins the oil), overloading (which squeezes the oil from the bearing face) and oil leakage or throw off from excessive bearing clearances, worn oil pump or high engine speeds all contribute to lubrication breakdown. Blocked oil passages will also starve a bearing and destroy it. When lack of lubrication is the cause of bearing failure, the bearing material is wiped or extruded from the steel backing of the bearing. Temperatures may increase to the pint where the steel backing and the journal turn blue from overheating.

6   Riding habits can have a definite effect on bearing life. Full throttle low speed operation, or lugging (labouring) the engine, puts very high loads on bearings, which tend to squeeze out the oil film. These loads cause the bearings to flex, which produces fine cracks in the bearing face (fatigue failure). Eventually the bearing material will loosen in pieces and tear away from the steel backing. Short trip riding leads to corrosion of bearings, as insufficient engine heat is produced to drive off the condensed water and corrosive gases produced. These products collect in the engine oil, forming acid and sludge. As the oil is carried to the engine bearings, the acid attacks and corrodes the bearing material.

7   Incorrect bearing installation during engine assembly will lead to bearing failure as well. Tight fitting bearings which leave insufficient bearing oil clearances result in oil starvation. Dirt or foreign particles trapped behind a bearing insert result in high spots on the bearing which lead to failure.

8   To avoid bearing problems, clean all parts thoroughly before reassembly, double check all bearing clearance measurements and lubricate the new bearings with clean engine oil during installation.

27.3 On removal make sure the bearing insert stays in the cap

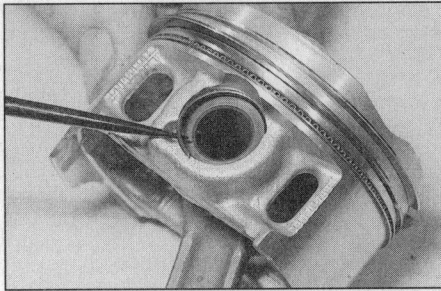

27.8 Using a scriber to pry out a piston pin snap-ring

27.11 Remove the piston rings with a ring removal and installation tool

27 Piston/connecting rod assemblies - removal, inspection, bearing selection, oil clearance check and installation

Removal

Refer to illustrations 27.3 and 27.8

1   Separate the crankcase halves as described in Section 24. Before removing the piston/connecting rods from the crankshaft measure the side clearance of each rod with a feeler gauge. If the clearance on any rod is greater than the service limit listed in this Chapter's Specifications, that rod will have to be replaced with a new one.

2   Using a center punch or paint, mark the relevant cylinder number on each connecting rod and bearing cap (No. 1 cylinder on left end).

3   Unscrew the bearing cap nuts and withdraw the cap, complete with the lower bearing insert, from each of the four connecting rods **(see illustration)**. Push the connecting rods up and off their crankpins, then remove the upper bearing insert. Keep the cap, nuts and (if they are to be reused) the bearing inserts together in their correct sequence.

4   Remove the ridge of carbon from the top of each cylinder bore. If there is a pronounced wear ridge on the top of each bore, remove with a ridge reamer.

5   Push each piston/connecting rod assembly up and remove it from the top of the bore making sure the connecting rod does not mark the cylinder bore walls. **Caution:** *Do not try to remove the piston/connecting rod from the bottom of the cylinder bore. The piston will not pass the crankcase main bearing webs. If the piston is pulled right to the bottom of the bore the oil control ring will expand and lock the piston in position. If this happens it is likely the ring will be broken.*

6   Immediately install the relevant bearing cap, inserts and nuts on each piston/connecting rod assembly so that they are all kept together as a matched set.

7   Using a sharp scriber, scratch the number of each piston into its crown (or use a suitable marker pen if the piston is clean enough).

8   Support the first piston and, using a small screwdriver or scriber, carefully pry out a snap-ring from the piston groove **(see illustration)**.

9   Push the piston pin out from the opposite end to free the piston from the rod. You may have to deburr the area around the groove to enable the pin to slide out (use a triangular file for this procedure). If the pin is tight, use a piston pin puller (see *Maintenance techniques, tools and working facilities*) or tap it out using a suitable hammer and punch, taking care not to damage the piston. Repeat the procedure for the other pistons.

Inspection

Pistons

Refer to illustrations 27.11, 27.18, 27.19 and 27.20

10   Before the inspection process can be carried out, the pistons must be cleaned and the old piston rings removed.

11   Using a piston ring removal and installation tool, carefully remove the rings from the pistons **(see illustration)**. Do not nick or gouge the pistons in the process.

12   Scrape all traces of carbon from the tops of the pistons. A hand-held wire brush or a piece of fine emery cloth can be used once most of the deposits have been scraped away. Do not, under any circumstances, use a wire brush mounted in a drill motor to remove deposits from the pistons; the piston material is soft and will be eroded away by the wire brush.

13   Use a piston ring groove cleaning tool to remove any carbon deposits from the ring grooves. If a tool is not available, a piece broken off an old ring will do the job. Be very careful to remove only the carbon deposits. Do not remove any metal and do not nick or gouge the sides of the ring grooves.

14   Once the deposits have been removed, clean the pistons with solvent and dry them thoroughly. Make sure the oil return holes below the oil ring grooves are clear.

15   If the pistons are not damaged or worn excessively and if the cylinders are not to be rebored, new pistons will not be necessary. Normal piston wear appears as even, vertical wear on the thrust surfaces of the piston and slight looseness of the top ring in its groove. New piston rings, on the other hand, should always be used when an engine is rebuilt.

16   Carefully inspect each piston for cracks around the skirt, at the pin bosses and at the ring lands.

17   Look for scoring and scuffing on the thrust faces of the skirt, holes in the piston crown and burned areas at the edge of the crown. If the skirt is scored or scuffed, the engine may have been suffering from overheating and/or abnormal combustion, which caused excessively high operating temperatures. The oil pump and oil cooler should be checked thoroughly. A hole in the piston crown, an extreme to be sure, is an indication that abnormal combustion (pre-ignition) was occurring. Burned areas at the edge of the piston crown are usually evidence of spark knock (detonation). If any of the above problems exist, the causes must be corrected or the damage will occur again.

18   Measure the piston ring-to-groove clearance by laying a new piston ring in the ring groove and slipping a feeler gauge in beside it **(see illustration)**. Check the clearance at three or four locations around the groove. Be sure to use the correct ring for each groove; they are different. If the clearance is greater than the service limit, new pistons will have to be used when the engine is reassembled.

27.18 Measuring piston ring-to-groove clearance

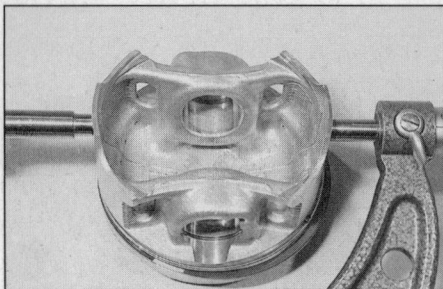

27.19 Measuring piston diameter

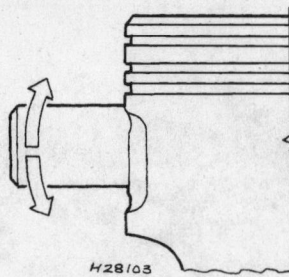

27.20 Slip the pin into the piston and try to wiggle it back-and-forth; if it's loose, replace the piston and pin

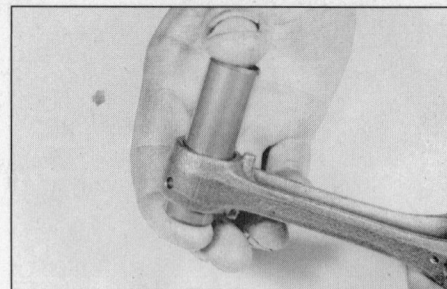

27.23 Slip the piston pin into the rod and rock it back-and-forth to check for looseness

19 Calculate the piston-to-bore clearance by measuring the bore (see Section 25) and the piston diameter. Make sure that the pistons and cylinders are correctly matched. Measure the piston across the skirt on the thrust faces at a 90° angle to the piston pin, 15 mm (0.6 inch) up from the bottom of the skirt **(see illustration)**. Subtract the piston diameter from the bore diameter to obtain the clearance. If it is greater than specified, the cylinders will have to be rebored and new oversized pistons and rings installed.

20 Apply clean engine oil to the pin, insert it into the piston and check for freeplay by rocking the pin back-and-forth **(see illustration)**. If the pin is loose, new pistons and pins must be installed. If the necessary measuring equipment is available measure the pin diameter and piston pin bore and check the readings obtained do not exceed the limits given in this Chapter's Specifications. Replace components that are worn beyond the specified limit.

21 If the pistons are to be replaced, ensure the correct size of piston is ordered. Honda produce two oversizes of piston as well as the standard piston. The piston oversizes available are: +0.50 mm and +1.0 mm. **Note:** *Oversize pistons have their relevant size stamped on top of the piston crown, eg. a 0.50 mm oversize piston will be marked 0.50.*

22 Install the rings on the pistons as described in Section 28.

Connecting rods

Refer to illustration 27.23

23 Check the connecting rods for cracks and other obvious damage Lubricate the piston pin for each rod, install it in its original rod and check for play **(see illustration)**. If it wobbles, replace the connecting rod and/or the pin. If the necessary measuring equipment is available measure the pin diameter and connecting rod bore and check the readings obtained do not exceed the limits given in this Chapter's Specifications. Replace components that are worn beyond the specified limit.

24 Refer to Section 26 and examine the connecting rod bearing inserts If they are scored, badly scuffed or appear to have been seized, new bearings must be installed. Always replace the bearings in the connecting rods as a set. If they are badly damaged, check the corresponding crankpin. Evidence of extreme heat, such as discol-

oration, indicates that lubrication failure has occurred. Be sure to thoroughly check the oil pump and pressure relief valve as well as all oil holes and passages before reassembling the engine.

25 Have the rods checked for twist and bending at a dealer service department or other motorcycle repair shop.

26 If a connecting rod is to replaced, it is essential that the new rod is of the correct weight group to minimise vibration. The weight is indicated by a letter stamped on the big-end cap of each rod. This letter together with the connecting rod size mark (see Step 29) should be quoted when purchasing new connecting rod(s). **Note:** *Honda state that the maximum difference between any two connecting rods in the engine must be only one weight group.*

Bearing selection

Refer to illustrations 27.27, 27.28 and 27.29

27 The connecting rod bearing running clearance is controlled in production by selecting one of five grades of bearing insert. The grades are indicated by a color-coding marked on the edge of each insert **(see illustration)**. In order, from the thickest to the thinnest, the insert grades are: Blue, Black, Brown, Green and Yellow. New bearing inserts are selected as follows using the crankpin and connecting rod size markings.

28 The standard crankpin journal diameter is divided into three size groups to allow for manufacturing tolerances. The size group of each crankpin can be determined by the letters which are stamped on the left end crank web **(see illustration)**. **Note:** *Ignore the numbers as these refer to the main bearing journals.* On the web will be an L followed by four letters, made up of the letters A, B and C, for example L ABAB. The letters indicate the diameter of each crankpin, starting with the left crankpin (No. 1 cylinder) and finishing with the right crankpin (no. 4 cylinder). If the equipment is available, these marks can be checked by direct measurement.

29 The connecting rods are also divided into three size groups to allow for manufacturing tolerances. The size group is in the form of numbers (either 1, 2 or 3) **(see illustration)**. **Note:** *Ignore the letter as this indicates the weight group of the connecting rod.* If the equipment is available, these marks can be checked by direct measurement.

30 Match the relevant connecting rod code with its crankshaft code

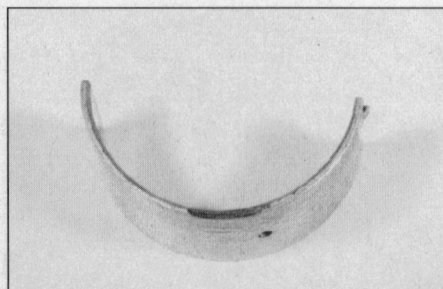

27.27 The color code is painted on the side of the bearing

27.28 Crankshaft crankpin journal identification marks (A) and main bearing journal identification marks (B)

27.29 Connecting rod bearing diameter number and weight group letter

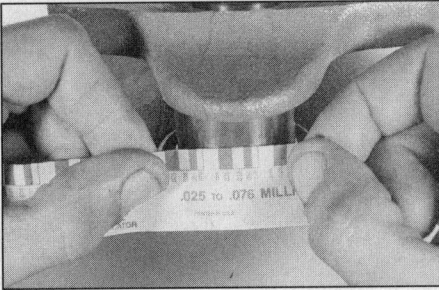

27.37 Place the Plastigage scale next to the flattened Plastigage to measure the bearing clearance

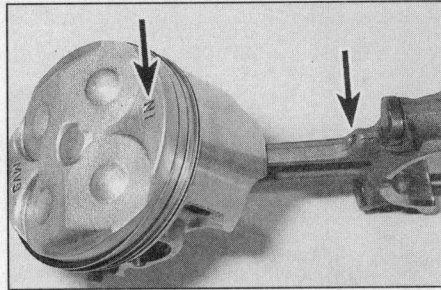

27.41 When coupling the rod and piston make sure the piston IN mark is on the same side as the rod oilway (arrows)

27.42 Make sure both piston pin snap-rings are securely seated in the piston grooves

and select a new set of bearing inserts using the following table.

Connecting rod mark	Crankshaft mark	Insert color
1	A	Yellow
1	B	Green
1	C	Brown
2	A	Green
2	B	Brown
2	C	Black
3	A	Brown
3	B	Black
3	C	Blue

Oil clearance check

Refer to illustration 27.37

31 Whether new bearing inserts are being fitted or the original ones are being re-used, the connecting rod bearing oil clearance should be checked prior to reassembly.

32 Clean the backs of the bearing inserts and the bearing locations in both the connecting rod and bearing cap.

33 Press the bearing inserts into their locations, ensuring that the tab on each insert engages in the notch in the connecting rod/bearing cap. Make sure the bearings are fitted in the correct locations and take care not to touch any insert's bearing surface with your fingers.

34 There are two possible ways of checking the oil clearance, the first method is by direct measurement (Steps 35 and 38) and the second by the use of a product known as Plastigage (Steps 36 to 38).

35 If the first method is to be used, fit the bearing cap to the connecting rod, with the bearing inserts in place. Make sure the cap is fitted the correct way around so the connecting rod and bearing cap weight/size markings are correctly aligned. Tighten the cap retaining nuts to the specified torque and measure the internal diameter of each assembled pair of bearing inserts. If the diameter of each corresponding crankpin journal is measured and then subtracted from the bearing internal diameter, the result will be the connecting rod bearing oil clearance.

36 If the second method is to be used, cut several lengths of the appropriate size Plastigage (they should be slightly shorter than the width of the crankpin). Place a strand of Plastigage on each (cleaned) crankpin journal and fit the (clean) piston/connecting rod assemblies, inserts and bearing caps. Make sure the cap is fitted the correct way around so the connecting rod and bearing cap weight/size markings are correctly aligned and tighten the bearing cap nuts to the specified torque wrench setting whilst ensuring that the connecting rod does not rotate. Take care not to disturb the Plastigage. Slacken the bearing cap nuts and remove the connecting rod assemblies, again taking great care not to rotate the crankshaft.

37 Compare the width of the crushed Plastigage on each crankpin to the scale printed on the Plastigage envelope to obtain the connecting rod bearing oil clearance **(see illustration)**.

38 If the clearance is not within the specified limits, the bearing inserts may be the wrong grade (or excessively worn if the original inserts are being reused). Before deciding that different grade inserts are needed, make sure that no dirt or oil was trapped between the bearing inserts and the connecting rod or bearing cap when the clearance was measured. If the clearance is excessive, even with new inserts (of the

correct size), the crankpin is worn and the crankshaft should be replaced.

39 On completion carefully scrape away all traces of the Plastigage material from the crankpin and bearing inserts using a fingernail or other object which is unlikely to score the inserts.

Installation

Refer to illustrations 27.41, 27.42, 27.44, 27.47, 27.48, 27.51a and 27.51b

40 Check that each piston has one new snap-ring fitted to it and insert the piston pin from the opposite side. If it is a tight fit, the piston should be warmed first. If the original pistons/connecting rods are being installed, use the marks made on disassembly to ensure each piston is fitted to its correct connecting rod.

41 Lubricate the piston pin and connecting rod bores with clean engine oil and fit each piston to its respective connecting rod making sure that the IN mark on the crown of the piston is on the same side as the connecting rod oilway **(see illustration)**.

42 Push the piston pin through both piston bosses and the connecting rod bore. If necessary the pin can be tapped carefully into position, using a hammer and suitable drift, whilst supporting the connecting rod and piston. Secure each piston pin in position with a second new snap-ring, making sure it is correctly seated in the piston groove **(see illustration)**.

43 Clean the backs of the bearing inserts and the bearing recesses in both the connecting rod and bearing cap. If new inserts are being fitted, ensure that all traces of the protective grease are cleaned off using kerosene (paraffin). Wipe dry the inserts and connecting rods with a lint-free cloth.

44 Press the bearing inserts into their locations. Ensure the tab on each insert engages in the notch in the connecting rod or bearing cap **(see illustration)**. Ensure the bearings are fitted in the correct locations and don't touch any insert's bearing surface with your fingers.

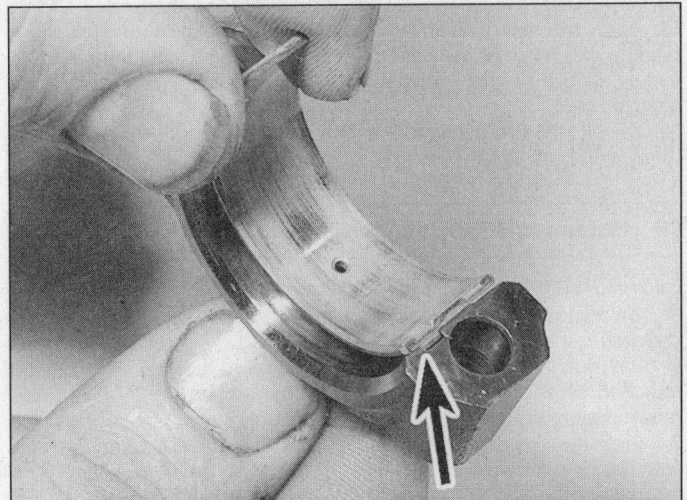

27.44 Install the bearings making sure each insert tab (arrow) is correctly engaged in its slot

27.47 Clamp the piston rings in position with a ring compressor . . .

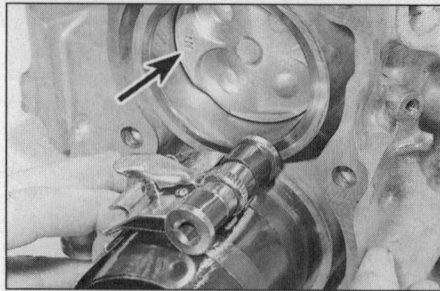

27.48 . . . insert the piston/connecting rod assembly; ensure each piston IN mark (arrow) is on the intake side of the bore

27.51a Lubricate the threads and heads of the connecting rod bearing cap nuts . . .

27.51b . . . and tighten the nuts to the specified torque as described in text

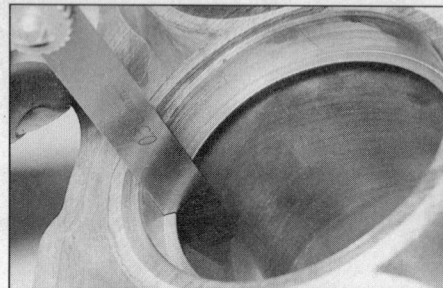

28.3 Measuring piston ring end gap

45 Lubricate the cylinder bores, the pistons and piston rings then lay out each piston/connecting rod assembly in its respective position.

46 Starting with assembly No. 1, position the top and second ring end gaps so they are 180° apart then position the oil control ring side rails so that their end gaps are 180° apart.

47 With the piston rings correctly positioned, clamp them in position with a piston ring compressor **(see illustration)**.

48 Insert the piston/connecting rod assembly into the top of its bore, taking care not to allow the connecting rod to mark the bore **(see illustration)**. Make sure the IN mark on the piston crown is on the intake side of the bore and push the piston into the bore until the piston crown is flush with the top of the bore.

49 Ensure that the connecting rod bearing insert is still correctly installed. Taking care not to mark the cylinder bores, liberally lubricate the crankpin and both bearing inserts, then pull the piston/connecting rod assembly down its bore and onto the crankpin.

50 Fit the bearing cap and insert to the connecting rod. Make sure the cap is fitted the correct way around so the connecting rod and bearing cap weight/size markings are correctly aligned **(see illustration 27.29)**.

51 Apply a smear of clean engine oil the threads and underside of the bearing cap nuts. Fit the nuts to the connecting rod and tighten them evenly, in two or three stages, to the specified torque setting **(see illustrations)**.

52 Check that the crankshaft is free to rotate easily, then install the three remaining assemblies in the same way.

28 Piston rings - installation

Refer to illustrations 28.3, 28.5, 28.9a, 28.9b and 28.11

1 Before installing the new piston rings, the ring end gaps must be checked.

2 Lay out the pistons and the new ring sets so the rings will be matched with the same piston and cylinder during the end gap measurement procedure and engine assembly.

3 Insert the top ring into the top of the first cylinder and square it up with the cylinder walls by pushing it in with the top of the piston. The ring should be about 25 mm below the top edge of the cylinder. To measure the end gap, slip a feeler gauge between the ends of the ring and compare the measurement to the Specifications **(see illustration)**.

4 If the gap is larger or smaller than specified, double check to make sure that you have the correct rings before proceeding.

5 If the gap is too small, it must be enlarged or the ring ends may come in contact with each other during engine operation, which can cause serious damage. The end gap can be increased by filing the ring ends very carefully with a fine file. When performing this operation, file only from the outside in **(see illustration)**.

6 Excess end gap is not critical unless it is greater than 1 mm. Again, double check to make sure you have the correct rings for your engine.

7 Repeat the procedure for each ring that will be installed in the first cylinder and for each ring in the remaining cylinders. Remember to keep the rings, pistons and cylinders matched up.

8 Once the ring end gaps have been checked/corrected, the rings can be installed on the pistons.

9 The oil control ring (lowest on the piston) is installed first. It is composed of three separate components. Slip the expander into the groove, then install the upper side rail. Do not use a piston ring

28.5 If the end gap is too small, clamp a file in a vise and file the ring ends (from outside in only) to enlarge the gap slightly

28.9a Installing the oil expander ring -
make sure the ends don't overlap

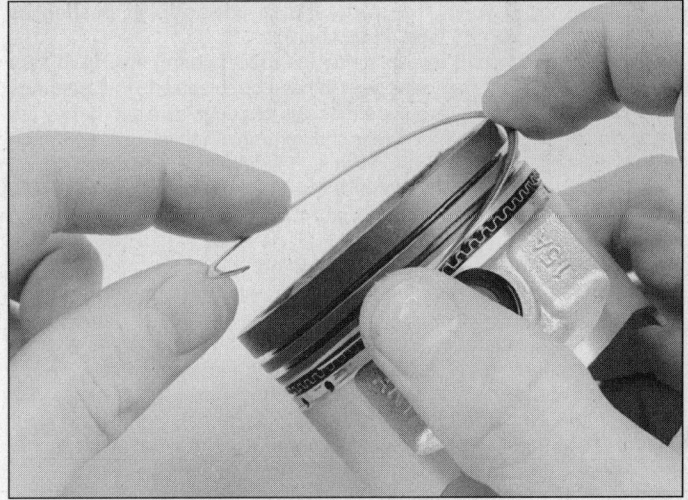

28.9b Installing an oil ring side rail - don't use a
ring installation tool to do this

installation tool on the oil ring side rails as they may be damaged. Instead, place one end of the side rail into the groove between the expander and the ring land. Hold it firmly in place and slide a finger around the piston while pushing the rail into the groove. Next, install the lower side rail in the same manner (see illustrations).

10 After the three oil ring components have been installed, check to make sure that both the upper and lower side rails can be turned smoothly in the ring groove.

11 Install the second (middle) ring next. **Note:** *The second ring and top rings are slightly different in profile.* To avoid breaking the ring, use a piston ring installation tool and make sure that the identification mark (the letters RN or T) is facing up (see illustration). Fit the ring into the middle groove on the piston. Do not expand the ring any more than is necessary to slide it into place.

12 Finally, install the top ring in the same manner. Make sure the identifying mark (the letter R or T) is facing up.

13 Repeat the procedure for the remaining pistons and rings.

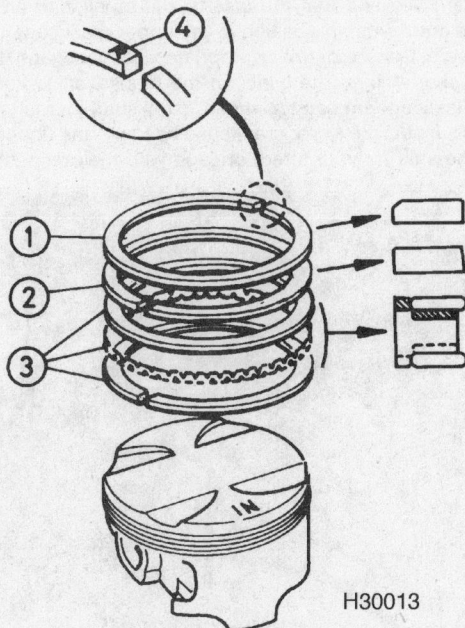

H30013

28.11 Piston ring profiles

1	Top ring	3	Oil ring
2	Second (middle) ring	4	Identification mark location

29 Crankshaft and main bearings - removal, inspection, bearing selection, oil clearance check and installation

Removal

Refer to illustration 29.3

1 Separate the crankcase halves as described in Section 24.

2 Remove the piston/connecting rod assemblies as described in Section 27. **Note:** *If no work is to be carried out on the piston/ connecting rod assemblies there is no need to remove them from the bores. The cylinder head can be left in position although the connecting rod bearing caps should be removed (see Section 27, Steps 1 through 3) and the pistons pushed up to the top of the bores so that the connecting rod ends are positioned clear of the crankshaft.*

3 Lift the crankshaft out of the upper crankcase half, taking care not to dislodge the bearing inserts (see illustration).

4 The main bearing inserts can be removed from the crankcase halves by pushing their centers to the side, then lifting them out. Keep the bearing inserts in order.

Inspection

Refer to illustration 29.5

5 Clean the crankshaft with solvent, using a rifle-cleaning brush to scrub out the oil passages. If available, blow the crank dry with

29.3 Lifting the crankshaft out of position

compressed air. Inspect the starter clutch needle roller bearing as described in Section 22 **(see illustration)**.

6   Refer to Section 26 and examine the main bearing inserts. If they are scored, badly scuffed or appear to have been seized, new bearings must be installed. Always replace the main bearings as a set. If they are badly damaged, check the corresponding crankshaft journal. Evidence of extreme heat, such as discoloration, indicates that lubrication failure has occurred. Be sure to thoroughly check the oil pump and pressure relief valve as well as all oil holes and passages before reassembling the engine.

7   The crankshaft journals should be given a close visual examination, paying particular attention where damaged bearing inserts have been discovered. If the journals are scored or pitted in any way a new crankshaft will be required. Note that undersizes are not available, precluding the option of re-grinding the crankshaft.

8   Set the crankshaft on V-blocks and check the runout with a dial indicator touching the center main bearing journal, comparing your findings with this Chapter's Specifications. If the runout exceeds the limit, replace the crank.

Bearing selection

Refer to illustration 29.11

9   The main bearing running clearance is controlled in production by selecting one of five grades of bearing insert. The grades are indicated by a color-coding marked on the edge of each insert **(see illustration 27.27)**. In order, from the thickest to the thinnest, the insert grades are: Black, Brown, Green, Yellow and Pink. New bearing inserts are selected as follows using the crankshaft journal and crankcase main bearing bore size markings.

10   The standard crankshaft journal diameter is divided into three size groups to allow for manufacturing tolerances. The size group of each journal can be determined by the numbers which are stamped on the left end crank web **(see illustration 27.28)**. Note: *Ignore the letters as these refer to the crankpin journals.* On the web will be an L followed by five numbers, made up of the numbers 1, 2 and 3, for example L 12121. The numbers indicate the diameter of each crankshaft journal, starting with the left journal and finishing with the right journal. If the equipment is available, these marks can be checked by direct measurement.

11   The crankcase main bearing bore diameters are divided into three size groups to allow for manufacturing tolerances. The size group of each main bearing bore can be determined using the five letters stamped on the left end of the upper crankcase half **(see illustration)**. These will be made up of the letters A, B or C. The first letter indicates the diameter of the left journal, and the last the diameter of the right journal. If the equipment is available, these marks can be checked by direct measurement.

12   Match the relevant crankcase code with its crankshaft code and select a new set of bearing inserts using the following table.

Crankshaft mark	Crankcase mark	Insert color
1	A	Pink
1	B	Yellow
1	C	Green
2	A	Yellow
2	B	Green
2	C	Brown
3	A	Green
3	B	Brown
3	C	Black

Oil clearance check

13   Whether new bearing inserts are being fitted or the original ones are being re-used, the main bearing oil clearance should be checked prior to reassembly.

14   Clean the backs of the bearing inserts and the bearing locations in both crankcase halves.

15   Press the bearing inserts into their locations, ensuring that the tab on each insert engages in the notch in the crankcase. Make sure the bearings are fitted in the correct locations and take care not to touch any insert's bearing surface with your fingers.

16   There are two possible ways of checking the oil clearance, the first method is by direct measurement (see Step 17 and 23) and the second by the use of a product known as Plastigage (see Steps 18 to 23).

17   If the first method is to be used, with the main bearing inserts in position, carefully lower the lower crankcase half onto the upper half. Make sure that the shift forks (if fitted) engage with their respective slots in the transmission gears as the halves are joined. Check that the lower crankcase half is correctly seated. **Note:** *Do not tighten the crankcase bolts if the casing is not correctly seated.* Install the ten 9 mm lower crankcase bolts in their original locations and, starting from the center and working outwards in a criss-cross pattern, tighten them to the specified torque setting. Measure the internal diameter of each assembled pair of bearing inserts. If the diameter of each corresponding crankshaft journal is measured and then subtracted from the bearing internal diameter, the result will be the connecting rod bearing oil clearance.

18   If the second method is to be used, ensure the main bearing inserts are correctly fitted and that the inserts and crankshaft are clean and dry. Lay the crankshaft in position in the upper crankcase.

19   Cut several lengths of the appropriate size Plastigage (they should be slightly shorter than the width of the crankshaft journal). Place a strand of Plastigage on each (cleaned) crankshaft journal.

20   Carefully install the lower crankcase half onto the upper half. Make sure that the shift forks (if fitted) engage with their respective slots in

29.5   Inspect the starter clutch needle roller bearing for signs of wear or damage

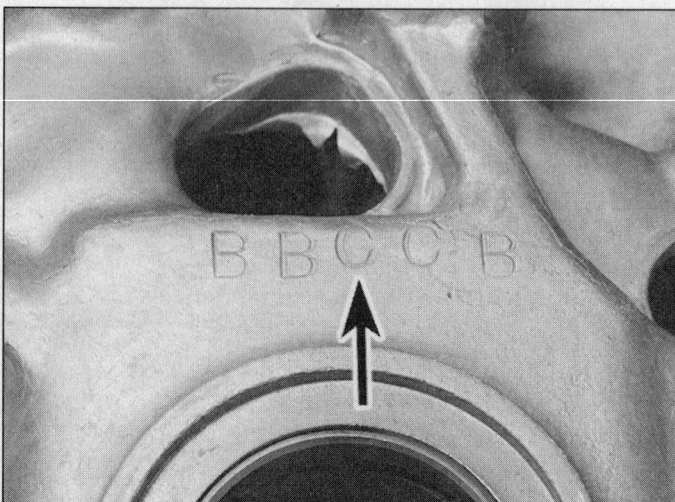

29.11   Crankcase main bearing bore diameter marks (arrow) are stamped on the left end of the upper crankcase half

the transmission gears as the halves are joined. Check that the lower crankcase half is correctly seated. **Note:** *Do not tighten the crankcase bolts if the casing is not correctly seated.* Install the ten 9 mm lower crankcase bolts in their original locations and, starting from the center and working outwards in a criss-cross pattern, tighten them to the specified torque setting. Make sure that the crankshaft is not rotated as the bolts are tightened.

21 Slacken and remove the crankcase bolts, working in a criss-cross pattern from the outside in, then carefully lift off the lower crankcase half, making sure the Plastigage is not disturbed.

22 Compare the width of the crushed Plastigage on each crankshaft journal to the scale printed on the Plastigage envelope to obtain the main bearing oil clearance **(see illustration 27.37)**.

23 If the clearance is not within the specified limits, the bearing inserts may be the wrong grade (or excessively worn if the original inserts are being reused). Before deciding that different grade inserts are needed, make sure that no dirt or oil was trapped between the bearing inserts and the crankcase halves when the clearance was measured. If the clearance is excessive, even with new inserts (of the correct size), the crankshaft journal is worn and the crankshaft should be replaced.

24 On completion carefully scrape away all traces of the Plastigage material from the crankshaft journal and bearing inserts; use a fingernail or other object which is unlikely to score the inserts.

Installation

Refer to illustrations 29.26 and 29.27

25 Clean the backs of the bearing inserts and the bearing recesses in both crankcase halves. If new inserts are being fitted, ensure that all traces of the protective grease are cleaned off using kerosene (paraffin). Wipe dry the inserts and crankcase halves with a lint-free cloth.

26 Press the bearing inserts into their locations. Make sure the tab on each insert engages in the notch in the casing **(see illustration)**. Make sure the bearings are fitted in the correct locations and take care not to touch any insert's bearing surface with your fingers.

27 Lubricate the bearing inserts in the upper crankcase with clean engine oil **(see illustration)**.

28 Lower the crankshaft into position in the upper crankcase.

29 Fit the piston/connecting rod assemblies to the crankshaft as described in Section 27 if they were disconnected.

30 Reassemble the crankcase halves as described in Section 24.

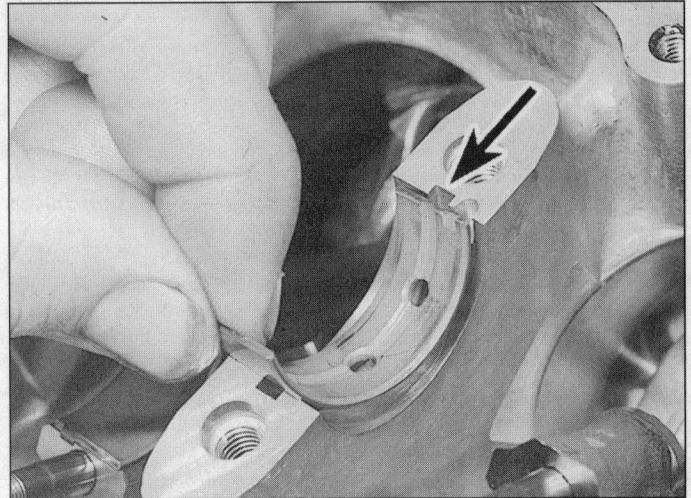

29.26 Install the main bearing inserts, making sure each insert tab (arrow) is correctly engaged in the crankcase slot

30 Transmission shafts - removal and installation

Removal

Refer to illustrations 30.2, 30.3a, 30.3b, 30.4a and 30.4b

1 Separate the crankcase halves as described in Section 24.

2 Unscrew the mainshaft bearing retaining plate bolts and remove the plate from the right side of the upper crankcase half, noting which way around it is fitted **(see illustration)**.

3 Lift the countershaft and mainshaft out of the crankcase **(see illustrations)**. Do not lose the end plate from the right end of the countershaft.

4 Recover the countershaft bearing half ring and dowel pin from the upper crankcase half and store them with the transmission shafts for safe-keeping **(see illustrations)**.

5 Remove the oil seal from the end of the countershaft and discard it, a new one must be used on installation.

29.27 Lubricate all bearing inserts with clean oil prior to installing the crankshaft

30.2 Undo the two bolts and remove the bearing retaining plate . . .

30.3a . . . then lift the countershaft . . .

30.3b . . . and mainshaft out of the crankcase

30.4a Recover the countershaft bearing half ring . . .

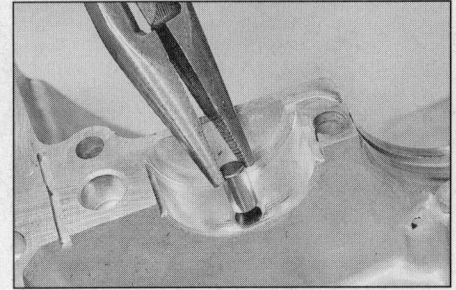

30.4b . . . and dowel pin from the upper crankcase half

30.8 Fit a new oil seal to the countershaft making sure that its sealing lip is facing inwards

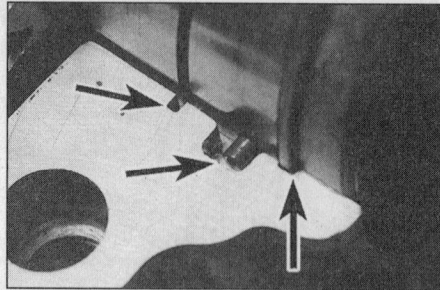

30.10a Lower the countershaft into position making sure the oil seal lip and bearing pin (arrows) . . .

30.10b . . . and the bearing race and dowel pin are correctly aligned

6   If necessary, the transmission shafts can be disassembled and inspected for wear or damage as described in Section 31.

Installation

Refer to illustrations 30.8, 30.10a, 30.10b and 30.13

7   Install the countershaft bearing half ring and dowel pin in the upper crankcase.

8   Slide a new oil seal on the end of the countershaft making sure it is fitted with its sealing lip (spring side) facing inwards **(see illustration)**.

9   Lower the mainshaft into position in the upper crankcase. Make sure the locating pin in the mainshaft bearing outer race is correctly located in the groove in the crankcase.

10   Lower the countershaft into position in the crankcase half. Make sure the oil seal lip, bearing pin and groove on the left end of the shaft are correctly engaged with the crankcase slots and bearing half ring, and the hole in the needle bearing race on the right end of the shaft is correctly engaged with the dowel pin **(see illustrations)**.

11   Make sure both transmission shafts are correctly seated. **Caution:** *If the mainshaft bearing locating pin and/or countershaft half ring or dowel pin are not correctly engaged, the crankcase halves will not seat correctly.*

12   Clean the mainshaft bearing plate bolts and apply a few drops of non-permanent locking compound to their threads.

13   Fit the bearing retaining plate to the crankcase making sure its OUTSIDE mark is facing outwards (away from the crankcase). Install the retaining bolts and tighten them securely **(see illustration)**.

14   Position the gears in the neutral position and check the shafts are free to rotate easily before proceeding further.

15   Reassembly the crankcase halves as described in Section 24.

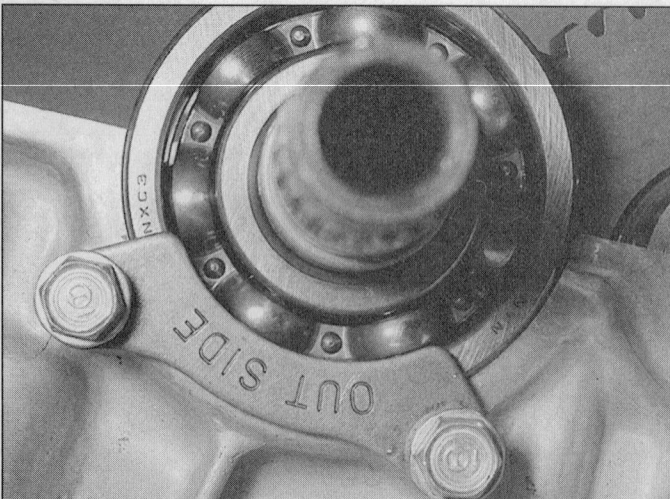

30.13 Ensure the mainshaft bearing retaining plate is fitted with its OUT SIDE mark facing outwards

31   Transmission shafts - disassembly, inspection and reassembly

Note: *When disassembling the transmission shafts, place the parts on a long rod or thread a wire through them to keep them in order and facing the proper direction.*

1   Remove the shafts from the casing as described in Section 30.

Mainshaft

Disassembly

Refer to illustration 31.2

2   Slide off the needle roller bearing and thrust washer from the left end of the shaft **(see illustration)**.

3   Slide off the 2nd gear.

4   Disengage the lock washer from the special splined washer and slide both off the mainshaft.

5   Remove the 6th gear followed by its splined bushing and thrust washer.

6   Remove the snap-ring using a suitable pair of snap-ring pliers.

7   Remove the 3rd/4th gear, noting which way around it is fitted.

8   Remove the second snap-ring.

9   Slide off the splined thrust washer followed by the 5th gear, 5th gear bushing and thrust washer.

Inspection

10   Wash all of the components in clean solvent and dry them off.

11   Check the gear teeth for cracking and other obvious damage. Check the gear bushings and the surface in the inner diameter of each gear for scoring or heat discoloration. If the gear or bushing is damaged, replace it.

12   Inspect the dogs and the dog holes in the gears for excessive wear. Replace the paired gears as a set if necessary.

13   The shafts is unlikely to sustain damage unless the engine has seized, placing an unusually high loading on the transmission, or the machine has covered a very high mileage. Check the surface of the shaft, especially where a pinion turns on it, and replace the shaft if it has scored or picked up. Inspect the threads of the shafts and check them for trueness by setting them up in V-blocks and measuring any runout with a dial gauge. Damage of any kind can only be cured by replacement.

14   Measure the internal diameter of all gears which run on bushings and the external diameter of the bushings which they run on. If either component has worn to beyond its service limit it must be replaced. Using the above measurements calculate the gear to bushing clearance, if this exceeds the specified limit replace the relevant gear and bushing as a pair.

15   Check that the outer race of the bearing fitted to the end of the shaft rotates freely and has no freeplay between its inner and outer races. If the bearing requires replacement, a bearing puller will be required to extract the bearing from its shaft. Note the position of the locating groove in the outer race of the bearing prior to removing it; ensure that the new bearing is fitted with the groove in the same position. Pull the bearing off the shaft and fit the new bearing using a hammer and tubular

31.2 Mainshaft components

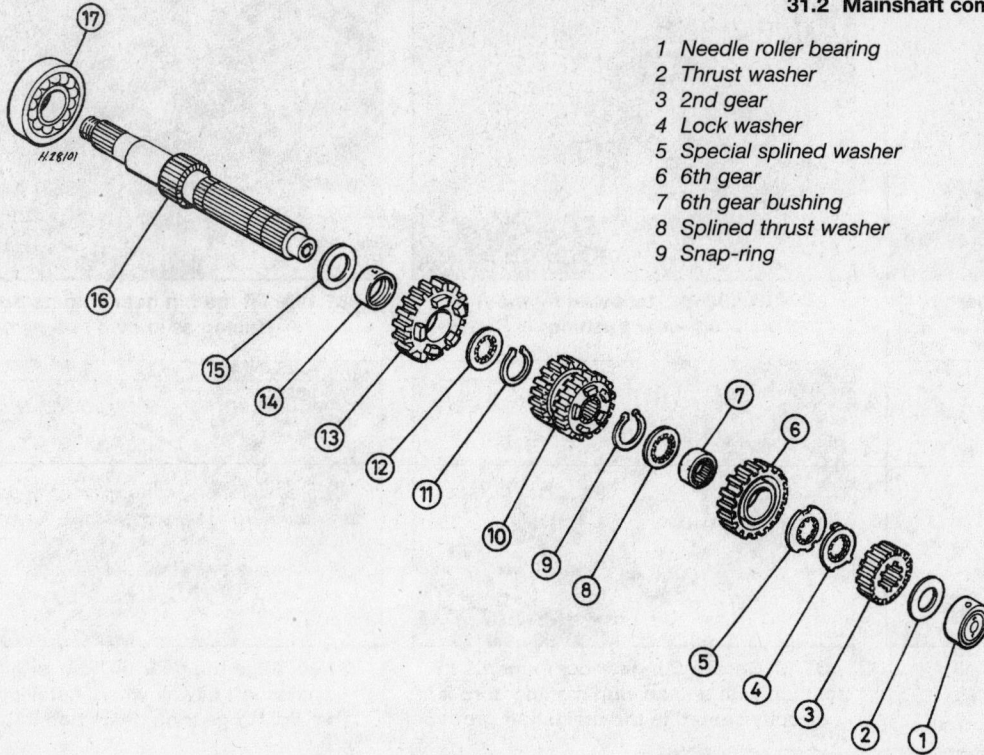

1 Needle roller bearing	10 3rd/4th gear
2 Thrust washer	11 Snap-ring
3 2nd gear	12 Splined thrust washer
4 Lock washer	13 5th gear
5 Special splined washer	14 5th gear bushing
6 6th gear	15 Thrust washer
7 6th gear bushing	16 Mainshaft
8 Splined thrust washer	17 Bearing
9 Snap-ring	

H.28101

drift which bears only on the inner race of the bearing.

Reassembly

Refer to illustrations 31.16, 31.17, 31.18a to 31.18c, 31.19, 31.20, 31.21, 31.22, 31.23a, 31.23b, 31.24a, 31.24b, 31.25a, 31.25b and 31.26

16 During reassembly, always use new snap-rings. Lubricate the components with engine oil before assembling them. **Note:** *If the thrust washers and snap-rings are examined closely it will be seen that they are chamfered on one side. During reassembly it is crucial that each thrust washer and snap-ring is fitted so its chamfer is on the correct side* **(see illustration)**.

17 Slide on the thrust washer making sure its chamfered edge is

2

31.16 Correct fitted orientation of mainshaft snap-rings and thrust washers

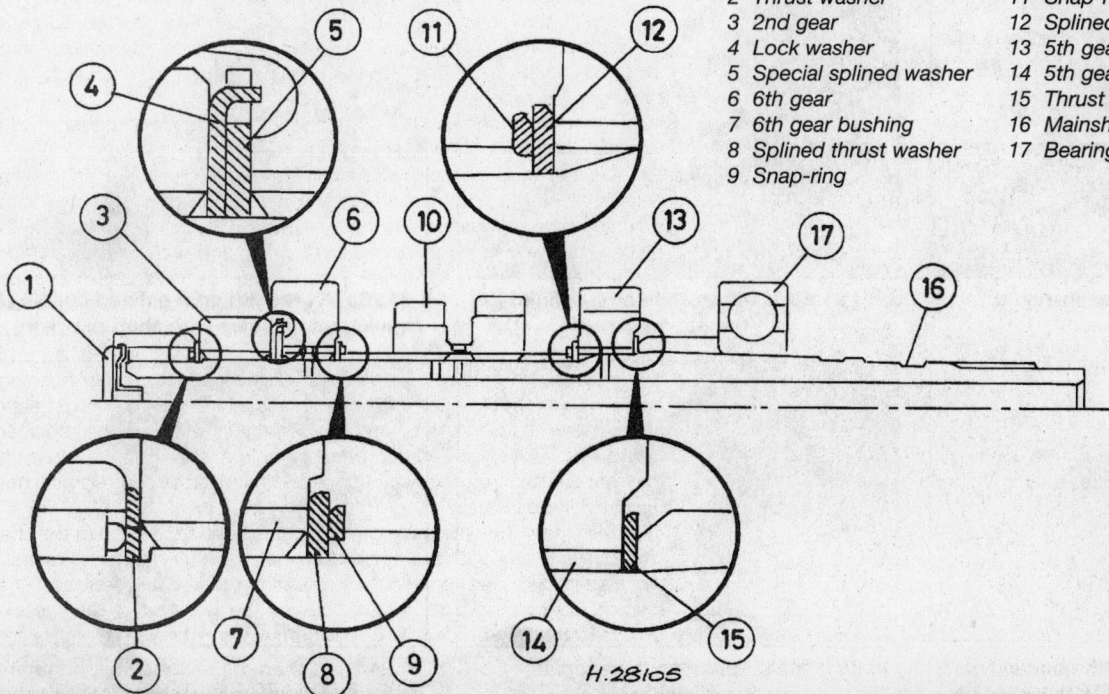

1 Needle roller bearing	10 3rd/4th gear
2 Thrust washer	11 Snap-ring
3 2nd gear	12 Splined thrust washer
4 Lock washer	13 5th gear
5 Special splined washer	14 5th gear bushing
6 6th gear	15 Thrust washer
7 6th gear bushing	16 Mainshaft
8 Splined thrust washer	17 Bearing
9 Snap-ring	

H.28105

31.17 Slide on the thrust washer . . .

31.18a . . . followed by the
5th gear bushing

31.18b Fit the 5th gear with its dogs
facing as shown . . .

31.18c . . . then fit the splined
thrust washer

31.19 Secure 5th gear components in
position with a snap-ring making sure it is
correctly seated in the mainshaft groove

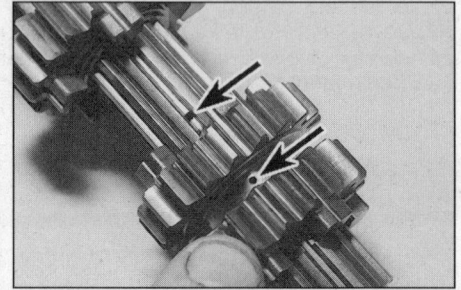

31.20 Align the gear oil hole with the
mainshaft oilway when installing
the 3rd/4th gear as described in text

facing away from the integral 1st gear **(see illustration)**.
18 Slide on the 5th gear bushing then fit the 5th gear with its dogs facing towards the left end of the shaft. Fit the splined thrust washer with its chamfered edge facing the 5th gear **(see illustrations)**.
19 Secure the 5th gear components in position with a new snap-ring; its chamfered edge must face the thrust washer **(see illustration)**. Check the snap-ring is correctly located in the mainshaft groove.
20 Install the 3rd/4th gear with its larger 4th gear facing the 5th gear. Engage the gear on the shaft splines ensuring that its oil holes are correctly aligned with the shaft oilways **(see illustration)**.
21 Fit a second new snap-ring to the shaft with its chamfered edge

facing away from the 3rd/4th gear pinion. Make sure the snap-ring is correctly located in the shaft groove **(see illustration)**.
22 Slide on the splined thrust washer with its chamfered edge facing away from the snap-ring **(see illustration)**.
23 Align the 6th gear splined bushing oil holes with the shaft oilways and slide it along the shaft. Fit the 6th gear so that its dogs are facing the 3rd/4th gear **(see illustrations)**.
24 Slide the special splined washer along until it abuts the 6th gear, followed by the lock washer. Rotate the splined washer until its cutouts align with the lock washer tabs, then engage the lock washer with the splined washer to lock it in position **(see illustrations)**.

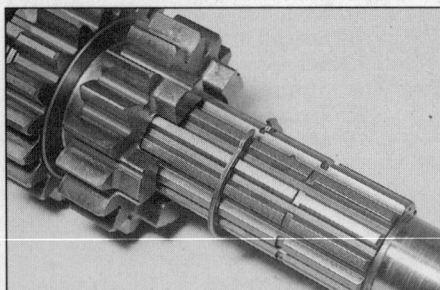

31.21 Fit a second snap-ring to
the next groove . . .

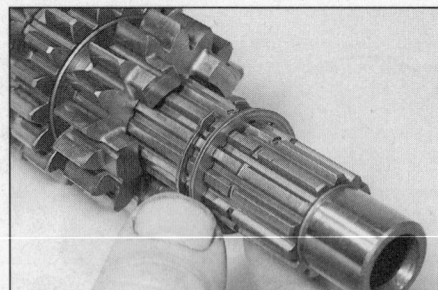

31.22 . . . then slide on a splined
thrust washer

31.23a Fit the 6th gear splined bushing,
aligning its oil holes with shaft oilways . . .

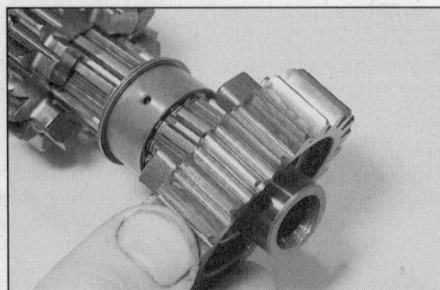

31.23b . . . and fit the 6th gear with its
dogs facing the 3rd/4th gear

31.24a Slide on the special
splined washer . . .

31.24b . . . then the lockwasher, engaging
its tabs with splined washer slots (arrows)

31.25a Fit the 2nd gear . . .

31.25b . . . followed by the thrust washer . . .

31.26 . . . then install the needle roller bearing

25 Fit the 2nd gear followed by the thrust washer, making sure the thrust washer chamfered edge faces the gear **(see illustrations)**.
26 Liberally oil the needle bearing and install it on the end of the shaft **(see illustration)**.

Countershaft

Disassembly
Refer to illustration 31.27
27 Remove the end plate from the right end of the countershaft and slide off the needle roller bearing **(see illustration)**.
28 Remove the thrust washer followed by the 1st gear, needle roller bearing and second thrust washer.
29 Remove the 5th gear noting which way around it is fitted.
30 Remove the snap-ring with a suitable pair of snap-ring pliers.
31 Slide off the splined washer then the 4th gear and splined bushing.
32 Disengage the lock washer from the special splined washer and slide both off the countershaft.
33 Remove the 3rd gear along with its splined bushing and thrust washer.
34 Remove the second snap-ring and slide off the 6th gear.
35 Remove the third snap-ring and slide off the splined thrust washer followed by the 2nd gear and 2nd gear bushing.

Inspection
36 Refer to Steps 10 through 15.

Reassembly
Refer to illustrations 31.37, 31.38a, 31.38b, 31.38c, 31.38d, 31.39, 31.40, 31.41, 31.42a, 31.42b, 31.43, 31.44a, 31.44b, 31.44c, 31.45, 31.46, 31.47a, 31.47b, 31.48a, 31.48b and 31.49
37 During reassembly, always use new snap-rings. Lubricate the components with engine oil before assembling them. **Note:** *If the thrust washers and snap-rings are examined closely it will be seen that they are chamfered on one side. During reassembly it is crucial that each thrust washer and snap-ring is fitted so its chamfer is on the correct side* **(see illustration)**.
38 Slide on the 2nd gear bushing and install the 2nd gear so that its plain surface abuts the countershaft bearing. Fit the splined thrust washer with its chamfered edge facing the 2nd gear and secure the 2nd gear components in position with a new snap-ring. Fit the snap-ring with its chamfered edge facing the thrust washer and make sure it is correctly located in the countershaft groove **(see illustrations)**.
39 Fit the 6th gear to the shaft so that its shift fork groove is facing away from the 2nd gear. Align the gear oil holes with the shaft oilways and slide the gear on the shaft **(see illustration)**.
40 Fit a second new snap-ring with its chamfered edge facing away from the 6th gear. Ensure the snap-ring is correctly located in the shaft groove **(see illustration)**.
41 Slide on the splined thrust washer with its chamfered edge facing away from the snap-ring **(see illustration)**.
42 Align the 3rd gear splined bushing oil holes with the shaft oilways

31.27 Countershaft components

1 Needle roller bearing	9 4th gear
2 1st gear needle roller bearing	10 4th gear bushing
3 Thrust washer	11 Lock washer
4 1st gear	12 Special splined washer
5 Thrust washer	13 3rd gear
6 5th gear	14 3rd gear bushing
7 Snap-ring	15 Splined thrust washer
8 Splined thrust washer	16 Snap-ring
	17 6th gear
	18 Snap-ring
	19 Splined thrust washer
	20 2nd gear
	21 2nd gear bushing
	22 End plate

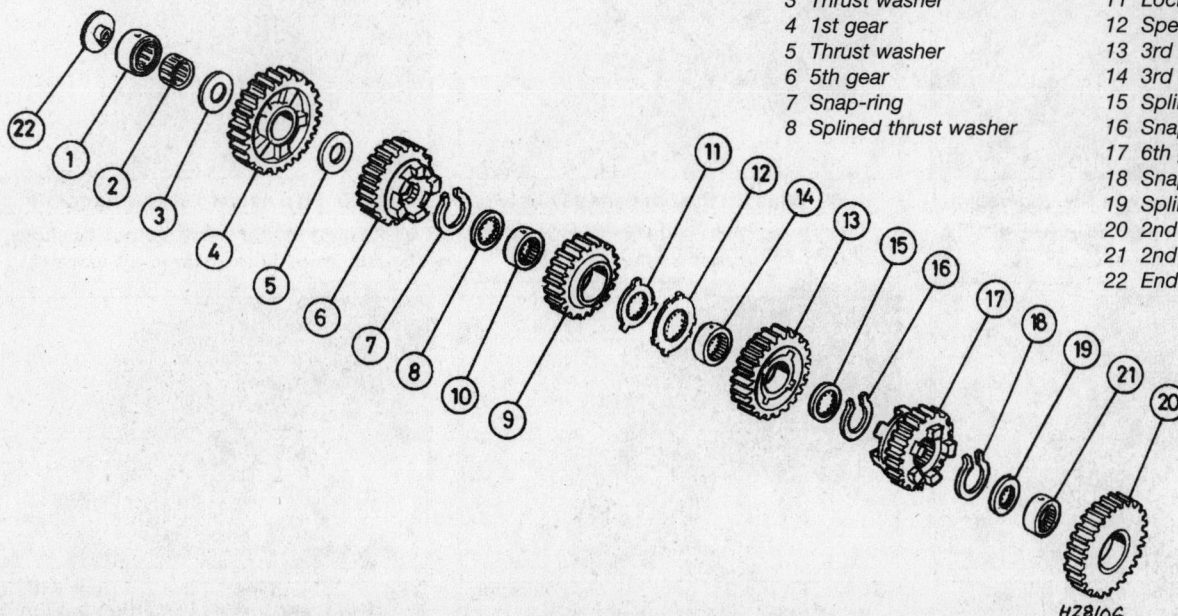

H28106

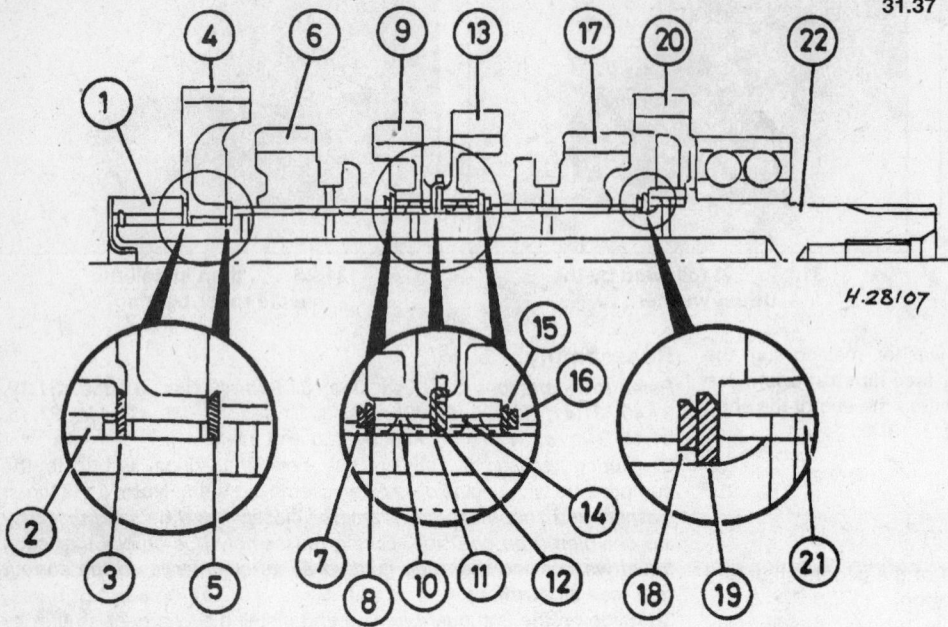

31.37 Correct fitted orientation of countershaft snap-rings and thrust washers

1 Needle roller bearing
2 1st gear needle roller bearing
3 Thrust washer
4 1st gear
5 Thrust washer
6 5th gear
7 Snap-ring
8 Splined thrust washer
9 4th gear
10 4th gear bushing
11 Lock washer
12 Special splined washer
13 3rd gear
14 3rd gear bushing
15 Splined thrust washer
16 Snap-ring
17 6th gear
18 Snap-ring
19 Splined thrust washer
20 2nd gear
21 2nd gear bushing
22 Countershaft

H.28107

31.38a Fit the 2nd gear bushing . . .

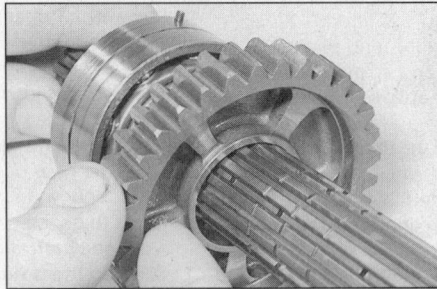

31.38b . . . and install the 2nd gear with its plain surface facing the bearing

31.38c Slide on a splined thrust washer . . .

31.38d . . . and secure the 2nd gear components in position with a snap-ring

31.39 Slide on the 6th gear as shown making sure its oil holes are correctly aligned with the shaft oilways (arrows)

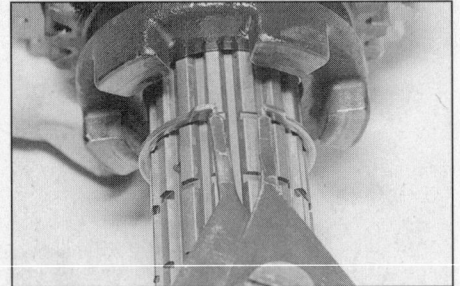

31.40 Fit a second snap-ring to the countershaft . . .

31.41 . . . and slide on a splined thrust washer

31.42a Fit the 3rd gear bushing, aligning its oil holes with the shaft oilways . . .

31.42b . . . and install the 3rd gear with its dog holes facing the 6th gear

31.43 Fit special splined washer and lock washer, engaging the lock washer tabs with the splined washer cutouts (arrows)

31.44a Slide on the 4th gear bushing, aligning its oil holes with the shaft oilways . . .

31.44b . . . and fit the 4th gear with its dog holes facing away from the 3rd gear

31.44c Slide on a splined thrust washer . . .

31.45 . . . and secure it in position with another snap-ring

31.46 Fit the 5th gear as shown

31.47a Install the thrust washer . . .

31.47b . . . followed by the 1st gear needle roller bearing . . .

31.48a . . . then fit the 1st gear with its dog holes facing the 5th gear

and slide it along the shaft. Fit the 3rd gear so that its dog holes are facing the 6th gear **(see illustrations)**.

43 Slide the special splined washer along until it abuts the 3rd gear, followed by the lock washer. Rotate the splined washer until its cutouts align with the lock washer tabs then engage the lock washer with the splined washer to lock it in position **(see illustration)**.

44 Align the 4th gear splined bushing oil holes with the shaft oilways and slide it along the shaft. Fit the 4th gear so that its dog holes are facing away from the 3rd gear, then slide on the splined thrust washer with its chamfered edge facing the 4th gear **(see illustrations)**.

45 Secure the 4th gear parts in position with a new snap-ring. Fit the snap-ring with its chamfered edge facing the thrust washer and make sure it is correctly located in the countershaft groove **(see illustration)**.

46 Fit the 5th gear to the shaft so that its shift fork groove is facing the 4th gear **(see illustration)**. Align the gear oil holes with the shaft oilways and slide the gear on the shaft.

47 Fit the thrust washer to the countershaft with its chamfered edge facing away from the 5th gear then install the 1st gear needle roller bearing **(see illustrations)**.

48 Liberally lubricate the bearing, then fit the 1st gear so that its dog holes are facing the 5th gear and slide on the second thrust washer **(see illustrations)**. The chamfered edge of the thrust washer should face the 1st gear.

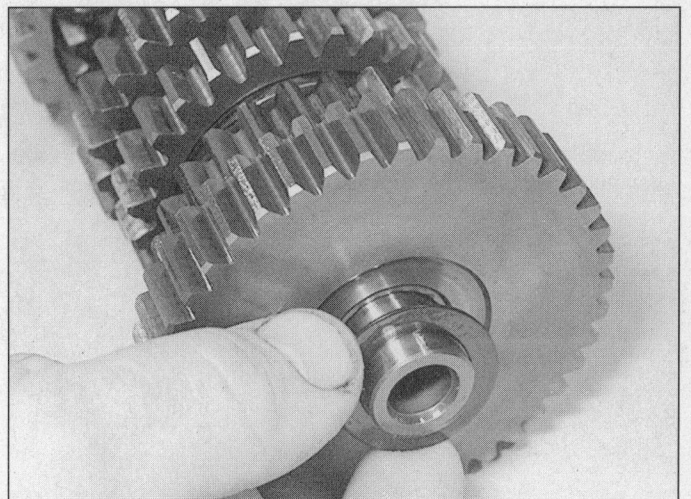

31.48b Slide on another thrust washer . . .

31.49 . . . then fit the needle roller bearing and end plate

49 Lubricate the needle roller bearing assembly and fit it to the end of the countershaft. Fit the end plate to the bearing **(see illustration)**.

32 Initial start-up after overhaul

1 Make sure the engine oil and coolant levels are correct (see Chapter 1), then remove the spark plugs from the engine. Place the engine kill switch in the OFF position.
2 Turn on the ignition switch and crank the engine over with the starter until the oil pressure indicator light goes off (which indicates that oil pressure exists). Reinstall the spark plugs, connect the HT leads and turn the kill switch to RUN.
3 Make sure there is fuel in the tank, then turn the fuel tap to the ON position and operate the choke.
4 Start the engine and allow it to run at a moderately fast idle until it reaches operating temperature. **Warning:** *If the oil pressure indicator light doesn't go off, or it comes on while the engine is running, stop the engine immediately.*
5 Check carefully for oil leaks and make sure the transmission and controls, especially the brakes, function properly before road testing the machine. See Section 33 for the recommended break-in procedure.
6 Upon completion of the road test, and after the engine has cooled down completely, recheck the valve clearances and check the engine oil and coolant levels (see Chapter 1).

33 Recommended break-in procedure

1 Treat the machine gently for the first few miles to make sure oil has circulated throughout the engine and any new parts installed have started to seat.
2 Even greater care is necessary if the engine has been rebored or a new crankshaft has been installed. In the case of a rebore, the engine will have to be broken in as if the machine were new. This means greater use of the transmission and a restraining hand on the throttle until at least 500 miles (800 km) have been covered. There's no point in keeping to any set speed limit - the main idea is to keep from lugging (labouring) the engine and to gradually increase performance until the 500 mile (800 km) mark is reached. These recommendations can be lessened to an extent when only a new crankshaft is installed. Experience is the best guide, since it's easy to tell when an engine is running freely. The following recommendations, which Honda provide for new motorcycles, can be used as a guide.

  a) *0 to 600 miles (0 to 1000 km): Keep engine speed below 5,000 rpm. Vary the engine speed and don't use full throttle.*
  b) *600 to 1000 miles (1,000 to 1,600 km): Keep engine speed below 7,000 rpm. Rev the engine freely through the gears, but don't use full throttle for prolonged periods.*
  c) *After 1000 miles (1,600 km): Full throttle can be used. Don't exceed maximum recommended engine speed (redline).*

3 If a lubrication failure is suspected, stop the engine immediately and try to find the cause. If an engine is run without oil, even for a short period of time, severe damage will occur.

Chapter 3 Cooling system

Contents

Specifications

Coolant
Mixture type .... 50% distilled water, 50% corrosion inhibited ethylene glycol antifreeze

Capacity
Radiator and engine .... 2.8 lit (3.0 US qt, 5.0 Imp pt)
Coolant reservoir .... 0.45 lit (0.48 US qt, 0.8 Imp pt)

Radiator
Cap valve opening pressure .... 16 to 20 psi (1.10 to 1.40 Bars)

Thermostat
Opening temperature .... 80 to 84°C (176 to 183°F)
Fully open .... 95°C (203°F)
Minimum valve lift .... 8 mm (0.32 in) @ 95°C (203°F)

Torque settings

	Nm	ft-lbs
Cooling fan thermostatic switch	18	13
Cooling fan nut	2.5	1.8
Temperature gauge sender unit	10	7
Water pump mounting and cover bolts	13	9

1 General information

The cooling system uses a water/antifreeze coolant to carry away excess energy in the form of heat. The cylinders are surrounded by a water jacket from which the heated coolant is circulated by thermo-syphonic action in conjunction with a water pump, driven off the oil pump. The hot coolant passes upwards to the thermostat and through to the radiator. The coolant then flows across the radiator core, where it is cooled by the passing air, down to the water pump and back up to the engine where the cycle is repeated. Air bleed hoses are fitted between the water pump and cylinder head, and between the thermostat housing and radiator filler neck.

A thermostat is fitted in the system to prevent the coolant flowing through the radiator when the engine is cold, therefore accelerating the speed at which the engine reaches normal operating temperature. A thermostatically-controlled cooling fan is also fitted to aid cooling in extreme conditions.

The cooling system is used to cool the engine oil. A short hose connects the coolant union on the front of the cylinder block to the oil cooler unit. Coolant circulates around the oil cooler body and is then routed back to the water pump via a rubber hose.

The complete cooling system is partially sealed and pressurised, the pressure being controlled by a valve contained in the spring-loaded radiator cap. By pressurising the coolant the boiling point is raised, preventing premature boiling in adverse conditions. The overflow pipe from the system is connected to a reservoir into which excess coolant is expelled under pressure. The discharged coolant automatically returns to the radiator when the engine cools.

Warning: *Do not allow antifreeze to come in contact with your skin or painted surfaces of the motorcycle. Rinse off any spills immediately with plenty of water. Antifreeze is highly toxic if ingested. Never leave antifreeze lying around in an open container or in puddles on the floor; children and pets are attracted by its sweet smell and may drink it. Check with the local authorities about disposing of used antifreeze. Many communities will have collection centers which will see that antifreeze is disposed of safely.*

Caution: *Do not remove the pressure cap from the radiator when the engine is hot. Scalding hot coolant and steam may be blown out under pressure, which could cause serious injury. To gain access to the pressure cap, remove the right lower fairing panel (see Chapter 8), then remove the fuse access panel from the upper fairing and the upper fairing right mounting screw; ease the edge of the fairing away from the fuel tank so that the cap can be reached between the fairing and tank. When the engine has cooled, place a thick rag, like a towel over the radiator cap; slowly rotate the cap counterclockwise (anti-clockwise) to the first stop. This procedure allows any residual pressure to escape. When the steam has stopped escaping, press down on the cap while turning it counterclockwise (anti-clockwise) and remove it.*

2   Radiator cap - check

If problems such as overheating or loss of coolant occur, check the entire system as described in Chapter 1. The radiator cap opening pressure should be checked by a Honda dealer or service station equipped with the special tester required to do the job. If the cap is defective, replace it with a new one.

3   Coolant reservoir - removal and installation

Removal

Refer to illustration 3.4

1   Remove the fuel tank as described in Chapter 4.
2   Remove the fuel pump (see Chapter 9). Remove the fuel pump mounting bracket bolt, leaving the bracket attached to the reservoir tank.
3   Free the hoses from the fuel filter mounting bracket slots and unbolt the bracket from the frame.
4   Maneuver the coolant reservoir out of the top of the frame. As the overflow hose becomes accessible, disconnect it and drain the coolant into a suitable container **(see illustration)**.

Installation

Refer to illustration 3.5

5   Installation is the reverse of removal, noting the following:

a) Engage the hook at the base of the tank with the hole in the frame, and the tab with the slot in the frame **(see illustration)**.
b) On completion refill the cooling system as described in Chapter 1.

4   Cooling fan and thermostatic switch - check and replacement

Check

Refer to illustrations 4.2 and 4.4

1   If the engine is overheating and the cooling fan isn't coming on, first check the cooling fan switch fuse. If the fuse is blown, check the fan circuit for a short to ground/earth (refer to the wiring diagrams).
2   If the fuse is sound, remove the fairing left middle section (Chapter 8) and disconnect the wire from the fan switch fitted to the left side of the radiator **(see illustration)**. Turn the ignition ON and ground (earth) the fan switch wire. As the wire is grounded (earthed) the fan should come on. If it does, the fan switch is defective and must be replaced.
3   If the fan does not come on, the fault lies in either the cooling fan motor or relevant wiring. To test the wiring, see Chapter 9.
4   To test the cooling fan motor, separate the two-pin connector behind the radiator **(see illustration)**. Using a 12 volt battery and two jumper wires, connect the battery across the connector terminals on the fan motor side of the connector. Once connected the fan should operate. If this is not the case the fan motor is faulty and must be replaced.

Replacement

Fan motor

Refer to illustrations 4.6 and 4.7

5   Remove the radiator as described in Section 7.
6   Unscrew the three bolts to separate the fan assembly from the radiator, noting the correct position of the fan motor ground (earth) lead **(see illustration)**.
7   Unscrew the retaining nut and remove the fan blade from the motor **(see illustration)**.
8   Release the clip and free the wiring from behind the fan motor shroud.
9   Unscrew the three nuts and separate the fan motor from the shroud **(see illustration 4.7)**.

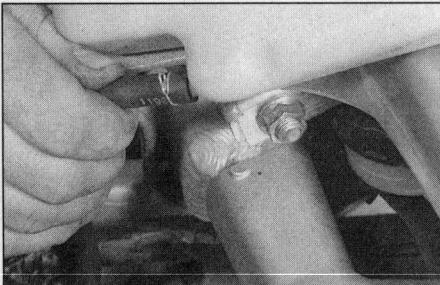

3.4 Have a container ready to catch the coolant as the overflow hose is removed

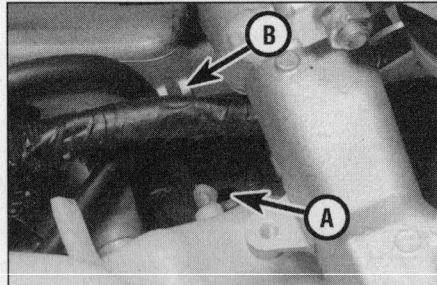

3.5 Coolant reservoir hook locates in frame hole (A) and tab in slot (B)

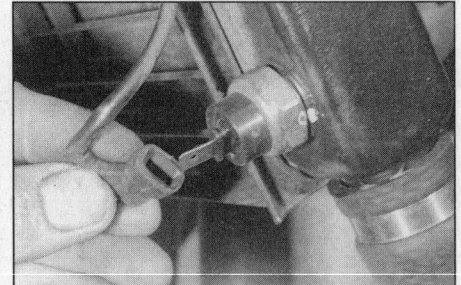

4.2 Disconnect wire from fan switch

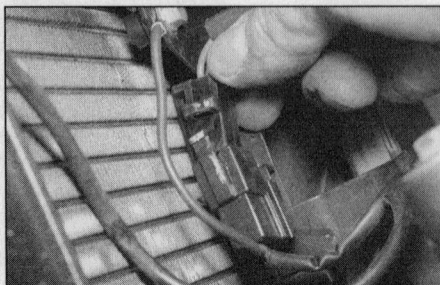

4.4 Fan motor two-pin wire connector is clamped behind the radiator

4.6 Remove three bolts (arrows) to separate fan assembly from radiator; note ground (earth) lead location

4.7 Fan blade is retained by a single nut (arrow)

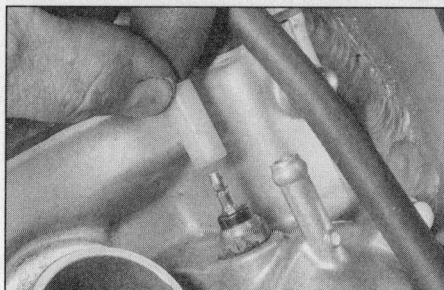

5.2 Disconnecting the temperature gauge sender unit wire

6.5 Disconnect the air bleed pipe from its union on the thermostat housing

6.6 Thermostat housing is retained by two bolts

10 Installation is the reverse of removal. On fitting the fan blade ensure its slot is correctly aligned with the motor shaft and tighten its nut to the specified torque setting. Make sure the motor ground (earth) lead is correctly positioned and that the wiring is secured by the clip.

11 Install the radiator as described in Section 7.

Thermostatic fan switch

Warning: *The engine must be completely cool before this procedure.*

12 Remove the fairing left side middle section (see Chapter 8).

13 Disconnect the wiring from the switch **(see illustration 4.2)**.

14 Unscrew the switch from the left side of the radiator and recover the O-ring. Plug the radiator opening to minimise coolant loss.

15 Fit a new O-ring to the switch and apply a smear of sealant to the switch threads.

16 Remove the plug and quickly install the new switch, tightening it to the specified torque setting.

17 Connect the wiring to the switch and install the fairing section.

18 Check the coolant level and top up as described in Chapter 1.

5   Coolant temperature gauge and sender unit - check and replacement

Check

Refer to illustration 5.2

1 The circuit consists of the sender unit mounted in the right end of the cylinder head and the gauge assembly in the instrument panel. If the system malfunctions check first that the battery is fully charged and that the fuse labelled METER, TAIL, HORN is in good condition.

2 If the gauge is not working, first remove the fuel tank (Chapter 4) to gain access to the sender unit. Turn the ignition switch ON and disconnect the wire from the temperature sender unit **(see illustration)**. Ground (earth) the sender unit wire on the engine. When the wire is grounded (earthed) the needle should swing immediately over to the H on the gauge. **Caution:** *Do not ground (earth) the wire for any longer than is necessary to take the reading, or the gauge may be damaged.* If the needle moves as described above, the sender unit is proven defective and must be replaced.

3 If the needle movement is still faulty, of if it does not move at all, the fault lies in the wiring or the gauge itself. Remove the instrument cluster as described in Chapter 9, and check the relevant wiring connectors. If all appears to be well, the gauge is defective and must be replaced.

Replacement

Temperature gauge sender unit

Warning: *The engine must be completely cool before this procedure.*

4 Remove the fuel tank as described in Chapter 4.

5 Disconnect the wiring from the switch **(see illustration 5.2)**.

6 Unscrew the sender unit from the cylinder head. Plug the head opening to minimise coolant loss.

7 Apply a smear of sealant to the switch threads.

8 Remove the plug and quickly install the new sender unit, tightening it to the specified torque setting.

9 Connect the wiring connector to the sender unit and install the fuel tank as described in Chapter 4.

10 Check the coolant level and top up as described in Chapter 1.

Temperature gauge

11 See Chapter 9.

6   Thermostat - removal, check and installation

Removal

Refer to illustrations 6.5 and 6.6

1 The thermostat is automatic in operation and should give many years service without requiring attention. In the event of a failure, the valve will probably jam open, in which case the engine will take much longer than normal to warm up. Conversely, if the valve jams shut, the coolant will be unable to circulate and the engine will overheat. Neither condition is acceptable, and the fault must be investigated promptly.

2 Partially drain the cooling system to prevent the escape of coolant when the thermostat housing is removed (see Chapter 1).

3 Remove the fairing right middle section as described in Chapter 8.

4 Remove the fuel tank as described in Chapter 4.

5 Release its spring clip and pull the small air bleed hose off its union on the thermostat housing **(see illustration)**.

6 Unscrew the two bolts and remove the thermostat housing; the large coolant hose can remain attached to the housing **(see illustration)**. Recover the housing O-ring.

7 Remove the thermostat from the cylinder head.

Check

Refer to illustration 6.9

8 Examine the thermostat before carrying out the test. If it remains in the open position at room temperature, it should be replaced.

9 Suspend the thermostat by a piece of wire in a container of cold water. Place a thermometer in the water so that the bulb is close to the thermostat **(see illustration)**. Heat the water, noting the temperature

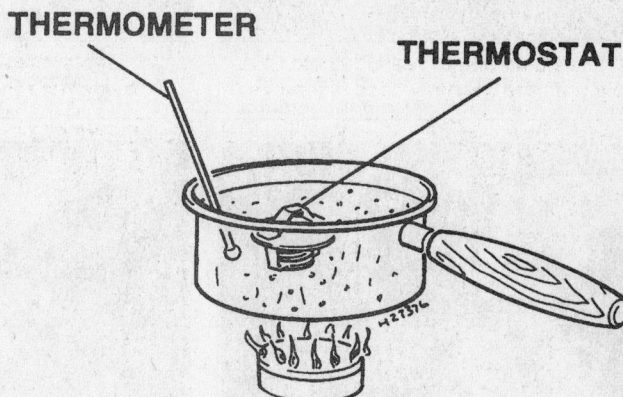

THERMOMETER

THERMOSTAT

6.9 Thermostat opening check

3

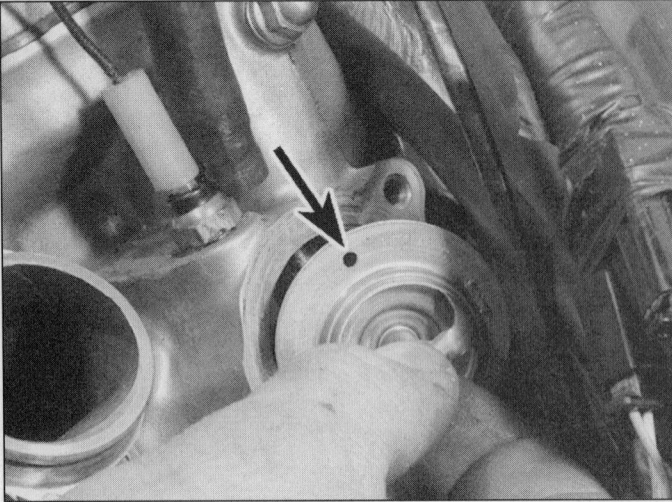

6.11 Thermostat bypass hole should be positioned at the top

6.12 Use a new O-ring in the thermostat housing groove

when the thermostat opens and how much valve lift it has when it is fully open, and compare the results with those given in the Specifications. If the readings obtained differ from those given, the thermostat is faulty and must be replaced.

10 In the event of thermostat failure, as an emergency measure only, it can be removed and the machine used without it. **Note:** *Take care when starting the engine from cold as it will take much longer than usual to warm up. A new unit should be installed as soon as possible.*

Installation

Refer to illustrations 6.11 and 6.12

11 Fit the thermostat to the cylinder head noting that its small bypass hole must be positioned at the top **(see illustration)**.

12 Fit a new O-ring to the housing cover and install the cover **(see illustration)**. Securely tighten the cover retaining bolts.

13 Reconnect the air bleed hose and secure with its retaining clip.

14 Refill the cooling system as described in Chapter 1.

15 Install the fuel tank as described in Chapter 4.

16 Fit the fairing middle section as described in Chapter 8.

7 Radiator - removal and installation

Removal

Refer to illustrations 7.3a, 7.3b, 7.3c, 7.5, 7.6a, 7.6b and 7.8

1 Remove both middle fairing sections as described in Chapter 8.

2 Drain the cooling system as described in Chapter 1.

3 Release their retaining clips and disconnect the overflow hose and air bleed hose from the top rear of the radiator **(see illustration)**. Slacken their clamps and pull the main coolant hoses off their unions at the top and bottom of the radiator **(see illustrations)**.

4 Disconnect the fan wiring connector behind the radiator **(see illustration 4.4)**.

5 Remove the radiator lower mounting bolt and collar **(see illustration)**.

6 Remove the radiator top mounting bolt and collar from the left side, then move the radiator to the right to disengage it from its mounting peg **(see illustrations)**.

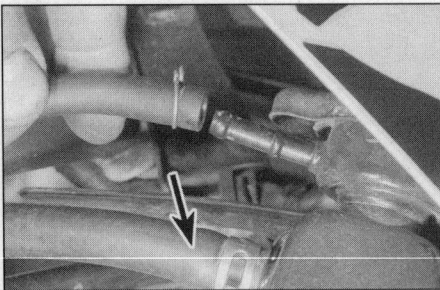

7.3a Disconnect air bleed hose and overflow hose (arrow) from the radiator

7.3b Disconnect the main radiator top hose . . .

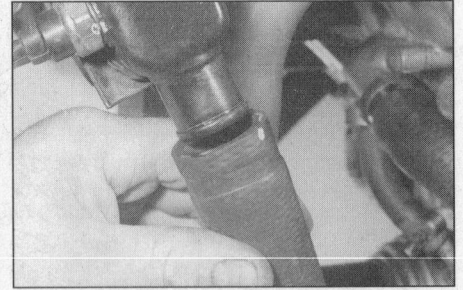

7.3c . . . and bottom hose

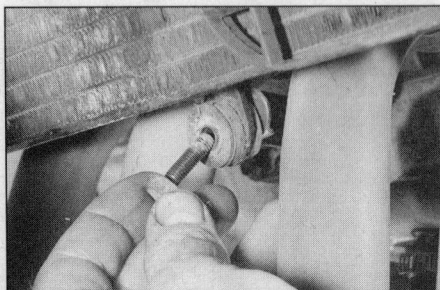

7.5 Radiator is retained to mounting bracket by a bolt and collar

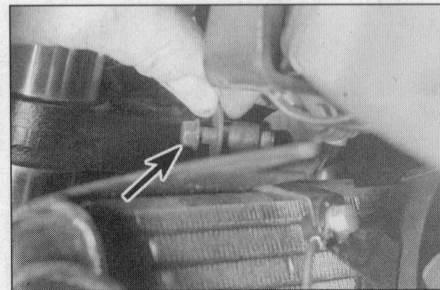

7.6a Radiator top mounting is by a bolt and collar (arrow) on the left side . . .

7.6b . . . and by a peg on right side

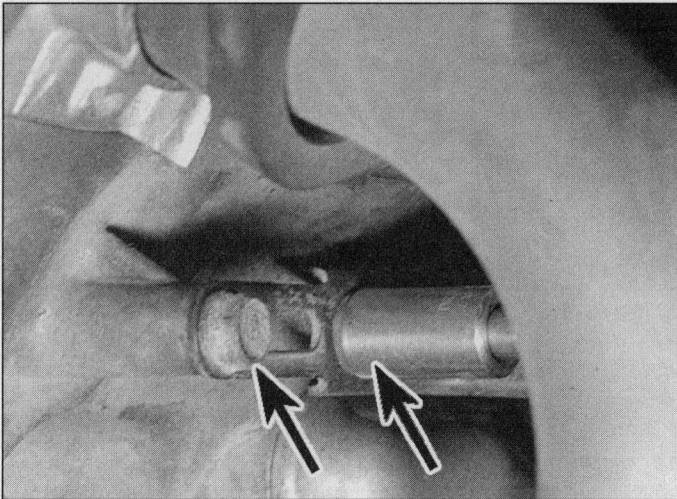

7.8 Radiator mounting bracket is retained by two bolts (arrows)

7   Check the radiator mounting rubbers for signs of damage or deterioration and replace if necessary.

8   The radiator mounting bracket can be detached from the cylinder block by removing its two bolts **(see illustration)**.

Installation

Refer to illustration 7.9

9   Installation is the reverse of the removal sequence noting the following **(see illustration)**.

a) *Locate the radiator on its mounting peg, fit the collar to each mounting bolt and securely tighten the mounting bolts.*

b) *The main wire harness must fit between the two raised plastic tangs on the radiator grille right side, next to the pressure cap.*

c) *Make sure that the fan wiring is correctly connected.*

d) *Ensure the coolant hoses are securely retained by their clips.*

e) *On completion refill the cooling system as described in Chapter 1.*

8   Water pump - check, removal and installation

Check

Refer to illustration 8.2

1   Visually check the area around the water pump for signs of leakage; if necessary remove the fairing lower section for improved access (see Chapter 8).

2   To prevent leakage of water or oil from the cooling system to the lubrication system and vice versa, two seals are fitted on the pump shaft. On the underside of the pump body there is also a drainage hole. If either seal fails, this hole should allow the coolant or oil to escape and prevent the oil and coolant mixing **(see illustration)**.

3   The seal on the water pump side is of the mechanical type which bears on the rear face of the impeller. The second seal, which is mounted behind the mechanical seal is of the normal feathered lip type. However, neither seal is available as a separate item as the pump is a sealed unit. Therefore, if on inspection the drainage hole shows signs of leakage, the pump must be removed and replaced.

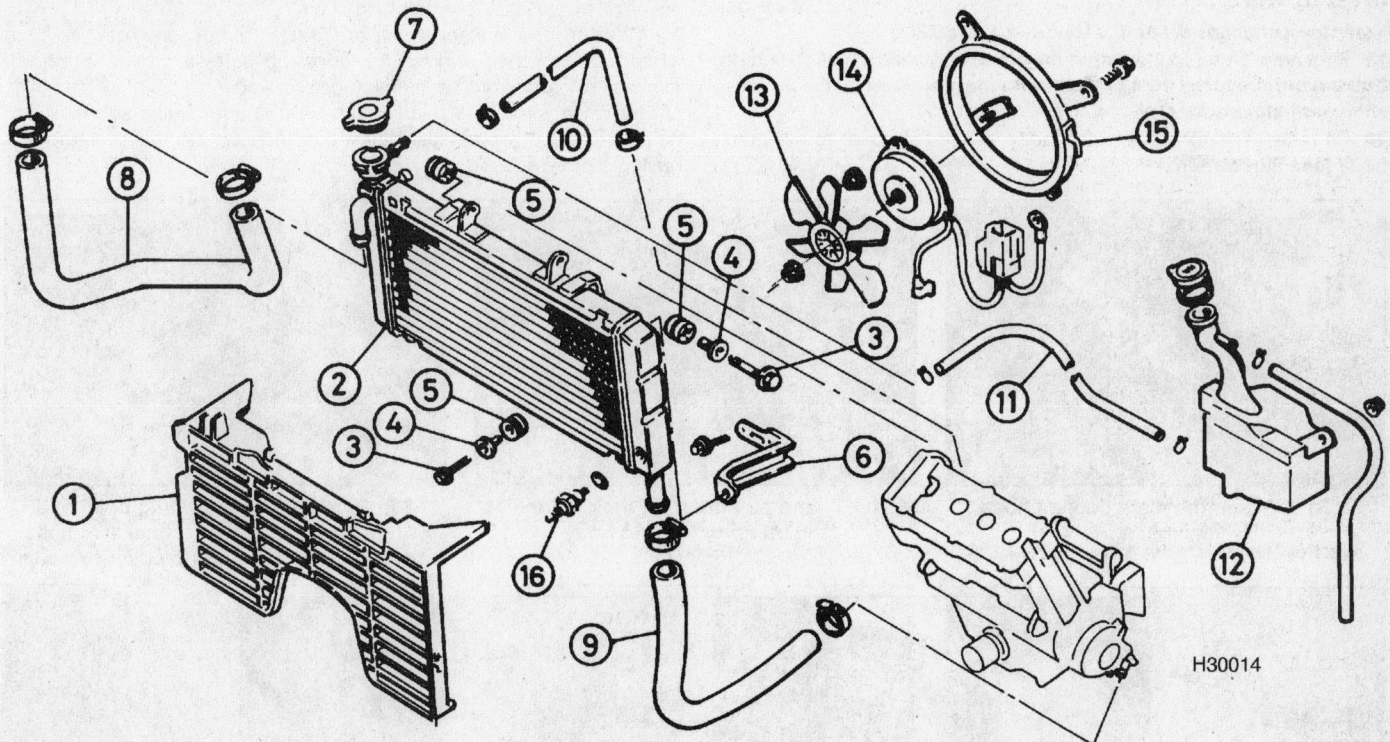

H30014

7.9 Radiator and fan detail

1 *Grille*	5 *Mounting rubbers*	9 *Bottom hose*	13 *Fan blade*
2 *Radiator*	6 *Mounting bracket*	10 *Air bleed hose*	14 *Fan motor*
3 *Mounting bolts*	7 *Pressure cap*	11 *Overflow hose*	15 *Fan shroud*
4 *Collars*	8 *Top hose*	12 *Reservoir tank*	16 *Thermostatic fan switch*

8.2 Leakage from the drainage hole (arrow) on the underside of the pump indicates mechanical seal failure

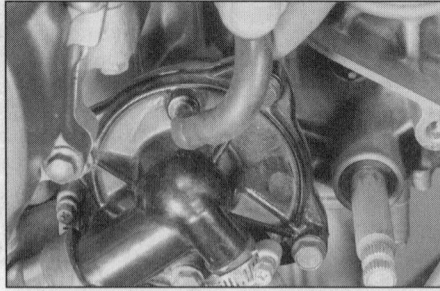

8.6a Disconnect the air bleed hose ...

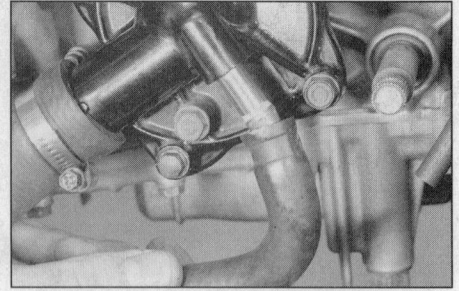

8.6b ... and oil cooler hose from the water pump cover

Removal

Refer to illustrations 8.6a, 8.6b, 8.7a, 8.7b and 8.8

4　Remove fairing lower section as described in Chapter 8.

5　Drain the cooling system as described in Chapter 1.

6　Release their retaining clips and disconnect the air bleed hose and oil cooler hose from the water pump cover **(see illustrations)**.

7　Slacken their retaining clamps and pull the main coolant hoses off their unions on the water pump body and cover **(see illustrations)**.

8　Unscrew the two mounting bolts and withdraw the water pump from the engine unit **(see illustration)**. Recover the O-ring from the water pump body.

9　Undo the remaining bolts and lift off the pump cover. Recover the sealing ring.

10　Wiggle the water pump impeller back-and-forth and in-and-out. If there is excessive movement the pump must be replaced.

Installation

Refer to illustrations 8.11a, 8.11b, 8.12a and 8.12b

11　Fit a new O-ring to the rear of the pump body and install the pump, aligning the slot in the impeller shaft with the projection on the oil pump shaft **(see illustrations)**.

12　Fit a new O-ring to the pump body and fit the cover to the pump body **(see illustrations)**.

13　Install the cover retaining bolts and the pump mounting bolts and tighten them to the specified torque setting.

14　Reconnect the coolant hoses to the pump and cover, making sure each one is securely retained by its clip or clamp.

15　Refill the cooling system as described in Chapter 1.

16　Install the fairing lower section as described in Chapter 8.

9　Coolant hoses and union - removal and installation

Coolant hoses

Removal

1　Before removing a hose, drain the coolant (see Chapter 1).

2　Use a screwdriver to slacken the hose clamps, then slide them back along the hose and clear of the union spigot. Many of the smaller-bore hoses are secured by spring clips which can be expanded by squeezing their ears together with pliers.

3　**Caution:** *The radiator unions are fragile. Do not use excessive force when attempting to remove the hoses.* If a hose proves stubborn, release it by rotating it on its union before working it off. If all else fails, cut the hose with a sharp knife then slit it at each union so that it can be peeled off in two pieces. Whilst this is expensive it is preferable to buying a new radiator.

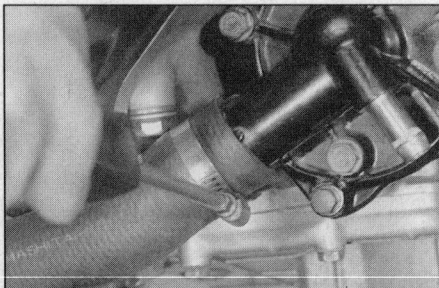

8.7a Slacken the main coolant hose clamps ...

8.7b ... and pull them off their unions on the pump cover and body

8.8 Remove two mounting bolts and withdraw water pump - other two bolts (arrows) retain the pump cover to body

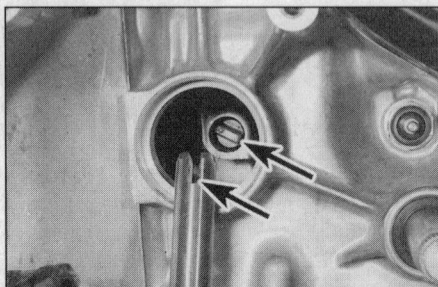

8.11a Align water pump impeller shaft slot with oil pump shaft projection (arrows) ...

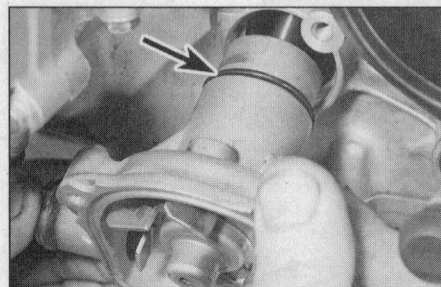

8.11b ... and insert pump in crankcase. Install a new pump O-ring (arrow)

8.12a Fit a new O-ring to the pump body groove . . .

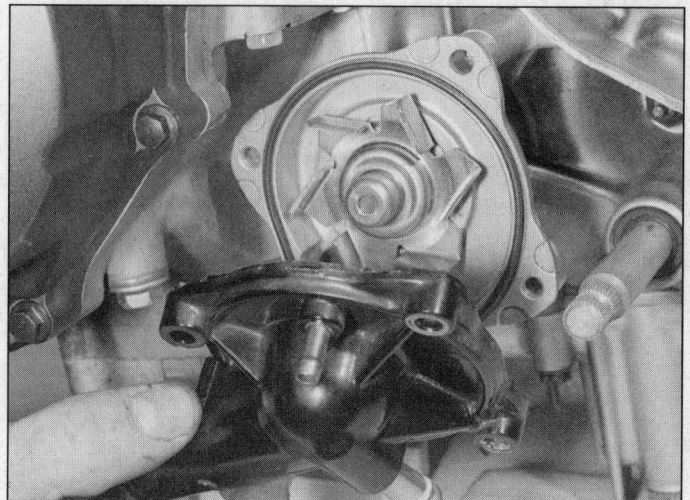

8.12b . . . before installing the pump cover

Installation

Refer to illustration 9.5

4   Slide the clips onto the hose and then work it on to its respective union. **Note:** *Do not use a lubricant of any kind. If necessary the hose can be softened by soaking it in hot water before installing, but take care to prevent the risk of personal injury whilst doing this.*

5   Rotate the hose on its unions to settle it in position before sliding the clamps into place and tightening them securely **(see illustration)**. Avoid overtightening the clamps, otherwise the union could distort.

Coolant union

Removal

Refer to illustrations 9.8 and 9.9

6   Remove the fairing lower section as described in Chapter 8.

7   Drain the cooling system as described in Chapter 1.

8   Slacken their clamps and disconnect the water pump hose and oil cooler hose from the coolant union **(see illustration)**.

9   Remove its two mounting bolts to free the union from the cylinder block **(see illustration)**. Recover the O-ring.

Installation

Refer to illustration 9.10

10   Ensure both mating faces are clean and install a new O-ring in the coolant union groove **(see illustration)**.

11   Install the two bolts and tighten them securely.

12   Reconnect the coolant hoses and secure with their clamps.

13   Refill the cooling system as described in Chapter 1.

14   Install the fairing lower section as described in Chapter 8.

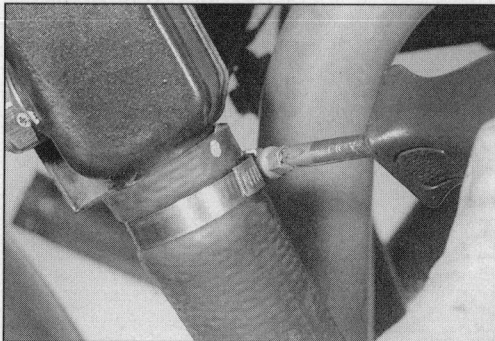

9.5 Ensure screw-type hose clamps are positioned correctly on the union before tightening

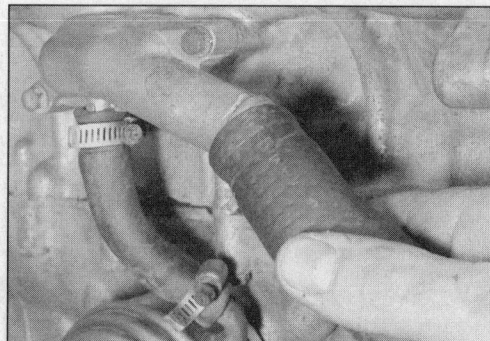

9.8 Disconnect the water pump hose and oil cooler hose from the coolant union

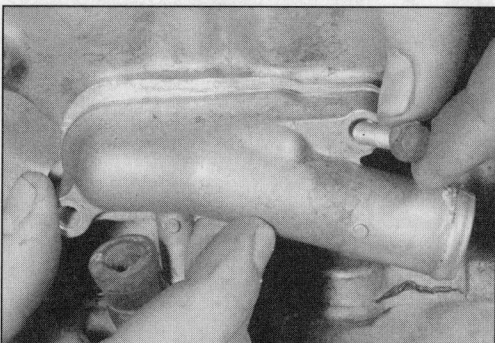

9.9 Coolant union is retained by two bolts

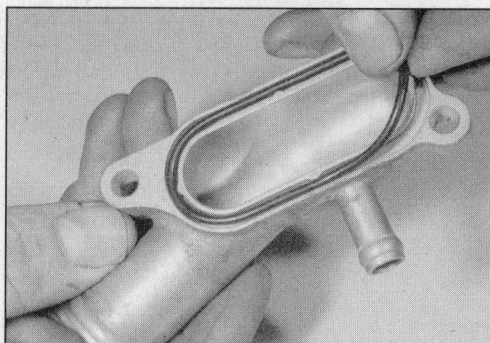

9.10 Fit a new O-ring to coolant union groove

Chapter 4 Fuel and exhaust systems

Contents

Specifications

Fuel

Grade	Unleaded or leaded (according to local regulations), minimum 91 octane (research method)
Fuel tank capacity	18 lit (4.8 US gal, 4.0 Imp gal)
Fuel tank reserve capacity	3.8 lit (1.0 US gal, 0.8 Imp gal)

Carburetor

Type	Keihin CV 38 mm
Identification code	
California models	VP84A
US models (except California)	VP81A
UK models	VP80B

Jet sizes

Main jet	115
Pilot jet	40
Pilot screw - initial setting (turns out)	
California models	2 3/4
US models (except California)	2 1/2
UK models	3

Carburetor adjustments

Float height (all models)	13.7 mm (0.54 in)
Idle speed	See Chapter 1

Torque settings

	Nm	ft-lbs
Carburetor joining bolts		
5 mm nut	5	3.6
6 mm nut	10	7
Exhaust system		
Front pipe nuts	12	9
Exhaust pipe and muffler (silencer) nuts	26	19

4

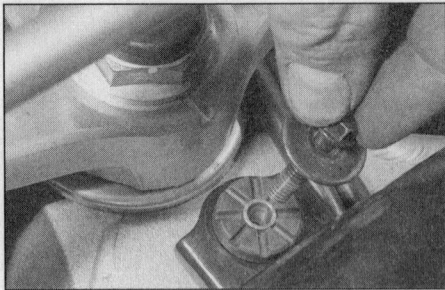

2.3a Fuel tank is mouted by a single bolt at the front . . .

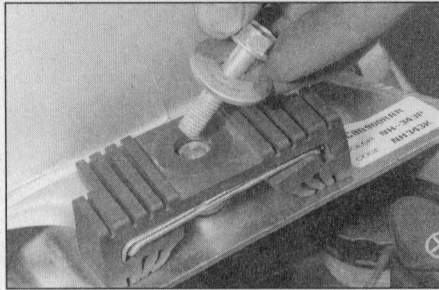

2.3b . . . and rear

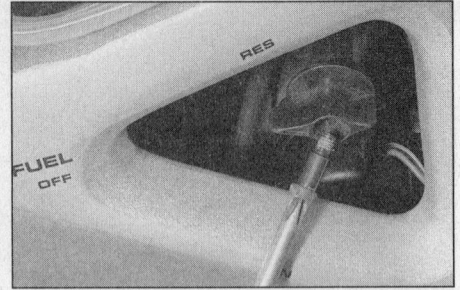

2.4 Remove screw to release fuel tap control knob

1   General information and precautions

General information

The fuel system consists of the fuel tank, fuel tap and filter, fuel pump, carburetors and the connecting lines, hoses and control cables.

The carburetors used on the CBR900RR are four 38 mm Keihin CV carburetors. For cold starting, an enrichment circuit is actuated by a cable and the choke knob on the top triple clamp.

Air is drawn to the carburetors from a moulded plastic air filter housing containing a pleated paper type element.

The exhaust system is a four-into-one design.

Many of the fuel system service procedures are considered routine maintenance items and for that reason are included in Chapter 1.

Precautions

Warning: *Gasoline (petrol) is extremely flammable, so take extra precautions when you work on any part of the fuel system. Don't smoke or allow open flames or bare light bulbs near the work area, and don't work in a garage where a natural gas-type appliance (such as a water heater or clothes dryer) is present. If you spill any fuel on your skin, rinse it off immediately with soap and water. When you perform any kind of work on the fuel system, wear safety glasses and have a fire extinguisher suitable for a class B type fire (flammable liquids) on hand.*

Always perform service procedures in a well-ventilated area to prevent a build-up of fumes.

Never work in a building containing a gas appliance with a pilot light, or any other form of naked flame. Ensure that there are no naked light bulbs or any sources of flame or sparks nearby.

Do not smoke (or allow anyone else to smoke) while in the vicinity of gasoline (petrol) or of components containing it. Remember the possible presence of vapor from these sources and move well clear before smoking.

Check all electrical equipment belonging to the house, garage or workshop where work is being undertaken (see the Safety first! section of this manual). Remember that electrical appliances such as drill, cutters etc create sparks in normal operation and must not be used near gasoline (petrol) or any component containing it. Again, remember the possible presence of fumes before using electrical equipment.

Always mop up any spilt fuel and safely dispose of the shop towel or rag used.

Any fuel that is drained off during servicing work, must be kept in a sealed container suitable for holding gasoline (petrol), and clearly marked as such; the container itself should be kept in a safe place. Note that this last point applies equally to the fuel tank, if it is removed from the machine; also remember to keep its cap closed at all times.

Note that the fuel system consists of the fuel tank, with its cap and related hoses, the fuel pump and filters. On US California models, this includes the Evaporative Emission Control (EVAP) System components.

Read the Safety first! section carefully before starting work.

Owners of machines used in the US, particularly California, should note that their machines must comply at all times with Federal or State legislation governing the permissible levels of noise and of pollutants such as unburnt hydrocarbons, carbon monoxide etc that can be emitted by those machines. All vehicles offered for sale must comply with legislation in force at the date of manufacture and must not subsequently be altered in any way which will affect their emission of noise or of pollutants.

In practice, this means that adjustments may not be made to any part of the fuel, ignition or exhaust systems by anyone who is not authorized or mechanically qualified to do so, or who does not have the tools, equipment and data necessary to properly carry out the task. Also if any part of these systems is to be replaced it must be replaced with only genuine Honda components or by components which are approved under the relevant legislation. The machine must never be used with any part of these systems removed, modified or damaged.

2   Fuel tank and tap - removal and installation

Warning: *Refer to the precautions in Section 1 before starting work.*

Fuel tank

Removal

Refer to illustrations 2.3a, 2.3b, 2.4, 2.5a and 2.5b

1   Set the bike on its side stand.

2   Remove the rider's seat as described in Chapter 8.

3   Unscrew the fuel tank front and rear mounting bolts together with their washers **(see illustrations)**.

4   Ensure the fuel tap is turned OFF. Using a cross-head screwdriver passed down through the center of the tap knob, unscrew the retaining screw and withdraw the knob **(see illustration)**.

5   Lift the tank at the rear to access the hoses on the tank underside. Release the spring clips and free the large-diameter drain hose and small-diameter breather hose from their unions **(see illustration)**. Disconnect the fuel hose from the tap union **(see illustration)**.

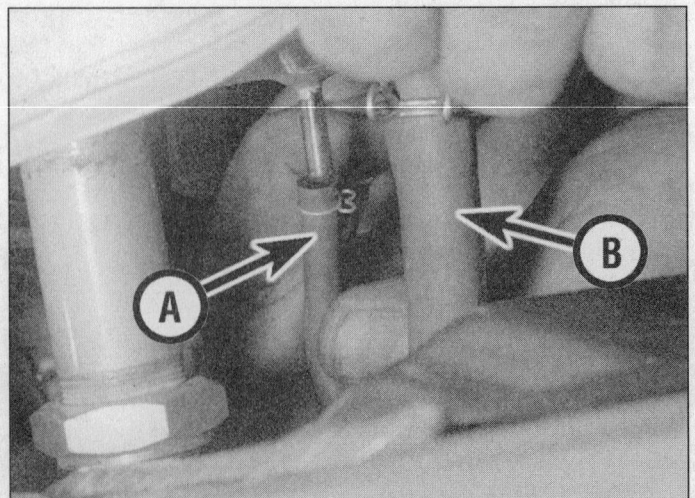

2.5a Disconnect breather (A) and drain (B) hoses from tank underside . . .

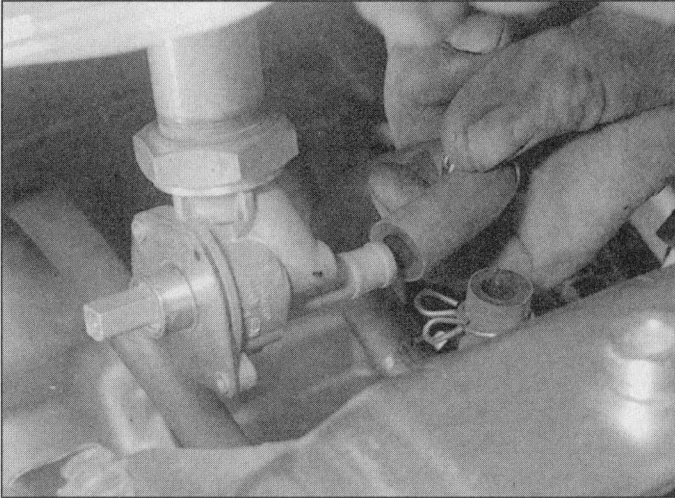

2.5b ... and disconnect fuel hose from fuel tap

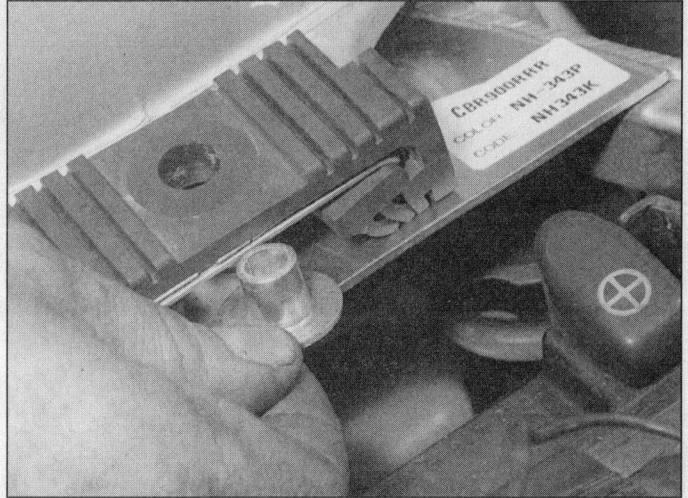

2.9 Tank mounting collars fit underneath mounting rubbers

6   Lift the fuel tank away from the machine.

7   Inspect the tank mounting rubbers for signs of damage or deterioration and replace if necessary.

Installation

Refer to illustrations 2.9 and 2.10

8   Lower the fuel tank into position, making sure the mounting rubbers remain in place, and reconnect the breather and drain hoses to their unions on the tank underside; secure the hoses with their clips. Fit the fuel hose to the tap and secure with its clip.

9   Ensure the collars are installed under the mounting rubbers and install the mounting bolts, tightening them securely **(see illustration)**.

10 Align the square hole of the tap knob with the corresponding square on the tap and secure it with the screw **(see illustration)**.

11 Install the rider's seat as described in Chapter 8.

12 Turn the fuel tap ON and check that there is no sign of fuel leakage.

Fuel tap

Removal

Refer to illustrations 2.14a and 2.14b

13 Remove the fuel tank as described above. Connect a drain hose to the fuel tap stub and insert its end in a container suitable for storing gasoline (petrol). Turn the fuel tap to RES and allow the tank to drain.

14 Unscrew the tap gland nut and withdraw the tap, O-ring and filter from the tank **(see illustrations)**.

15 Clean the gauze filter to remove all traces of dirt and fuel sediment. If it is holed or torn replace it with a new one.

16 The tap is a sealed unit - if it fails, a new tap must be fitted.

Installation

17 Installation is a reverse of the removal procedure, noting that a new O-ring should be used at the tap joint.

3   Fuel tank - cleaning and repair

1   All repairs to the fuel tank should be carried out by a professional who has experience in this critical and potentially dangerous work. Even after cleaning and flushing of the fuel system, explosive fumes can remain and ignite during repair of the tank.

2   If the fuel tank is removed from the vehicle, it should not be placed in an area where sparks or open flames could ignite the fumes coming out of the tank. Be especially careful inside garages where a natural gas-type appliance is located, because the pilot light could cause an explosion.

4   Idle fuel/air mixture adjustment - general information

1   Due to the increased emphasis on controlling motorcycle exhaust emissions, certain governmental regulations have been formulated which directly affect the carburation of this machine. In order to comply with the regulations, the carburetors on some models have a plastic limiter cap stuck onto the end of the pilot screw (which controls the idle fuel/air mixture) on each carburetor, so they can't be tampered with. These should only be removed in the event of a complete carburetor overhaul, and even then the screws should be returned to their original settings. The pilot screws on other models are accessible, but the use of an exhaust gas analyzer is the only accurate way to adjust the idle fuel/air mixture and be sure the machine doesn't exceed the emissions regulations.

2   If the engine runs extremely rough at idle or continually stalls, and if a carburetor overhaul does not cure the problem, take the motorcycle to a Honda dealer or other repair shop equipped with an exhaust gas analyzer. They will be able to properly adjust the idle fuel/air mixture to achieve a smooth idle and restore low speed performance.

4

2.10 Tap control knob end locates over square on the tap body

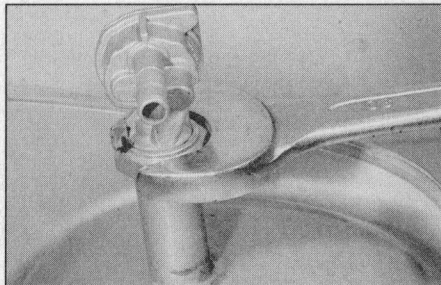

2.14a Unscrew gland nut to free fuel tap from the tank

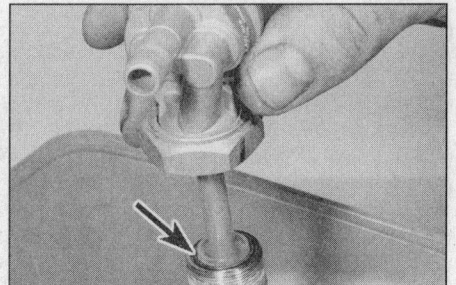

2.14b Note the O-ring between the tap and tank (arrow)

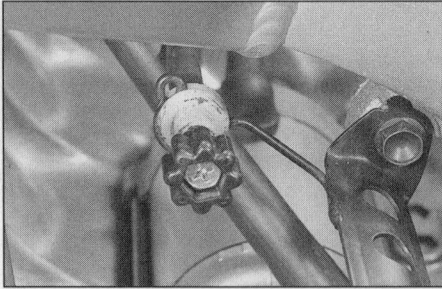

6.2 Slip the idle speed screw out of its wire holder

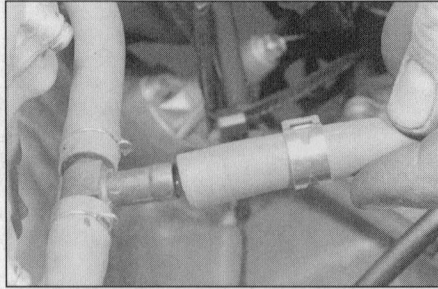

6.4a Release its clip and disconnect the fuel hose from the T-piece

6.4b Free the carburetor vent hose from its wire guide

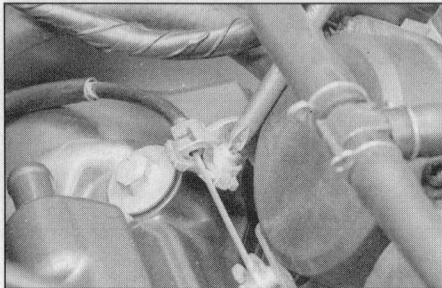

6.5a Slacken the choke cable clamp screw to free the outer cable . . .

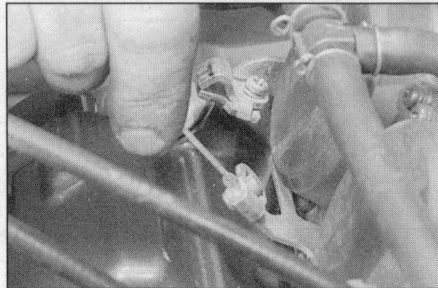

6.5b . . . and disconnect the inner cable from the choke linkage shaft

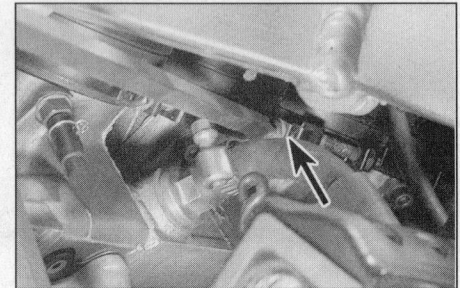

6.6 Slacken all four intake manifold joint clamps

5 Carburetor overhaul - general information

1   Poor performance, hesitation, hard starting, stalling, flooding and backfiring are signs that major carburetor maintenance may be required.

2   Keep in mind that many so-called carburetor problems are really not carburetor problems at all, but mechanical problems within the engine or ignition system. Try to establish for certain that the carburetors are in need of maintenance before beginning a major overhaul.

3   Check the fuel filter, fuel lines, tank breather hose (except California models), intake manifold joint clamps, fuel pump, air filter element, cylinder compression, spark plugs and carburetor synchronization before assuming a carburetor overhaul is required.

4   Most carburetor problems are caused by dirt particles, varnish and other deposits which build up in and block the fuel and air passages. Also, in time, gaskets and O-rings shrink or deteriorate and cause fuel and air leaks which lead to poor performance.

5   When the carburetor is overhauled, it is generally disassembled completely and the parts are cleaned thoroughly with a carburetor cleaning solvent and dried with filtered, unlubricated compressed air. The fuel and air passages are also blown through with compressed air to force out any dirt that may have been loosened but not removed by the solvent. Once the cleaning process is complete, the carburetor is reassembled using new gaskets and O-rings.

6   Before disassembling the carburetors, make sure you have a carburetor rebuild kit (which will include all necessary O-rings and other parts), some carburetor cleaner, a supply of rags, some means of blowing out the carburetor passages and a clean place to work. It is recommended that only one carburetor be overhauled at a time to avoid mixing up parts.

6 Carburetors - removal and installation

Warning: *Refer to the precautions in Section 1 before starting work*

Removal

Refer to illustrations 6.2, 6.4a, 6.4b, 6.5a, 6.5b, 6.6, 6.7a, 6.7b and 6.7c

1   Remove the air filter housing as described in Section 12.

2   Remove both middle fairing sections (Chapter 8). Free the idle speed screw from its wire holder on the left fairing bracket **(see illustration)**.

3   On California models, disconnect the emission (EVAP) system hoses from the carburetors noting the correct fitted position of each hose.

4   Free the fuel hose from T-piece at the rear of the carburetors **(see illustration)**. Release the vent hose (all models except California) from the wire guide on the right side of the frame **(see illustration)**.

5   Slacken the choke outer cable clamp screw and detach the inner cable from the choke linkage shaft **(see illustrations)**.

6   Slacken the four retaining clamps securing the carburetor intake rubbers to the cylinder head **(see illustration)**.

7   Ease the carburetors off the cylinder head stubs; tilt the assembly to access the throttle cam **(see illustration)**. Slacken the adjusters on the throttle cable bracket to get slack in the cables, then hook each cable out of the slot in the cam **(see illustrations)**. **Note:** *Keep the*

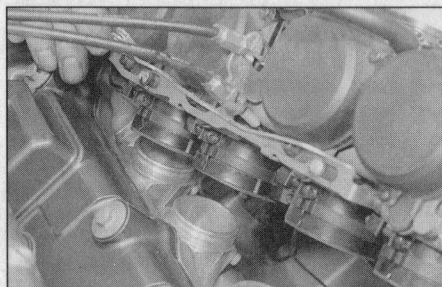

6.7a Ease the carburetors rearwards, off the cylinder head

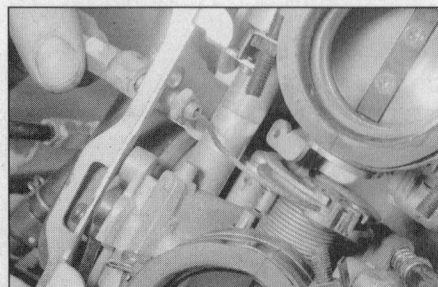

6.7b Slacken the throttle cable adjusters to provide enough slack in the cables . . .

6.7c . . . to slip their inner cables out of the throttle cam

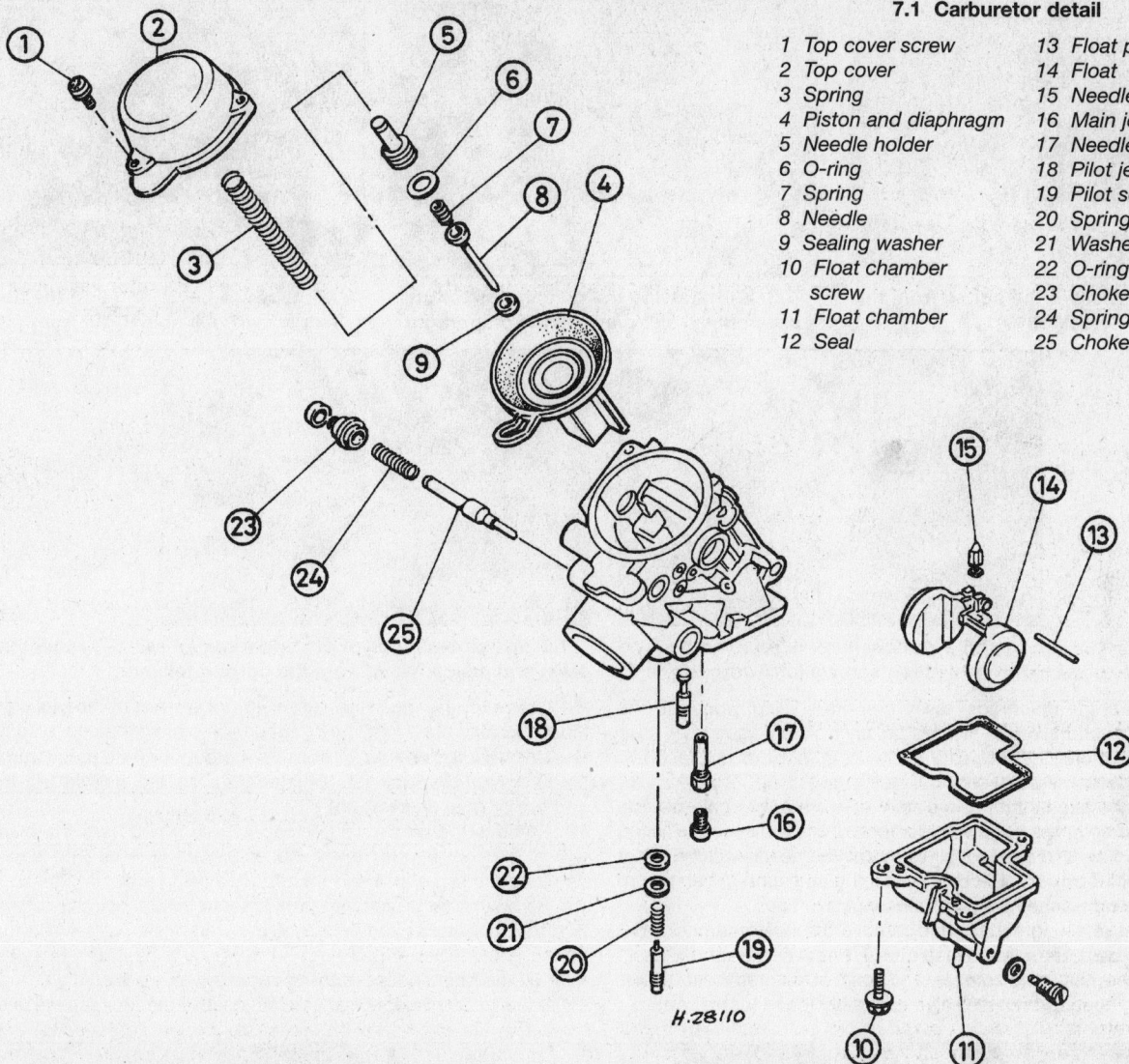

7.1 Carburetor detail

1 Top cover screw	13 Float pivot pin
2 Top cover	14 Float
3 Spring	15 Needle valve
4 Piston and diaphragm	16 Main jet
5 Needle holder	17 Needle jet
6 O-ring	18 Pilot jet
7 Spring	19 Pilot screw
8 Needle	20 Spring
9 Sealing washer	21 Washer
10 Float chamber	22 O-ring
screw	23 Choke plunger nut
11 Float chamber	24 Spring
12 Seal	25 Choke plunger

carburetors the right way up to prevent fuel spillage from the float chambers and the possibility of the piston diaphragms being damaged.

8 With the carburetors removed, place a suitable container below their float chambers then slacken the drain screws and drain all the fuel from the carburetors. Once all the fuel has been drained, tighten all the drain screws securely. **Caution:** *Honda advise against resting the carburetors on the air funnels.*

Installation

9 Installation is the reverse of removal making sure the carburetor intake rubbers are fully engaged with the cylinder head and their retaining clips are securely tightened. Prior to installing the air filter housing, adjust the throttle and choke cables (see Chapter 1).

7 Carburetors - disassembly, cleaning and inspection

Warning: *Refer to the precautions given in Section 1 before proceeding*

Disassembly

Refer to illustrations 7.1, 7.2a, 7.2b, 7.3, 7.4, 7.5, 7.6, 7.7a, 7.7b, 7.8a, 7.8b, 7.9 and 7.12

1 Remove the carburetors from the machine as described in the previous Section. **Note:** *Do not separate the carburetors unless absolutely necessary; each carburetor can be dismantled sufficiently for all normal cleaning and adjustments while in place on the mounting*

brackets. Dismantle the carburetors separately to avoid interchanging parts. Note that it is necessary to separate the carburetors to remove the choke plungers **(see illustration)**.

2 Slacken and remove the top cover retaining screws. Lift off the cover and remove the spring from inside the piston **(see illustrations)**.

7.2a Undo the three screws and remove the top cover . . .

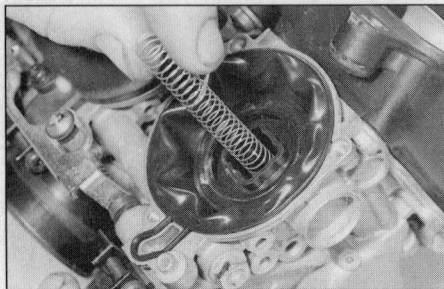
7.2b . . . then recover the spring from the piston

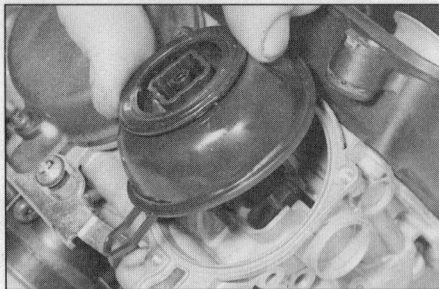
7.3 Withdrawing the diaphragm/piston assembly. Don't damage the diaphragm

7.4 Screw a cover bolt into the needle holder and pull holder out from piston

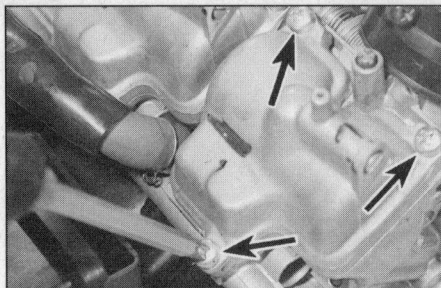
7.5 Float chamber is retained by three screws (arrows)

7.6 Withdraw the pivot pin and remove float and needle valve from the carburetor

3   Carefully peel the diaphragm away from its sealing groove in the carburetor and withdraw the diaphragm and piston assembly **(see illustration)**. **Caution:** *Do not use a sharp instrument to displace the diaphragm as it is easily damaged.*

4   To remove the needle from the piston, screw a 4 mm bolt into the thread in the centre of the needle holder (one of the top cover retaining screws will do), then use a pair of pliers to pull the needle holder out of the piston **(see illustration)**. Recover the O-ring and spring, then tip the needle and sealing washer out of the piston.

5   Remove the retaining screws and remove the float chamber from the base of the carburetor **(see illustration)**. Recover the rubber seal.

6   Withdraw the float pivot pin, using a pair of pointed-nose pliers, and remove the float and needle valve assembly **(see illustration)**.

7   Unscrew the main jet from the base of the needle jet **(see illustrations)**.

8   Unscrew the needle jet from the carburetor **(see illustrations)**.

9   Unscrew the pilot jet, situated next to the needle jet, from the carburetor **(see illustration)**.

10   Remove the plastic limiter cap (where fitted) from the pilot screw which is screwed into the base of the carburetor. The cap will be cemented in place and can be removed using a pair of pliers.

11   Screw the pilot screw in until it seats lightly, counting the number of turns necessary to achieve this, then remove the screw along with its spring, flat washer and O-ring. If the screw is bent or damaged in any way, all the pilot screws must be replaced as a set.

12   If the air funnels require removal, release the four screws and lift the

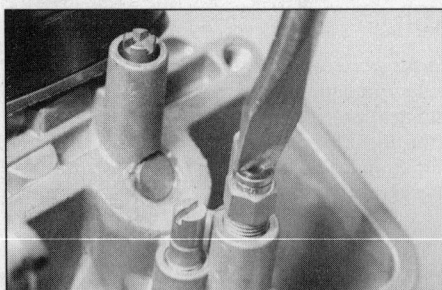
7.7a Unscrew the main jet with a close-fitting screwdriver . . .

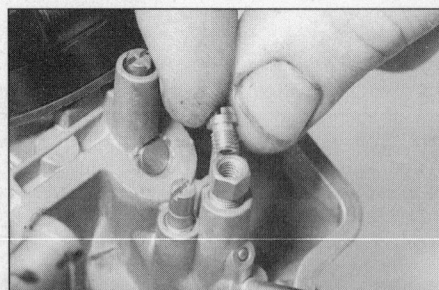
7.7b . . . and remove it from the base of the needle jet

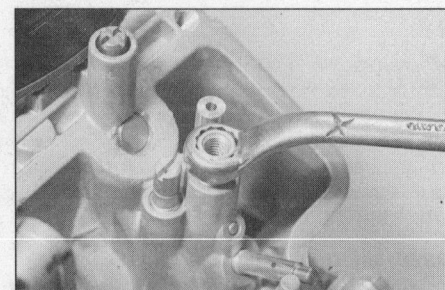
7.8a Unscrew the needle jet . . .

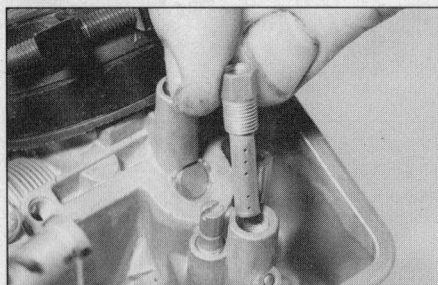
7.8b . . . and remove it from the carburetor

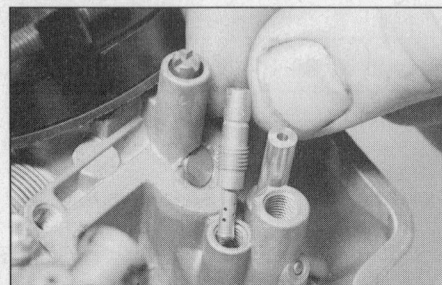
7.9 Pilot jet is also a screw fit in the carburetor

7.12 Air funnel holders are retained by four screws

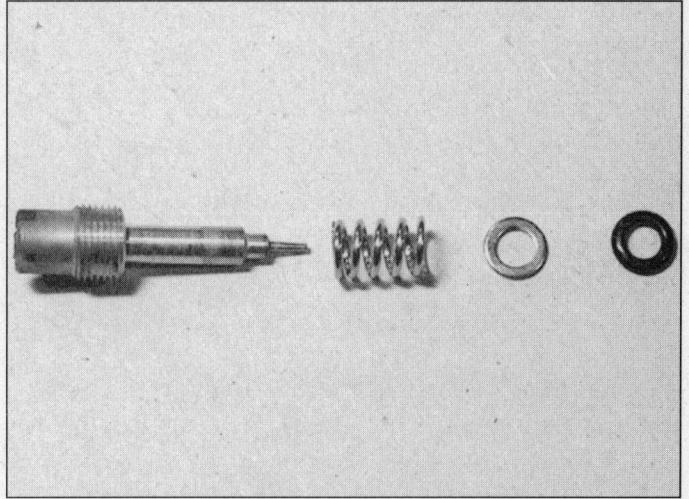

7.18 Pilot screw components. Ensure the tapered portion of the screw is not bent or damaged

holders and air funnels free **(see illustration)**. Recover the O-rings.
13 If the carburetors have been separated, unscrew the valve nut and remove the choke plunger and spring from the carburetor.

Cleaning

Caution: *Use only a petroleum based solvent for carburetor cleaning. Don't use caustic cleaners.*
14 Submerge the metal components in the solvent for approximately thirty minutes (or longer, if the directions recommend it).
15 After the carburetor has soaked long enough for the cleaner to loosen and dissolve most of the varnish and other deposits, use a brush to remove the stubborn deposits. Rinse it again, then dry it with compressed air. Blow out all of the fuel and air passages in the main and upper body. **Caution:** *Never clean the jets or passages with a piece of wire or a drill bit, as they will be enlarged, causing the fuel and air metering rates to be upset.*
16 Use a jet of compressed air directed in the fuel inlet valve needle seat to clear the gauze fuel filter inside the carburetor body.

Inspection

Refer to illustrations 7.18 and 7.20
17 Check the operation of the choke plunger. If it doesn't move smoothly, replace it, along with its spring. Inspect the needle on the end of the choke plunger and replace the plunger if it's worn or bent.
18 Check the tapered portion of the pilot screw for wear or damage. Replace the pilot screw if necessary **(see illustration)**.
19 Check the carburetor body, float chamber and top cover for cracks, distorted sealing surfaces and other damage. If any defects are found, replace the faulty component, although replacement of the entire carburetor will probably be necessary (check with your parts supplier for the availability of separate components).
20 Check the diaphragm for splits, holes and general deterioration. Holding it up to a light will help to reveal problems of this nature **(see illustration)**.
21 Insert the diaphragm piston in the carburetor body and check that it moves up-and-down smoothly. Check the surface of the piston for wear. If it's worn excessively or doesn't move smoothly in the bore, replace the carburetor.
22 Check the jet needle for straightness by rolling it on a flat surface (such as a piece of glass). Replace it if it's bent or if the tip is worn.
23 Check the tip of the float needle valve. If it has grooves or scratches in it, it must be replaced. Push in on the rod in the other end of the needle, then release it - if it doesn't spring back, replace the needle valve. If the needle valve seat is damaged the carburettor assembly must be replaced; it is not possible to replace the seat individually.
24 Check the float chamber gasket and replace it if it's damaged.

25 Operate the throttle shaft to make sure the throttle butterfly valve opens and closes smoothly. If it doesn't, replace the carburetor.
26 Check the floats for damage. This will usually be apparent by the presence of fuel inside one of the floats. If the floats are damaged, they must be replaced.

8 Carburetors - separation and joining

Warning: *Refer to the precautions given in Section 1 before proceeding*

Separation

Refer to illustrations 8.1, 8.5a, 8.5b, 8.5c, 8.5d, 8.5e and 8.7
1 The carburetors do not need to be separated for normal overhaul. If you need to separate them (to replace a carburetor body, for example), refer to the following procedure **(see illustration)**.
2 Remove the carburetors from the machine as described in Section 6. Mark the body of each carburetor with its cylinder number to ensure that it is positioned correctly on reassembly.
3 To prevent damage to the air funnels, it is advised that they are removed. Remove the four screws and lift the holders and air funnels off the carburetors. Recover the O-rings.
4 Release the spring clips and pull the fuel hose off of the unions between carburetor Nos. 1 and 2 and carburetor Nos. 3 and 4. On all

4

7.20 Check the piston diaphragm for signs of splitting and replace if necessary

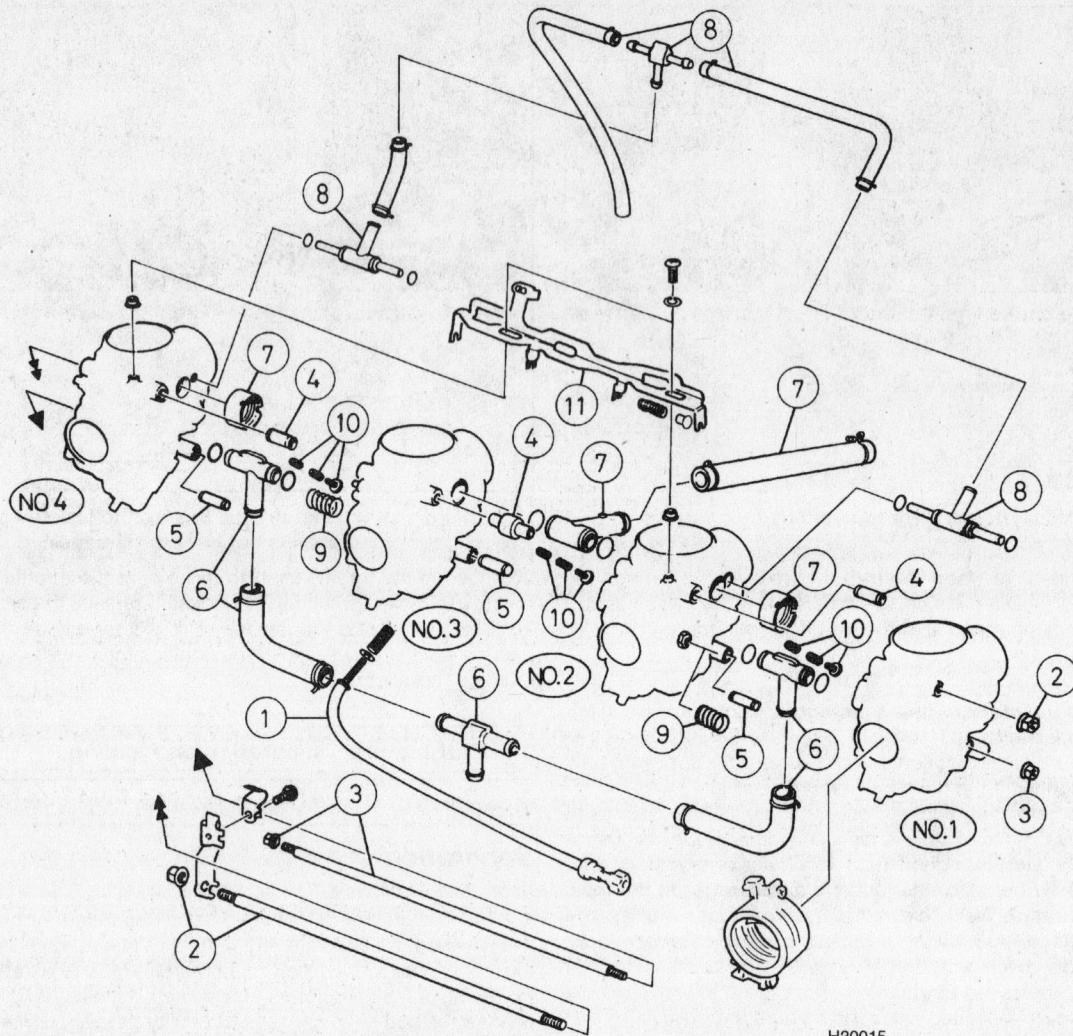

H30015

8.1 Carburetor linkage components

1 Idle speed screw	4 Joining bolt dowel - 6 mm	7 Breather hose and unions	10 Synchronization screws
2 Joining bolt and nuts - 6 mm	5 Joining bolt dowel - 5 mm	8 Air vent hose and unions	11 Choke linkage shaft
3 Joining bolt and nuts - 5 mm	6 Fuel hoses and unions	9 Throttle linkage springs	

models except California, disconnect the air vent hoses from the unions between carburetor Nos. 1 and 2 and carburetor Nos. 3 and 4.

5 Unhook the choke linkage return spring and washer from between Nos. 1 and 2 carburetors. Undo the two screws and washers securing the linkage shaft to the top of the carburetors then lift off the linkage shaft and recover the spacers from underneath the shaft **(see illustrations)**.

6 Make a note of how the throttle linkage spring and No. 2 carburetor

synchronization spring are arranged to ensure that they are fitted correctly on reassembly.

7 Evenly unscrew the two retaining nuts from the No. 1 carburetor end **(see illustration)**. Withdraw the two joining bolts from the other end of the carburetors.

8 Carefully separate carburetor Nos. 2 and 3, leaving Nos. 1 and 2, and Nos. 3 and 4 still joined. Retrieve the synchronization springs, the two dowels and the breather hose T-piece as they are separated.

8.5a Remove the choke linkage return spring and washer (arrow) . . .

8.5b . . . then undo the two retaining screws . . .

8.5c . . . and recover the washers

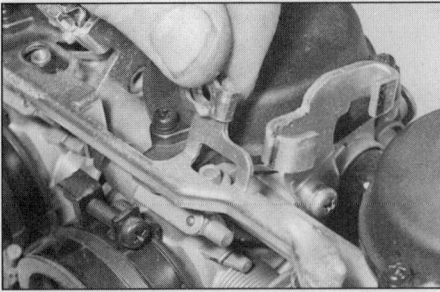
8.5d Lift off the choke linkage shaft . . .

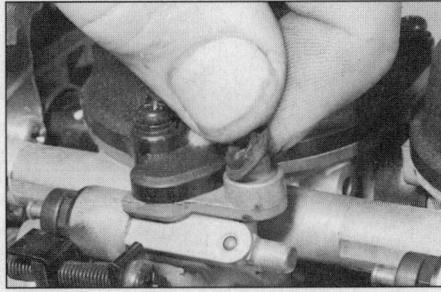
8.5e . . . recover the spacers underneath

8.7 Unscrew the carburetor nuts (arrows)

8.9a On reassembly renew all O-rings (arrows) on the unions

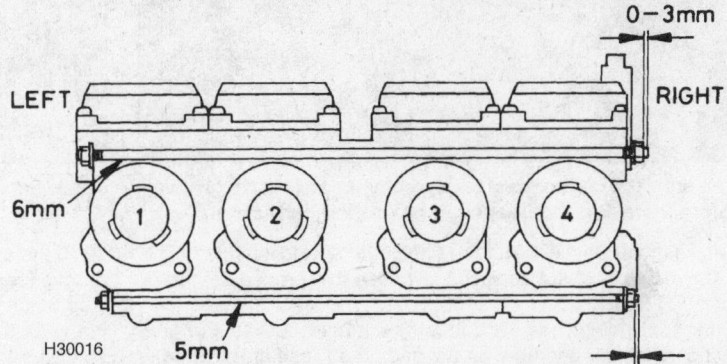
8.9b The joining bolt threads must not extend more than 3 mm from the face of the nut

9   Separate carburetors 1 and 2 by gently pulling them apart. Retrieve the throttle linkage spring, synchronization springs, both dowels, air vent T-piece, fuel T-piece and breather joint. Carry out the same procedure to separate carburetor Nos. 3 and 4.

Joining

Refer to illustrations 8.9a and 8.9b

9   Assembly is the reverse of the disassembly procedure, noting the following.

a) Use new O-rings on all the T-piece unions **(see illustration)**.
b) The carburetor joining bolts are different diameter; the front bolt is 6 mm and must have the choke cable clamp fitted to its right end, whereas the rear bolt is 5 mm diameter. Similarly, the dowels between each carburetor are sized accordingly. Tighten the nuts evenly to the specified torque setting, ensuring correct alignment. No more than 3 mm of thread should extend from the outer face of any nut **(see illustration)**.
c) If the intake manifold joints were removed, ensure that the edge with the CARB faces the carburetor and align the groove with the carburetor lug.
d) Check the operation of both the choke and throttle linkages ensuring that both operate smoothly and return quickly under spring pressure before installing the carburetors on the machine.
e) Check carburetor synchronization (see Chapter 1).

9   Carburetors - reassembly and float height check

Refer to illustrations 9.2, 9.6, 9.7, 9.8, 9.9a, 9.9b, 9.10, 9.11, 9.12a, 9.12b and 9.12c

Note: When reassembling the carburetors, be sure to use the new O-rings, gaskets and other parts supplied in the rebuild kit. Do not overtighten the carburetor jets and screws as they are easily damaged.

1   Install the choke plunger in its bore, followed by its spring and nut. Tighten the nut securely and install the cap.
2   Install the pilot screw (if removed) along with its spring, washer and O-ring, turning it in until it seats lightly **(see illustration)**. Now, turn the screw out the number of turns previously recorded. Where fitted, install a new limiter cap on the screw, applying a little bonding agent to hold it in position.
3   Screw the needle jet into position in the carburetor.
4   Screw the main jet into the end of the needle jet.
5   Screw the pilot jet into position.
6   Hook the needle valve over the float, then install the float and secure it with the pivot pin **(see illustration)**.
7   To check the float height, hold the carburetor so the float hangs down, then tilt it back until the needle valve is just seated, but not so far that the needle's spring-loaded tip is compressed. Measure the distance between the gasket face and the bottom of the float with an accurate ruler **(see illustration)**. The correct setting should be as given

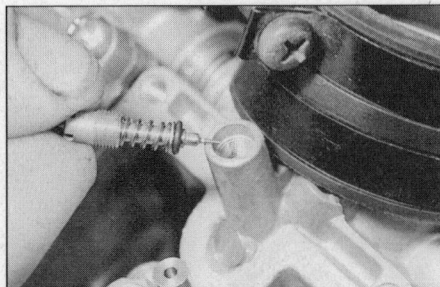
9.2 Assemble pilot screw components on the screw and install the screw as described in text

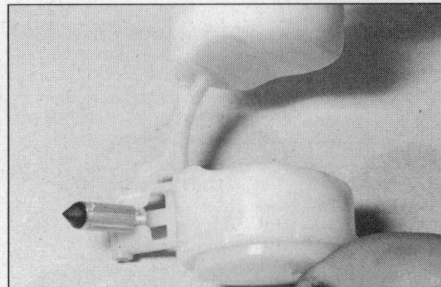
9.6 Make sure that the needle valve is correctly slotted into the float

9.7 Measuring the float height

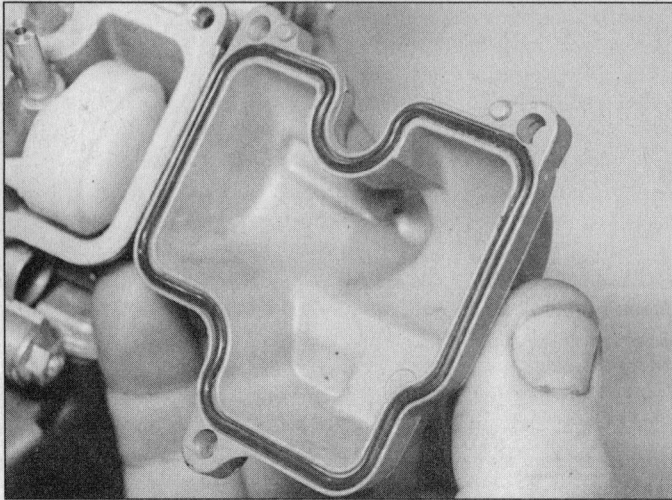

9.8 Make sure the rubber seal is correctly located in its groove before installing the float chamber on the carburetor

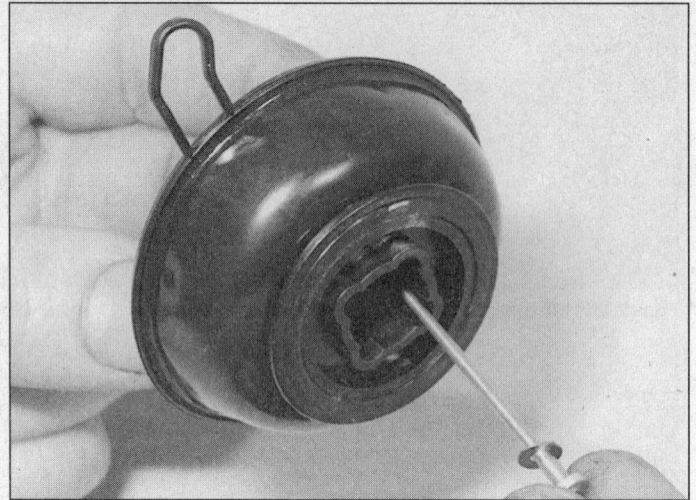

9.9a Insert the jet needle and sealing washer into the piston . . .

in the Specifications Section. The float height is not adjustable; if it is incorrect the float must be replaced. Repeat the procedure for all carburetors.

8   With the float height checked, fit a new rubber seal to the float chamber and install the chamber on the carburetor **(see illustration)**.

9   Fit the washer to the jet needle and insert the needle into the piston **(see illustration)**. Position a new O-ring in the groove of the needle jet holder and fit the spring inside the holder **(see illustration)**. Insert the needle holder into the center of the piston and press it into position until the O-ring is heard to click into the groove in the base of the piston.

10 Insert the piston assembly into the carburetor body and lightly push it down, ensuring the needle is correctly aligned with the needle jet. Press the diaphragm outer edge into its groove, ensuring the diaphragm tongue is correctly seated in the cutout on the carburetor **(see illustration)**. Check the diaphragm is not creased, and that the piston moves smoothly up and down the bore.

11 Insert the spring and fit the top cover to the carburetor **(see illustration)**.

12 If the air funnels where removed, install the O-ring in the carburetor groove **(see illustration)**. Insert the air funnel into the holder and rotate it so that its tabs are locked in place **(see illustration)**. Align the cutout in the air funnel with the carburetor lugs and install it; tighten the four screws securely **(see illustration)**.

10 Throttle cables - removal and installation

Warning: *Refer to the precautions given in Section 1 before proceeding*

Removal

Refer to illustrations 10.3a and 10.3b

1   Remove the air filter housing as described in Section 12.

2   Slacken the throttle cable locknuts then free each outer cable from its mounting bracket and detach the inner cables from the throttle cam **(see illustrations 6.7b and 6.7c)**. If necessary to improve access to throttle cam, slacken the four retaining clips securing the carburetor

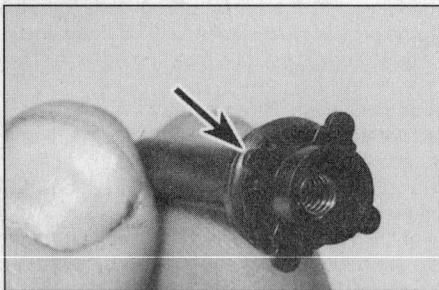

9.9b . . . and fit the needle holder and spring. Make sure the O-ring (arrow) is correctly positioned

9.10 Insert the piston and locate the diaphragm in the carburetor groove

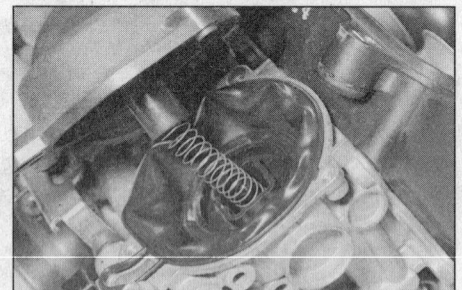

9.11 Install the spring and cover whilst making sure the diaphragm remains correctly seated

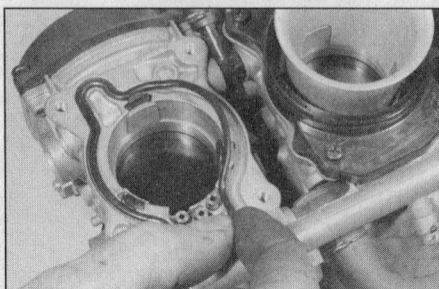

9.12a Install the O-ring in the carburetor groove

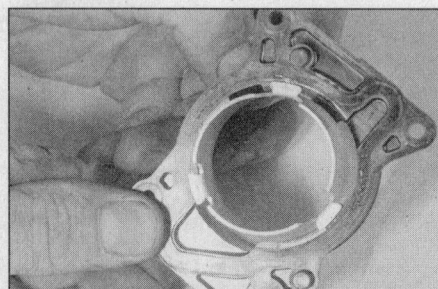

9.12b Insert air funnel in its holder and rotate it so it tabs are locked in place

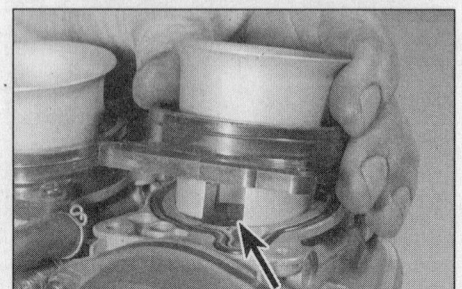

9.12c Align the cutout in the air funnel with the carburetor lug (arrow)

10.3a Release the two screws to free the throttle grip
housing cover . . .

10.3b . . . and access the throttle cables

intake rubbers to the cylinder head and disengage the carburetors from the cylinder head (see illustration 6.7a). Keep the carburetors upright to prevent fuel spillage.

3   Remove the two screws to separate the throttle grip pulley housing on the handlebar (see illustration). Hook the cable ends out of the pulley and unscrew the cable elbows from the rear of the housing (see illustration). Mark each cable to ensure it is connected correctly on installation.

4   Remove the cables from the machine noting the correct routing of each cable.

Installation

5   Install the cables making sure they are correctly routed. The cables must not interfere with any other component and should not be kinked or bent sharply.

6   Screw the cables into the lower half of the throttle grip pulley housing, making sure they are correctly connected. Lubricate the end of each cable with multi-purpose grease and attach the cables to the pulley. Assemble the throttle housing, aligning its locating pin with the hole in the handlebar. Install the retaining screws and tighten the upper screws first, followed by the lower.

7   Lubricate the end of each cable with multi-purpose grease and attach them to the carburetor throttle cam.

8   Make sure the cables are correctly connected and locate the outer cable adjusters in the mounting bracket.

9   Where necessary, fit the carburetors to the cylinder head and securely tighten the intake rubber clips.

10 Adjust the cables (see Chapter 1). Turn the handlebars back and forth to make sure the cables don't cause the steering to bind.

11 Install the air filter housing as described in Chapter 12.

12 Start the engine and check that the idle speed does not rise as the handlebars are turned. If it does, correct the problem before riding.

11 Choke cable - removal and installation

Removal

Refer to illustration 11.3

1   Remove the air filter housing as described in Section 12.

2   Slacken the screw then free the choke outer cable from its retaining clamp and detach the inner cable from the carburetor choke linkage (see illustrations 6.5a and 6.5b).

3   Peel the rubber cover up to reveal the knurled ring under the choke knob. Unscrew the ring and the nut beneath the bracket, then slip the cable out of the bracket slot (see illustration).

4   Remove the cable from the machine noting its correct routing.

Installation

5   Install the cable, making sure that it is correctly routed, as noted during removal. It is important to ensure that the cable does not interfere with any other component, and it should also not be kinked or bent sharply.

6   Slip the cable into the bracket on the triple clamp, aligning its flat with that on the bracket. Secure with the nut and knurled ring. Fit the rubber cover back into place.

7   Lubricate the cable end with multi-purpose grease and attach it to the choke linkage.

8   Locate the outer cable in the retaining clamp and adjust the cable as described in Chapter 1.

9   Install the air filter housing as described in Section 12.

12 Air filter housing and carburetor breather filter - removal and installation

Air filter housing

Removal

Refer to illustrations 12.2, 12.3, 12.4a and 12.4b

1   Remove the air filter element as described in Chapter 1.

4

11.3 The choke cable can be slipped out of its bracket
after releasing the knurled ring and nut

12.2 Air filter housing is retained by a single bolt to ignition HT coil bracket

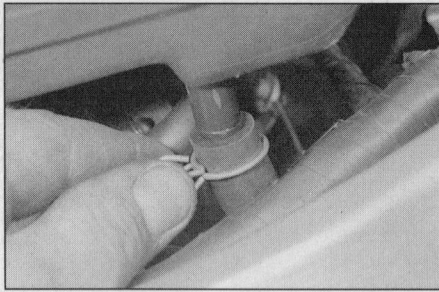

12.3 Disconnect the crankcase breather hose from the base of the housing

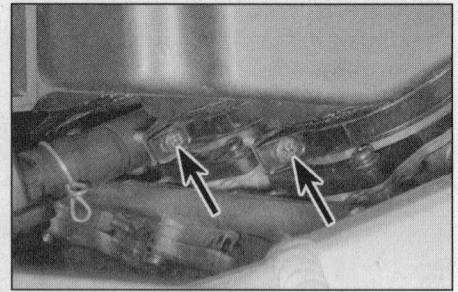

12.4a Slacken the two joint clamp screws on each side (arrows) . . .

12.4b . . . lift housing off the carburetors to enable carburetor breather hose to be disconnected from its filter (arrow)

12.7 Carburetor breather cover unclips for access to filter

2   Undo the bolt securing the filter housing base in position **(see illustration)**.
3   Free the crankcase breather pipe from the front right corner of the housing **(see illustration)**. On California models, disconnect the pulse secondary air injection (PAIR) system hose from the rear of the housing.
4   Slacken all four carburetor joint clamp screws and ease the housing off the air funnels **(see illustration)**. Disconnect the carburetor breather pipe from the filter on the base of the air filter housing **(see illustration)**.

Installation

5   Installation is the reverse of removal, noting that the air filter housing has a guide for the main wire harness.

Carburetor breather filter

Removal

Refer to illustration 12.7
6   Remove the fuel tank as described in Section 2 of this Chapter.
7   Unclip the cover to free the filter from the base of its chamber on the air filter housing **(see illustration)**. If the filter is excessively dirty it must be replaced with a new one.

Installation

8   Install the filter and clip the cover back into place.
9   Install the fuel tank as described in Section 2 of this Chapter.

13   Exhaust system - removal and installation

Muffler (silencer)

Removal

Refer to illustrations 13.1 and 13.2
1   Slacken and remove the muffler mounting nut and bolt and recover the washer and collar from the rubber mounting **(see illustration)**.
2   Loosen the muffler clamp bolts, then release the muffler from the exhaust front pipe using a twisting motion **(see illustration)**. Remove the muffler and recover the sealing ring.

Installation

Refer to illustration 13.3
3   Fit a new sealing ring to the front pipe **(see illustration)**.
4   Ensure the collar is in place in the mounting rubber (inserted with its flange on the inside), then fit the muffler with its mounting inside of the footpeg bracket. Install the bolt, washers and nut and tighten to the specified torque setting. Tighten the three muffler clamp bolts securely.

Complete system

Removal

Refer to illustrations 13.7a, 13.7b and 13.9
5   Remove the fairing lower and middle sections (see Chapter 8).

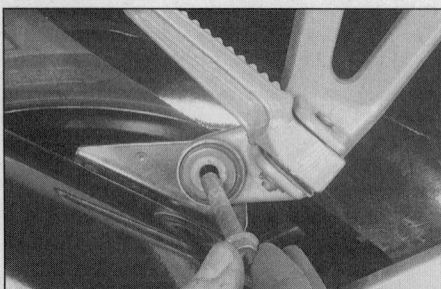

13.1 Muffler (silencer) bolts to pillion footpeg bracket

13.2 Three bolts retain muffler (silencer) to exhaust pipe

13.3 Use a new gasket at the muffler-to-exhaust pipe joint

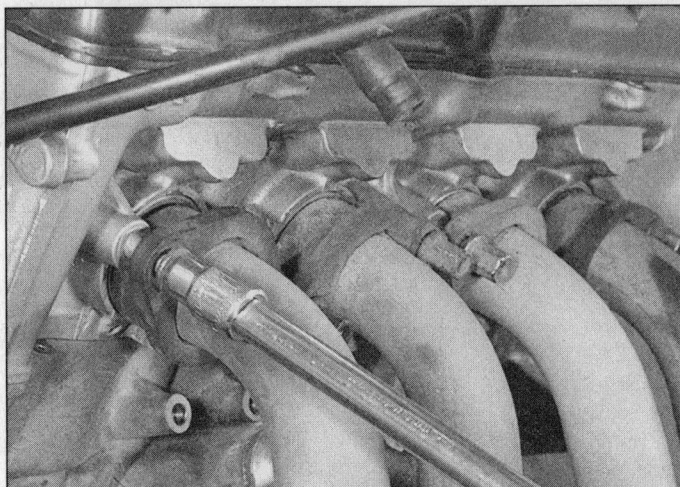

13.7a Remove the eight joint nuts . . .

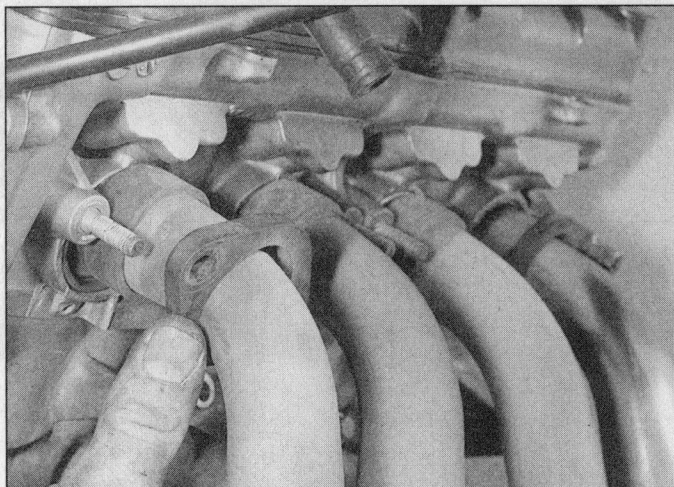

13.7b . . . and slide the joint plates off the cylinder head studs

6   Drain the cooling system (see Chapter 1) and remove the radiator (see Chapter 3).
7   Remove the eight nuts and slip the joint plates off the cylinder head studs **(see illustrations)**.
8   Slacken and remove the muffler mounting nut and bolt and recover the washer and collar from the rubber mounting **(see illustration 13.1)**.
9   Slacken and remove the nut and bolt securing the exhaust system to the base of the frame, taking exact note of the washer and collar positions in the rubber mounting **(see illustration)**. Remove the exhaust assembly from the bike and recover the gaskets from the cylinder head ports.

Installation

Refer to illustration 13.12
10   Inspect the mounting rubbers for damage and replace if necessary.
11   Fit the collar to the exhaust front pipe mounting rubber so that its flange is on the outside. Insert the collar into the muffler mounting rubber so that its flange is on the inside.
12   Position a new gasket in each of the cylinder head ports **(see illustration)**.
13   Fit the exhaust front pipe assembly, aligning the pipes with the cylinder head ports, and fit the bolt, washers and nut to the frame mounting.
14   Slide the joint plates onto the studs and fit the nuts. Tighten the exhaust front pipe nuts to the specified torque setting.
15   Securely tighten the frame mounting bolt nut to the specified torque setting.

16   Insert the muffler mounting bolt, washers and nut and tighten to the specified torque setting.
17   Install the radiator (see Chapter 3) and fill the cooling system (see Chapter 1).
18   Install the fairing sections as described in Chapter 8.

14   Pulse secondary air injection (PAIR) system (California models) - general information, removal and installation

General information

1   On California models to reduce the amount of unburnt hydrocarbons released in the exhaust gases, a Pulse secondary air injection (PAIR) system is fitted **(see illustration 15.1)**. The system consists of the pulse air control valve assembly, which is mounted on the front of the engine unit, and the air feed pipes which link the control valve to the cylinder head. The control valve is linked to the No. 3 intake manifold by a vacuum hose and to the air filter housing by an air suction hose.
2   When the engine is running, depression present in the intake manifold acts on the diaphragm in the control valve and opens the valve.
3   With the valve open, whenever there is a negative pulse in the exhaust system (ie on the overrun), filtered air is drawn from the air filter housing through the control valve and into the exhaust ports in the cylinder head. This fresh air promotes the burning of any excess fuel present in the exhaust gases, so reducing the amount of harmful hydrocarbons reduced into the atmosphere via the exhaust gases.

4

13.9 Remove the exhaust pipe-to-frame mounting bolt

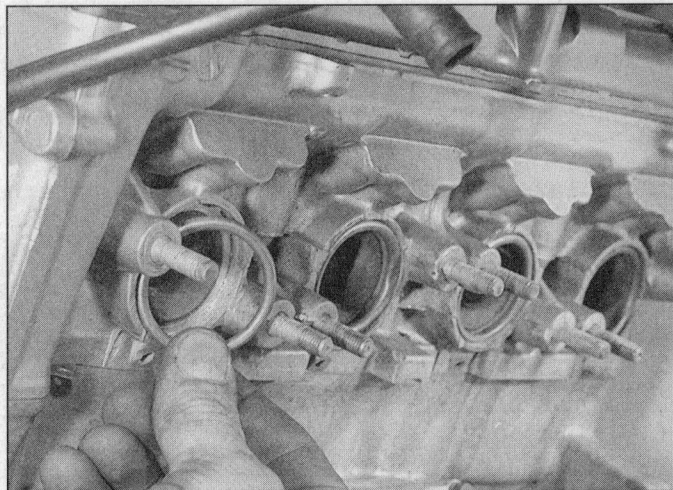

13.12 Use new gaskets in the cylinder head ports

14.8 Pulse secondary air injection (PAIR) system components

1 Control valve
2 Control valve bracket
3 Air feed pipes
4 Gaskets
5 Hoses

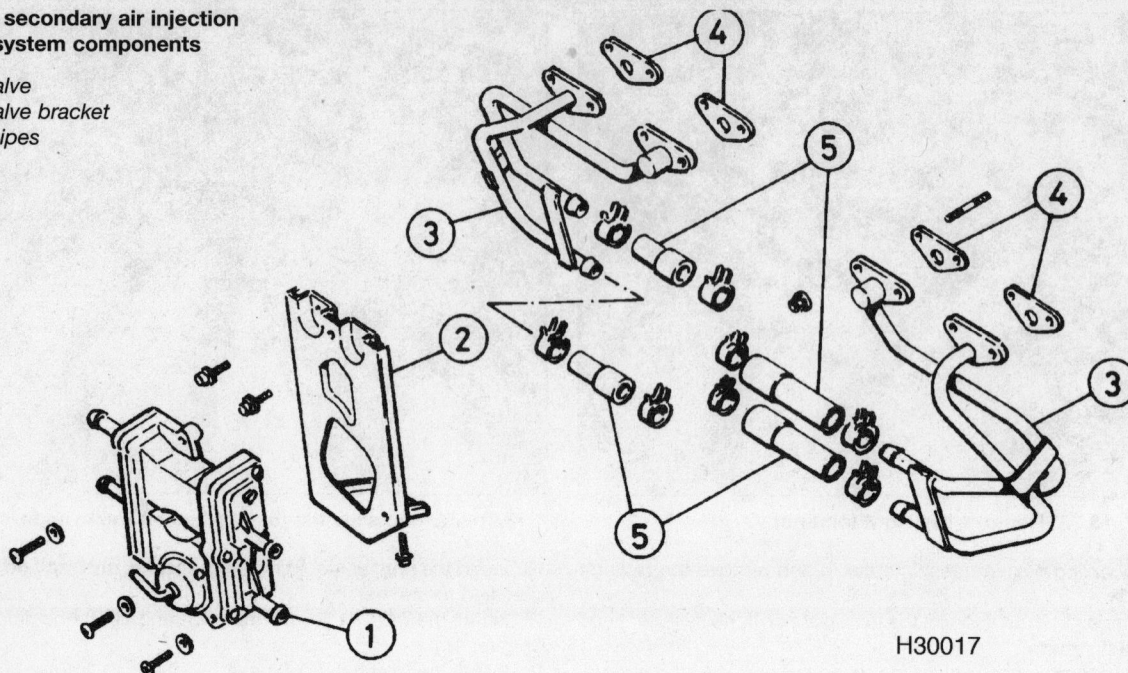

H30017

4   The control valve assembly is fitted with a pair of one-way check valves to prevent the exhaust gases passing through the control valve and into the air filter housing.

5   The system is not adjustable, but can be tested by a Honda dealer. Checks which can be performed by the owner are given in Chapter 1.

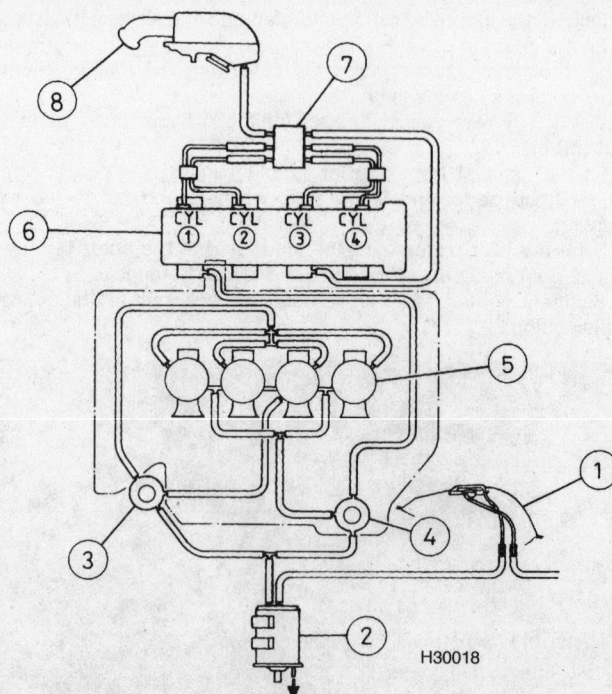

Removal

Refer to illustration 14.8

6   Remove the fairing middle sections as described in Chapter 8.

7   Release their retaining clips and disconnect the vacuum hose and air suction hose from the control valve.

8   Undo the three screws and washers securing the control valve to the front of the crankcase **(see illustration)**.

9   Unscrew the four nuts and release the air feed pipes from the front of the cylinder head.

10 Remove the control valve assembly, complete with pipes and hoses, from the front of the engine unit.

11 Recover the gaskets from the front of the head and discard them.

12 If necessary, unbolt the control valve mounting bracket from the front of the crankcase.

13 Inspect the pipes and hoses for signs of cracks and splits and replace damaged components.

Installation

14 Ensure the air feed pipe and cylinder head mating surfaces are clean and dry. Install the mounting bracket (where removed).

15 Fit a new gasket to each of the cylinder head unions, making sure each gasket is fitted the correct way around.

16 Install the control valve assembly locating the feed pipes on the cylinder head studs. Fit the air feed pipe nuts and control valve retaining screws and tighten them securely.

17 Connect the vacuum hose and air suction hose to the control valve and secure in position with the retaining clips.

18 Fit the fairing sections as described in Chapter 8.

15 Evaporative emission control (EVAP) system (California models only) - general information

Refer to illustration 15.1

1   On all California models, an evaporative emission control (EVAP) system is fitted **(see illustration)**. This system prevents the escape of fuel vapors into the atmosphere and functions as follows.

2   When the engine is stopped, fuel vapor from the tank is directed into a charcoal canister where it is absorbed and stored whilst the motorcycle is standing. When the engine is started, inlet manifold depression opens the purge control valve diaphragm. The vapors which are stored in the canister are then drawn into the engine to be burned during the normal combustion process.

3   The system is not adjustable, but can be tested by a Honda dealer. Checks which can be performed by the owner are given in Chapter 1.

15.1 Evaporative emission control (EVAP) system and pulse secondary air injection (PAIR) system connections

H30018

1 Fuel tank	5 Carburetors
2 EVAP charcoal canister	6 Cylinder head
3 EVAP purge control valve	7 PAIR system control valve
4 EVAP air vent control valve	8 Air filter housing

Chapter 5 Ignition system

Contents

Specifications

General information

Firing order	1-2-4-3
Cylinder identification	1-2-3-4 left to right
Spark plugs	See Chapter 1

Ignition timing

Initial	10°BTDC @ idle
Full advance	
US models and UK CBR900RR-R and RR-S	36°BTDC @ 4300 ± 100 rpm
UK CBR900RR-N and RR-P models	36°BTDC @ 6000 ± 100 rpm

Pulse generator

Resistance	460 to 580 ohms @ 20°C (68°F)

Ignition HT coils

Primary winding resistance	2.5 to 3.2 ohms @ 20°C (68°F)
Secondary winding resistance	
With plug caps	21 to 27 K ohms @ 20°C (68°F)
Without plug caps	11 to 17 K ohms @ 20°C (68°F)

Torque settings

	Nm	ft-lbs
Crankshaft right end cover		
Retaining bolts	12	9
Center cap	18	13

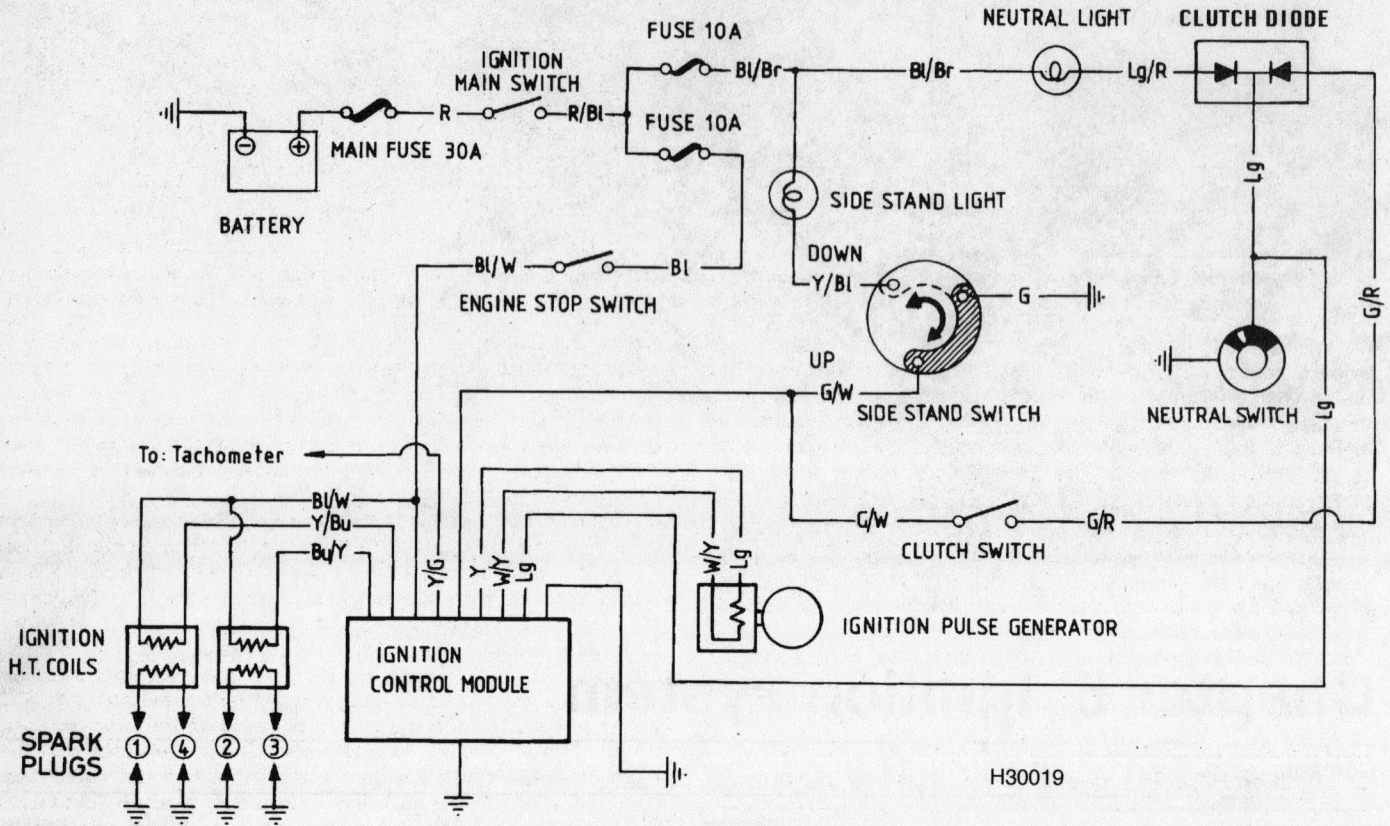

1.1 Ignition circuit diagram

Bl	Black	G	Green	Lg/R	Light green and red	Y	Yellow
Bl/Br	Black and brown	G/R	Green and red	R	Red	Y/Bl	Yellow and black
Bl/W	Black and white	G/W	Green and white	R/Bl	Red and black	Y/G	Yellow and green
Bu/Y	Blue and yellow	Lg	Light green	W/Y	White and yellow		

1 General information

Refer to illustration 1.1

These models are fitted with a fully transistorized digital ignition system, which due to its lack of mechanical parts is totally maintenance free. The system comprises a rotor, pulse generator, control module and two ignition HT coils **(see illustration)**.

The raised triggers on the rotor, which is fitted to the right end of the crankshaft, magnetically operate the pulse generator as the crankshaft rotates. The pulse generator sends a signal to the ignition control module which then supplies the ignition HT coils with the power necessary to produce a spark at the plugs. Each coil supplies two spark plugs.

Cylinders 1 and 4 operate off one coil and cylinders 2 and 3 off the other. For any given cylinder, the plug is fired twice for every engine cycle, but one of the sparks occurs during the exhaust stroke and therefore performs no useful function. This arrangement is usually known as a "spare spark" or "wasted spark" system.

Because of their nature, the individual ignition system components can be checked but not repaired. If ignition system troubles occur, and the faulty component can be isolated, the only cure for the problem is to replace the part with a new one. Keep in mind that most electrical parts, once purchased, can't be returned. To avoid unnecessary expense, make very sure the faulty component has been positively identified before buying a replacement part.

2 Ignition system - check

Caution: *The energy levels in electronic systems can be very high. On no account should the ignition be switched on whilst the plugs or plug caps are being held. Shocks from the HT circuit can be unpleasant. Secondly, it is vital that the plugs are soundly grounded (earthed) when the system is checked for sparking. The ignition system components can be seriously damaged if the HT circuit becomes isolated.*

1   As no means of adjustment is available, any failure of the system can be traced to failure of a system component or a simple wiring fault. Of the two possibilities, the latter is by far the most likely. In the event of failure, check the system in a logical fashion, as described below.

2   Disconnect the HT leads from No. 1 and No. 2 cylinder spark plugs and connect each lead to a spare spark plug. Lay each plug on the engine with the threads contacting the engine. If necessary, hold each spark plug with an insulated tool. **Warning:** *Don't remove one of the spark plugs from the engine to perform this check - atomized fuel being pumped out of the open spark plug hole could ignite, causing severe injury!*

3   Having observed the above precautions, check that the kill switch is in the RUN position, turn the ignition switch to ON and turn the engine over on the starter motor. If the system is in good condition a regular, fat blue spark should be evident at each plug electrode. If the spark appears thin or yellowish, or is non-existent, further investigation will be necessary. Before proceeding further, turn the ignition off and remove the key as a safety measure.

3.3 Ignition HT coils are mounted on frame cross-member

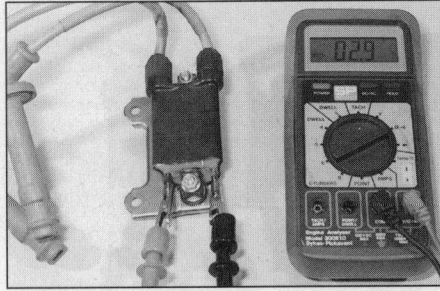
3.5 Measuring ignition HT coil primary winding resistance

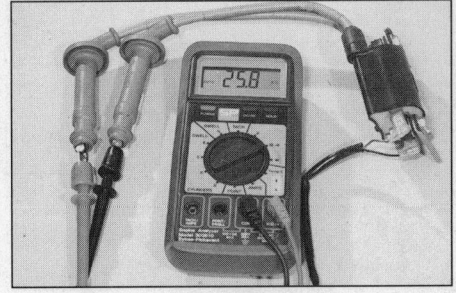
3.6 Measuring ignition HT coil secondary winding resistance (plug caps connected)

4   Ignition faults can be divided into two categories, namely those where the ignition system has failed completely, and those which are due to a partial failure. The likely faults are listed below, starting with the most probable source of failure. Work through the list systematically, referring to the subsequent sections for full details of the necessary checks and tests. **Note:** *Before checking the following items ensure that the battery is fully charged and that all fuses are in good condition.*

   a) *Loose, corroded or damaged wiring connections, broken or shorted wiring between any of the component parts of the ignition system (see Chapter 9).*
   b) *Faulty ignition or engine kill switch (see Chapter 9).*
   c) *Faulty neutral or side stand switch (see Chapter 9).*
   d) *Faulty pulse generator or damaged rotor.*
   e) *Faulty ignition HT coil(s).*
   f) *Faulty ignition control module.*

3   Ignition HT coils - check, removal and installation

Check

Refer to illustrations 3.3, 3.5 and 3.6

1   In order to determine conclusively that the ignition HT coils are defective, they should be tested by a Honda dealer equipped with the special electrical tester required for this check.
2   The coils can be checked visually (for cracks and other damage) and the primary and secondary coil resistances can be measured with an ohmmeter. If the coils are undamaged, and if the resistances are as specified, they are probably capable of proper operation.
3   To gain access to the coils, remove the air filter housing as described in Chapter 4 **(see illustration)**.
4   Disconnect the primary circuit electrical connectors from the coil(s) and disconnect the HT leads from the plugs that are connected to the

coil being checked. Gently pull the HT leads and plug caps through the dust/heat deflector sheet. Mark the locations of all wires before disconnecting them.
5   Set the meter to the ohms x 1 scale and measure the resistance between the low tension terminals **(see illustration)**. This will give a resistance reading of the primary windings and should be within the limits given in the Specifications.
6   To check the condition of the secondary windings, set the meter to the K ohm scale and connect the meter probes to the spark plug caps, noting the reading obtained **(see illustration)**. If this reading is not within the range in the Specifications, unscrew the plug caps from the HT leads then measure the resistance between the HT lead ends. If both values obtained differ greatly from those specified it is likely that the coil is defective. **Note:** *If only the first reading obtained is suspect, then the fault lies in the spark plug caps rather than the coil or its leads.*
7   Should any of the above checks not produce the expected result, the coil should be taken to a Honda dealer or auto-electrician for a thorough check. If the coil is confirmed to be faulty, it must be replaced; the coil is a sealed unit and cannot therefore be repaired. Note that the HT leads can be unscrewed from the coils and replaced separately.

Removal

Refer to illustrations 3.10 and 3.11

8   Remove the air filter housing as described in Chapter 4.
9   Disconnect the HT leads from the spark plugs that are connected to the coil(s) being removed. The cylinder number should be marked on the HT leads; if not, mark all leads before disconnecting them. Gently pull the HT leads and plug caps through the rubber dust cover.
10   Disconnect the primary circuit electrical connectors from the coils, having made note of their positions **(see illustration)**.
11   Remove the two bolts to release each coil from the frame cross-member **(see illustration)**. The coils can be separated from their mounting brackets by removing the throughbolts, washers and nuts.

5

3.10 Label the coil terminals or make written note of the wire connections before disconnecting primary circuit wires

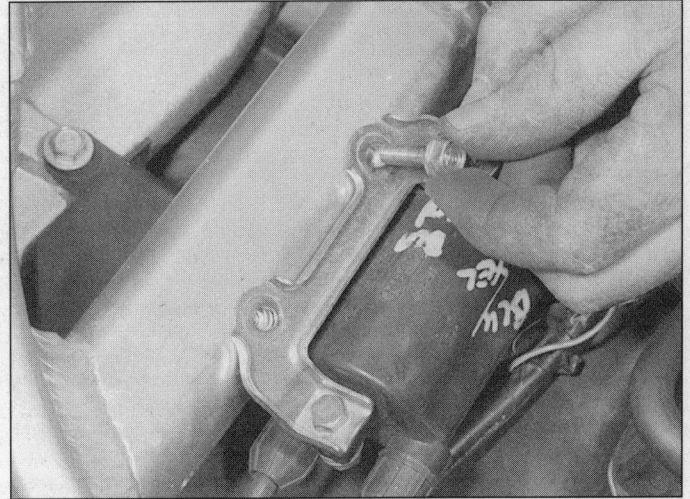
3.11 Remove the two bolts to free the coil mounting bracket from the frame cross-member

4.2 Pulse generator wire connector is located on inside of the frame

4.11 Pulse generator coil is retained by two bolts

4.15a Install the locating dowels (arrows) . . .

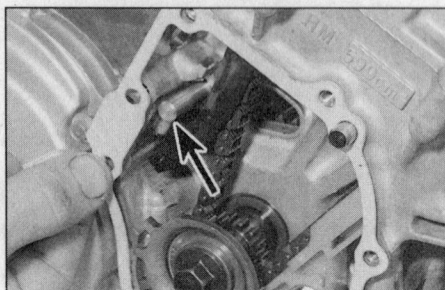

4.15b . . . and fit a new gasket to the crankcase. Ensure the oil gallery plug (arrow) is in position

4.17a Fit the cover to the crankcase and install the bolts . . .

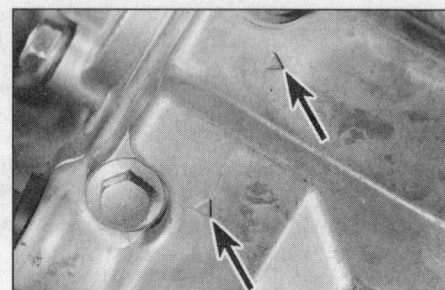

4.17b . . . the two bolts installed in the holes with the triangular marks (arrows) should have thread lock applied to them

Installation

12 Installation is the reverse of removal making sure the wiring connectors and HT leads are securely connected.

4   Pulse generator coil - check, removal and installation

Check

Refer to illustration 4.2

1   Remove the fuel tank as described in Chapter 4.

2   Trace the pulse generator wiring up from the crankshaft end cover to the black two-pin connector at the top of the frame **(see illustration)**. Disconnect the connector and using a multimeter set to the ohms x 100 scale, measure the resistance between the white/yellow and yellow wires on the pulse generator side of the connector.

3   Compare the reading obtained with that given in the Specifications at the start of this Chapter. The pulse generator must be replaced if the reading obtained differs greatly from that given, particularly if the meter indicates a short circuit (no measurable resistance) or an open circuit (infinite, or very high resistance).

4   If the generator is thought to be faulty, first check that this is not due to a damaged or broken wire from the coil to the connector; pinched or broken wires can usually be repaired.

Removal

Refer to illustration 4.11

5   Remove the fuel tank as described in Chapter 4.

6   Remove the fairing lower section as described in Chapter 8.

7   Drain the engine oil as described in Chapter 1.

8   Trace the pulse generator wiring up from the crankshaft end cover to the black two-pin connector at the top of the frame. Disconnect the connector and free the wiring from any relevant ties or clips.

9   Unscrew the cover retaining bolts and remove the cover squarely from the engine unit. Note the position of the oil gallery plug fitted behind the cover and take care to ensure that it stays in position in the crankcase.

10 Remove the cover locating dowels from the crankcase and discard the gasket.

11 Remove the bolts which secure the pulse generator to the cover and remove the generator **(see illustration)**.

12 Examine the rotor triggers for signs of damage such as chipped or missing teeth and replace if necessary (See Chapter 2, Section 10).

Installation

Refer to illustration 4.15a, 4.15b, 4.17a and 4.17b

13 Fit the pulse generator to the cover and securely tighten its retaining bolts.

14 Remove all traces of old gasket from the crankcase and cover mating surfaces. Make sure the oil gallery plug is in position in the crankcase.

15 Install the locating dowels and fit a new gasket to the crankcase **(see illustrations)**.

16 Apply a smear of sealant to the right crankshaft end cover wiring grommet.

17 Fit the cover to the engine noting that the bolts which are installed in the two holes next to the cam chain tensioner should have a drop of non-permanent thread locking compound applied to their threads **(see illustration)**. These holes are indicated by triangular marks cast in the cover **(see illustration)**. Tighten the cover bolts to the specified torque setting.

18 Route the wiring up to the wire harness and reconnect the two-pin connector. Secure the wiring in position with all relevant clips and ties.

19 Fill the engine with the correct type and amount of oil as described in Chapter 1.

20 Install the fuel tank and fairing section as described in Chapters 4 and 8 respectively.

5   Ignition control module - removal, check and installation

Removal

Refer to illustrations 5.2 and 5.3

1   Remove the seat (see Chapter 8).

2   Free the control module from its two retaining tabs **(see illustration)**.

3   Disconnect the wiring connector from the control module and remove it from the motorcycle **(see illustration)**.

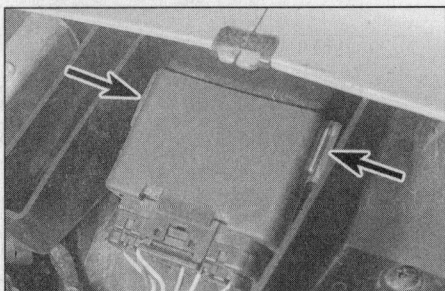

5.2 Ignition control module is mounted on two tabs (arrows)

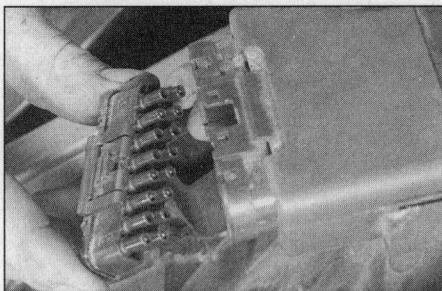

5.3 Release its tab and pull the wire connector free of the control module

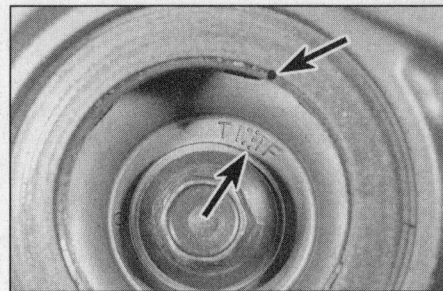

6.9 Three dots on the rotor should align with the notch at idle if the ignition timing is correct (arrows)

Check

4   If the tests shown in the preceding Sections have failed to isolate the cause of an ignition fault, it is likely that the control module itself is faulty. No test details are available with which the unit can be tested; refer to a Honda dealer for advice.

Installation

5   Installation is the reverse of removal ensuring the wiring connector is securely connected.

6   Ignition timing - general information and check

General information

1   Since no provision exists for adjusting the ignition timing and since no component is subject to mechanical wear, there is no need for regular checks; only if investigating a fault such as a loss of power or a misfire, should the ignition timing be checked.
2   The ignition timing can only be checked whilst the engine is running using a stroboscopic (timing) lamp. The inexpensive neon lamps should be adequate in theory, but in practice may produce a pulse of such low intensity that the timing mark remains indistinct. If possible, one of the more precise xenon tube lamps should be used, powered by an external source of the appropriate voltage. **Note:** *Do not use the machine's own battery as an incorrect reading may result from stray impulses within the machine's electrical system.*

Check

Refer to illustrations 6.9 and 6.12
3   Warm the engine up to normal operating temperature then stop it.
4   Remove the fairing lower section and middle sections as described in Chapter 8.
5   Unscrew the center cap from the right crankshaft end cover to reveal the pulse generator rotor. Recover the cap O-ring.
6   The timing mark is stamped on the center of the rotor and is represented by a line of three punch marks next to the F mark. The static index mark, with which the three dots should align is a notch in the cover.

7   Connect the timing light to the No. 1 cylinder HT lead as described in its manufacturer's instructions.
8   Start the engine and aim the light at the inspection hole.
9   With the machine idling at the specified speed, the timing mark (three dots) should align with the cover notch **(see illustration)**.
10 Slowly increase the engine speed whilst observing the timing mark. Starting at approximately 1600 rpm, the timing mark should move counterclockwise (anti-clockwise), increasing in relation to the engine speed.
11 As already stated, there is no means of adjustment of the ignition timing on these machines. If the ignition timing is incorrect one of the ignition system components is at fault, and the system must be tested as described in the preceding Sections of this Chapter.
12 When the check is complete, fit a new O-ring to the center cap and lubricate it with a smear of clean engine oil **(see illustration)**. Install the cap and tighten it to the specified torque setting.
13 Install the fairing sections as described in Chapter 8.

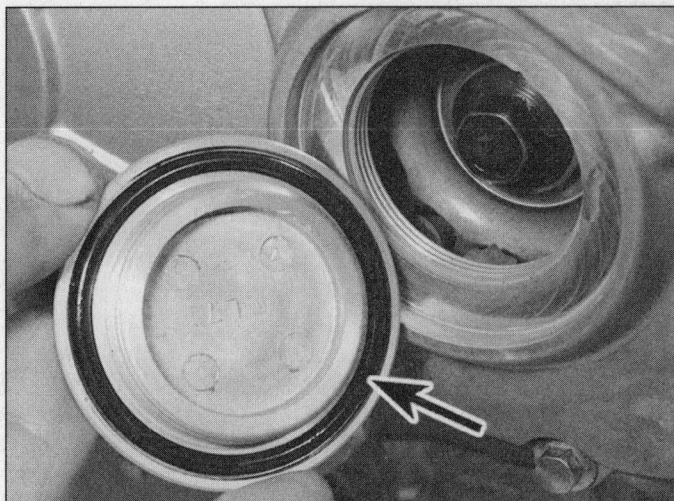

6.12 On installation, fit a new O-ring (arrow) to the cover center cap

Chapter 6 Frame, suspension and final drive

Contents

Specifications

Front forks

Spring free length
US 1993 and 1994 models, UK CBR900RR-N and RR-P
New............................. 306.7 mm (12.07 in)
Service limit .... 300.6 mm (11.83 in)
US 1995 model, UK CBR900RR-R and RR-S
New............................. 243.8 mm (9.60 in)
Service limit .... 238.9 mm (9.41 in)
Oil capacity - per leg
US 1993 and 1994 models, UK CBR900RR-N and RR-P............ 509 cc (17.2 US fl oz, 17.9 Imp fl oz)
US 1995 model, UK CBR900RR-R and RR-S .... 564 cc (19.1 US fl oz, 19.9 Imp fl oz)
Oil level
US 1993 and 1994 models, UK CBR900RR-N and RR-P............ 110 mm (4.3 in)*
US 1995 model, UK CBR900RR-R and RR-S .... 112 mm (4.4 in)*
Fork oil type
US models............ Pro Honda Suspension Fluid SS-8
UK models .... SAE 10W fork oil

*Oil level is measured from the top of the tube with the fork spring removed and the leg fully compressed.

Rear shock absorber

Spring free length
New............ 163.5 mm (6.44 in)
Service limit .... 160.2 mm (6.31 in)

Torque settings

	Nm	ft-lbs
Handlebar clamp bolt	23	17
Triple clamp pinch bolts		
Top triple clamp	23	17
Bottom triple clamp	50	36
Bottom triple clamp brake hose union assembly bolts	17	12
Front forks		
Top bolts	23	17
Damper rod Allen bolt	20	14
Steering stem adjuster nut (see text)	31	22
Steering stem top nut	105	76
Shock absorber mounting bolt nuts	45	33
Rear suspension linkage pivot bolt nuts	45	33
Swingarm pivot adjusting bolt	15	11
Swingarm pivot locknut	65	47
Swingarm pivot nut	95	69
Front sprocket bolt	54	39
Rear sprocket nuts	88	64

6

1 General information

The models covered in this manual use a diamond type twin-spar frame made of extruded aluminum, which uses the engine as a stressed member.

Front suspension is by a pair of Showa oil-damped, coil spring, telescopic forks with 45 mm stanchions and forged aluminum sliders.

Rear suspension is by Honda's "Pro-Link" system in which the swingarm acts on a Showa gas-charged, hydraulically-damped suspension unit via a two piece linkage.

The final drive uses an O-ring drive chain with a riveted master link. A rubber damper (often called a "cush drive") is installed between the rear wheel coupling and the wheel.

2 Frame - inspection and repair

1 The frame should not require attention unless accident damage has occurred. In most cases, frame replacement is the only satisfactory remedy for such damage. A few frame specialists have the jigs and other equipment necessary for straightening the frame to the required standard of accuracy, but even then there is no simple way of assessing to what extent the frame may have been over stressed.

2 After the machine has accumulated a lot of miles, the frame should be examined closely for signs of cracking or splitting at the welded joints. Loose engine mount bolts can cause ovaling or fracturing of the mounting tabs. Minor damage can often be repaired by welding, depending on the extent and nature of the damage.

3 Remember that a frame which is out of alignment will cause handling problems. If misalignment is suspected as the result of an accident, it will be necessary to strip the machine completely so the frame can be thoroughly checked.

3 Footpegs and brackets - removal and installation

Rider's footpegs

Removal

1 Remove the cotter pin (split pin) and washer. Slide out the pivot pin and remove the footpeg from the bracket along with its return spring.

2 If necessary, undo the retaining screws and separate the rubber from the footpeg.

Installation

3 Installation is a reverse of removal using a new cotter pin (split pin).

Rider's left footpeg bracket

Removal

Refer to illustrations 3.4 and 3.5

4 Remove the pivot bolt securing the gearshift lever (models with gear linkage) and footpeg to the footpeg bracket and recover the washer from behind the lever **(see illustration)**.

5 Unscrew the two footpeg bracket mounting bolts and remove the bracket assembly from the frame **(see illustration)**.

Installation

6 Install the bracket and securely tighten its mounting bolts.

7 Apply a smear of grease to the bearing surface of the gearshift lever pivot and ensure that the washers are in place (models with gear linkage). Install the pivot bolt and tighten it securely.

Rider's right footpeg bracket

Removal

Refer to illustration 3.9

8 Remove the cotter pin (split pin) and the clevis pin from the clevis joining the rear brake pedal to the rear brake master cylinder pushrod.

9 Unscrew the step guard and master cylinder mounting bolts. Remove the step guard and support the master cylinder so that no strain is put on the hoses **(see illustration)**.

10 Unhook the brake pedal return spring and the brake light switch return spring and keep them safe. Release the brake light switch from its holder on the footpeg bracket.

11 Unscrew the pivot bolt securing the brake pedal and footpeg to the footpeg bracket and recover the washer from behind the pedal.

12 Unscrew the two footpeg bracket mounting bolts and remove the bracket from the frame.

Installation

13 Install the bracket and securely tighten its mounting bolts.

14 Apply a smear of grease to the bearing surface of the brake pedal pivot. Position the washer behind the brake pedal then fit the pivot bolt, tightening it securely.

15 Install the rear brake light switch into its holder on the footpeg bracket, and hook up its return spring and the brake pedal return spring.

16 Install the rear brake master cylinder and the step guard on the footpeg bracket and tighten the mounting bolts securely.

17 Align the brake pedal with the master cylinder pushrod clevis and install the clevis pin. Fit a new cotter pin (split pin) to the clevis pin.

Passenger footpegs

Removal

18 Remove the cotter pin (split pin) and washer, then slide out the pivot pin and remove the footpeg from the bracket. As the footpeg is removed, recover the footpeg detent plate, spring and ball bearing, noting the correct fitted positions of each component.

Installation

19 Installation is the reverse of removal using a new cotter pin (split pin).

Left passenger footpeg bracket

Removal

Refer to illustration 3.21

20 Remove the left side cover (see Chapter 8, if necessary).

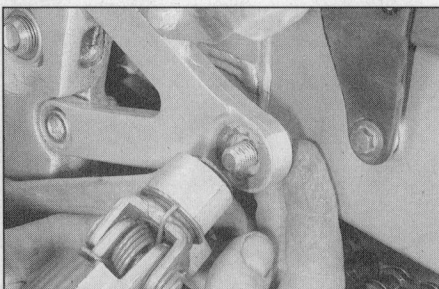

3.4 Remove the retaining bolt to release the gearshift lever and footpeg

3.5 Unscrew the two footpeg bracket mounting bolts

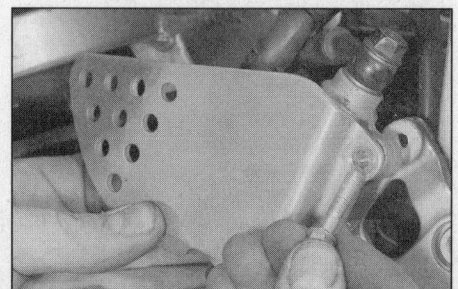

3.9 Unscrew the two step guard and master cylinder mounting bolts

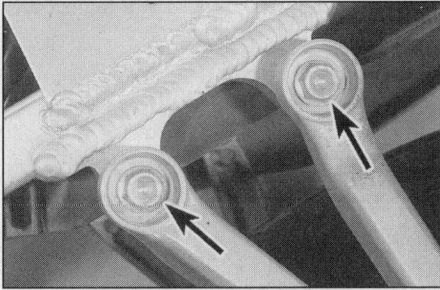

3.21 Passenger footpeg brackets are retained by two bolts (arrows)

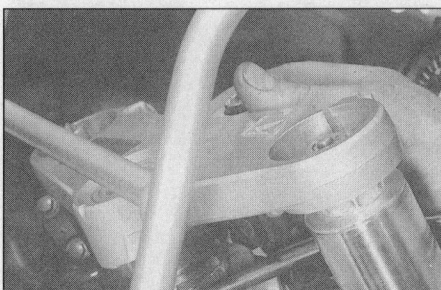

5.5a Note the brackets linking the top triple clamp to the handlebars (arrow) . . .

5.5b . . . and the choke knob bracket on the left handlebar (arrow)

21 Unscrew the footpeg bracket retaining bolts and remove the bracket from the motorcycle (**see illustration**).

Installation
22 Installation is the reverse of removal.

Right passenger footpeg bracket

Removal
23 Remove the right side cover (see Chapter 8, if necessary).
24 Slacken and remove the mounting bolt securing the exhaust system muffler (silencer) to the footpeg bracket. Tie the muffler (silencer) to the frame to prevent strain on the exhaust system.
25 Unscrew the footpeg bracket retaining bolts and remove the bracket from the motorcycle.

Installation
26 Installation is the reverse of removal.

4   Sidestand - maintenance

1   The sidestand is attached to a bracket on the frame. An extension spring anchored to the bracket ensures that the stand is held in the retracted position. The sidestand incorporates a rotary switch which cuts out the ignition if the sidestand is down when the engine is running and in gear.
2   Make sure the pivot bolt is tight and the extension spring is in good condition and not over-stretched. An accident is almost certain to occur if the stand extends while the machine is in motion. For check and replacement of the sidestand switch see Chapter 9.

5   Handlebars - removal and installation

Right handlebar

Removal
Refer to illustrations 5.5a, 5.5b and 5.5c
1   Refer to Chapter 4 *"Throttle cables - removal and installation"* and release the throttle cables from the throttle grip. Position the throttle housing away from the handlebar.
2   Disconnect the brake light switch wires from the master cylinder assembly. Refer to Chapter 9 and unscrew the two handlebar switch retaining screws and free the switch from the handlebar.
3   Unscrew the two master cylinder assembly clamp bolts and position the assembly clear of the handlebar, ensuring no strain is placed on the hydraulic hose. Keep the master cylinder reservoir upright to prevent possible fluid leakage.
4   Remove the steering stem nut cap, then unscrew the steering stem nut.
5   Unscrew the top triple clamp pinch bolts on each side, noting the position of the choke knob and the brackets between the triple clamp and the handlebars, then slide the triple clamp off the forks and position it away from the handelbars (**see illustrations**).
6   Slacken the handlebar clamp pinch bolt and slide the handlebar off the fork.
7   If necessary, unscrew the handlebar weight retaining screw, then remove the weight from the end of the handlebar and slide off the throttle twistgrip.

Installation
Refer to illustrations 5.10a, 5.10b, 5.11 and 5.14
8   Where removed, apply a smear of grease to the throttle twistgrip and slide the grip on the handlebar. Apply locking compound to the handlebar weight retaining screw, then install the weight and securely tighten the screw.
9   Install the handlebar onto the fork, but do not yet tighten the clamp pinch bolt.
10   Install the top triple clamp on the forks and install the steering stem nut, tightening it to the torque setting specified at the beginning of the Chapter. Install the steering stem nut cap. Now position the handlebar so that the lug on the handlebar clamp fits into the slot in the underside of the triple clamp, and the bracket on the handlebar clamp aligns with the pinch bolt hole in the triple clamp. Tighten the handlebar clamp bolt to the specified torque setting (**see illustrations**).
11   Install the choke knob bracket to the left side of the triple clamp then tighten both triple clamp pinch bolts to the specified torque setting (**see illustration**).
12   Reconnect the throttle cables to the twistgrip (see Chapter 4).

5.5c Slide the top triple clamp off the forks

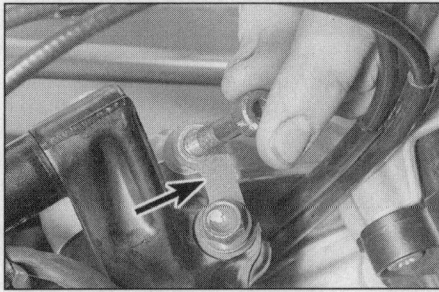

5.10a Make sure the lug on the handlebar clamp fits into the slot in the top triple clamp (arrow)

5.10b Tighten the handlebar clamp pinch bolt to the specified torque setting

6

5.11 Tighten the top triple clamp pinch bolts to the specified torque setting

5.14 Align mating surfaces of master cylinder clamp with punch mark on handlebar (arrow)

13 Install the handlebar switch as described in Chapter 9.
14 Install the master cylinder assembly, making sure the "UP" mark on the clamp is upwards. Align the mating surfaces of the clamp with the punch mark on the handlebar, then tighten the upper clamp bolt first, followed by the lower **(see illustration)**.
15 Reconnect the front brake light switch wiring.

Left handlebar

Removal

16 Disconnect the clutch switch wires from the clutch lever assembly. Free the handlebar switch as described in Chapter 9.
17 Unscrew the two clutch lever bracket mounting bolts, and position the lever assembly away from the handlebar.
18 Remove the steering stem nut cap. Unscrew the steering stem nut.
19 Unscrew the top triple clamp pinch bolts on each side, noting the position of the choke knob and the brackets between the triple clamp and the handlebars, then slide the triple clamp off the forks and place it away from the handlebars **(see illustrations 5.5a, 5.5b and 5.5c)**.
20 Slacken the handlebar clamp pinch bolt and slide the handlebar off the fork.
21 If necessary, undo the retaining screw then remove the weight from the end of the handlebar and peel off the handlebar grip.

Installation

22 Where removed, apply a suitable adhesive to the inside of the throttle grip and slide it on the handlebar. Apply locking compound to the handlebar weight retaining screw then install the weight and securely tighten the screw.
23 Install the handlebar on the fork, but do not yet tighten the clamp pinch bolt.
24 Install the top triple clamp on the forks and install the steering stem nut, tightening it to the specified torque setting. Install the steering

stem nut cap. Now position the handlebar so that the lug on the handlebar clamp fits into the slot in the underside of the triple clamp **(see illustration 5.10a)**, and the bracket on the handlebar clamp aligns with the pinch bolt hole in the triple clamp. Tighten the handlebar clamp bolt to the specified torque setting **(see illustration 5.10b)**.
25 Install the choke knob bracket to the triple clamp then tighten both triple clamp pinch bolts to the specified torque **(see illustration 5.11)**.
26 Install the handlebar switch as described in Chapter 9.
27 Install the clutch lever assembly, ensuring the "UP" mark on the clamp is upwards. Align the mating surfaces of the clamp with the punch mark on the handlebar. Tighten the upper bolt first, then the lower.
28 Reconnect the clutch switch wiring.

6 Forks - removal and installation

Removal

Refer to illustrations 6.5 and 6.6
Note: *If the fork legs are to be dismantled it is preferable to adjust the preload setting to minimum as described in Section 12, then slacken the top bolts whilst the forks are still held in the triple clamps.*
1   Remove the front wheel as described in Chapter 7.
2   Remove the front fender/mudguard as described in Chapter 8, and pry off the fender/mudguard bracket from the top of the fork slider.
3   Slacken but do not remove the top triple clamp pinch bolts **(see illustration 5.11)**.
4   Slacken but do not remove both left and right handlebar clamp bolts **(see illustration 5.10b)**.
5   Slacken but do not remove the bottom triple clamp pinch bolts **(see illustration)**.
6   Remove the forks by twisting them and pulling them downwards **(see illustration)**. If the fork legs are seized in the triple clamps, spray

6.5 Slacken the bottom triple clamp pinch bolts

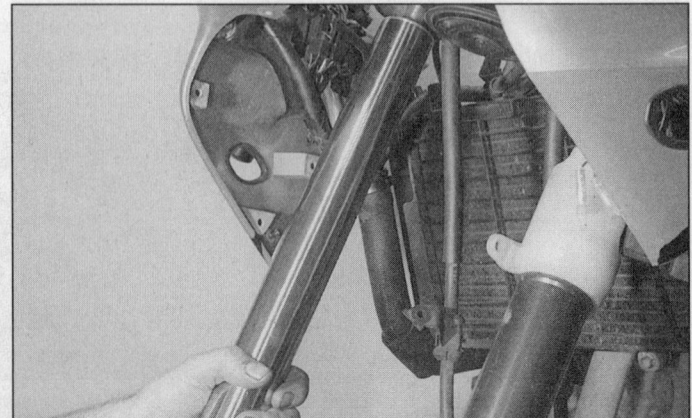

6.6 Remove the forks by twisting and pulling them down

6.7 Slide the forks into position through the bottom triple clamp

6.8 Align top of fork with top surface of top triple clamp

the area with penetrating oil and allow time for it to soak in before trying again. Once the forks are removed, make sure the right handlebar is supported so that the master cylinder reservoir is upright and no strain is placed on the hoses.

Installation
Refer to illustrations 6.7 and 6.8

7 Remove all traces of corrosion from the fork tubes and the triple clamps and slide the fork legs back into place **(see illustration)**.

8 Position each leg so that the top surface on the fork tube aligns with the top surface of the top triple clamp, then tighten the top and bottom triple clamp pinch bolts to the torque settings specified at the beginning of the Chapter **(see illustration)**.

9 Tighten both handlebar clamp pinch bolts to the specified torque.

10 If the fork legs have been dismantled, the fork tube top bolts should now be tightened to the specified torque setting.

11 Install the fender/mudguard bracket to the top of each fork slider, then install the front fender/mudguard as described in Chapter 8.

12 Install the front wheel as described in Chapter 7.

13 Adjust the fork settings as described in Section 12. Check the operation of the front forks and brake before taking the machine out on the road.

7 Forks - disassembly, inspection and reassembly

Disassembly
Refer to illustrations 7.1, 7.5, 7.9, 7.10 and 7.13

1 Always dismantle the fork legs separately to avoid interchanging parts and thus causing an accelerated rate of wear. Store all components in separate, clearly marked containers **(see illustration)**.

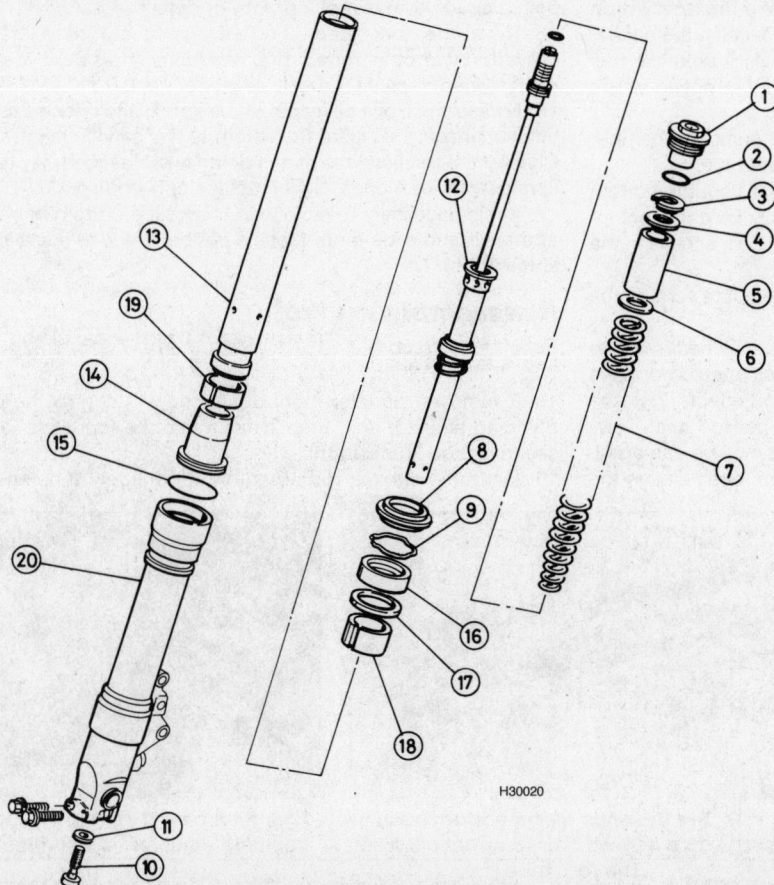

7.1 Exploded view of a front fork

1 Top bolt
2 O-ring
3 Slotted spring collar
4 Washer (later models only)
5 Spacer
6 Spring seat
7 Fork spring
8 Dust seal
9 Snap-ring
10 Damper rod Allen bolt
11 Allen bolt washer
12 Damper rod assembly
13 Fork tube
14 Damper rod seat
15 O-ring (later models only)
16 Oil seal
17 Washer
18 Top bush
19 Bottom bush
20 Fork slider

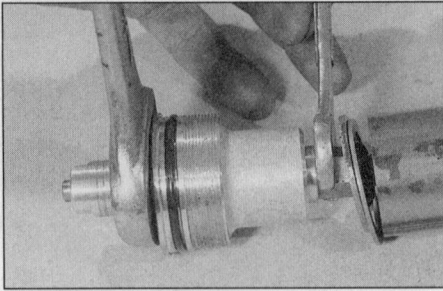

7.5 Unscrew the fork top bolt whilst counter-holding the locknut

7.9 Pry the dust seal out of the fork lower leg using a flat-bladed screwdriver

7.10 Carefully remove the snap-ring

2   Before dismantling the fork, it is advised that the damper rod Allen bolt be slackened. Compress the fork leg so that the spring exterts pressure on the damper rod head, then have an assitant unscrew the damper rod bolt from the base of the fork leg.

3   If the fork top bolt was not slackened with the fork in situ, carefully clamp the fork tube in a vise, taking care not to overtighten or score its surface, then fully unscrew the preload adjuster as described in Section 12 and slacken the fork top bolt.

4   Unscrew the fork top bolt from the top of the fork tube.

5   Carefully clamp the fork slider in a vise and slide the fork tube down into the slider a little way (wrap a rag around the spring and the top of the tube to minimize oil spillage) while, with the aid of an assistant if necessary, keeping the damper rod and top cap fully extended. Counter-hold the locknut immediately below the top cap with a wrench, whilst the top bolt is unscrewed **(see illustration)**.

6   Remove the slotted spring collar by holding down the fork spring (keeping the damper rod fully extended) and slipping the collar out to the side. **Warning:** *The fork spring may be exerting considerable pressure, making this a potentially dangerous operation. Wipe off as much oil as possible to minimize the risk of your hands slipping on oily components and enlist the help of an assistant.* Keep the restraint on the fork spring and remove the washer (later models only), the spacer and the spring seat. Slowly release the spring until all pressure has been relieved, then withdraw the spring from the tube, noting which way up it fits.

7   Invert the fork leg over a suitable container and pump the damper rod piston vigorously to expel as much fork oil as possible.

8   Remove the damper rod Allen bolt and its washer from the bottom of the slider, then withdraw the damper rod assembly from the fork.

9   Pry out the dust seal from the top of the slider to gain access to the oil seal retaining snap ring **(see illustration)**.

10   Carefully remove the snap ring whilst taking care not to scratch the surface of the tube **(see illustration)**.

11   In order to separate the tube from the slider it will be necessary to displace the top bush and oil seal. The bottom bush should not pass through the top bush, and this can be used to good effect. Push the tube gently inwards until it stops against the damper rod seat. Take great care not to do this forcibly, otherwise the seat may be damaged. Then pull the tube sharply outwards until the bottom bush strikes the

top bush. Repeat this operation until the top bush and seal are tapped out of the slider.

12   With the tube removed, slide off the oil seal and washer, noting which way up they fit. The top bush can then also be slid off its upper end. **Caution:** *Do not remove the bottom bush from the tube unless it is to be replaced.*

13   Tip the damper rod seat out of the slider. If necessary, on later models, unscrew the compression damping adjuster unit using a suitable wrench **(see illustration)**.

Inspection

Refer to illustration 7.17

14   Clean all parts in solvent and blow them dry with compressed air, if available. Check the fork tube and slider, the bushes and the damper rod assembly for score marks, scratches, flaking of the chrome and excessive or abnormal wear. Look for dents in the tubes and replace the tubes if any are found. Check the fork seal seat for nicks, gouges and scratches. If damage is evident, leaks will occur. If either bush is worn so badly that the copper base metal appears through the Teflon coating over more than 3/4 of the bush's surface area, that bush must be replaced. Always replace worn or defective parts with new ones.

15   Have the fork tube checked for runout at a dealer service department or other repair shop. **Warning:** *If it is bent, it should not be straightened; replace it with a new one.*

16   Measure the overall length of the spring and check it for cracks and other damage. Compare the length to the service length listed in this Chapter's Specifications. If it is defective or sagged, replace both fork springs with new ones. Never replace only one spring.

17   If it is necessary to replace the fork tube bottom bush, pry it apart at the slit and slide it off. Make sure the new one seats properly **(see illustration)**.

Reassembly

Refer to illustrations 7.18, 7.19, 7.20, 7.21a, 7.21b, 7.22, 7.23, 7.24, 7.25, 7.27a, 7.27b, 7.28a, 7.28b and 7.29

18   If removed on later models, fit a new O-ring to the compression damping adjuster unit, then install it into the fork slider and tighten it securely **(see illustration)**.

19   Insert the damper rod assembly into the fork tube and slide it into

7.13 The compression damper unit can be removed if necessary using a suitable spanner

7.17 Pry the ends of the bottom bush apart with a screwdriver to remove it from the fork tube

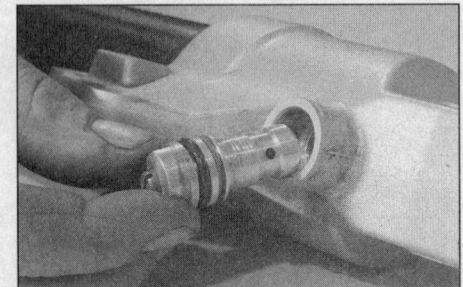

7.18 Fit a new O-ring to the compression damping adjuster unit and install it in the fork slider

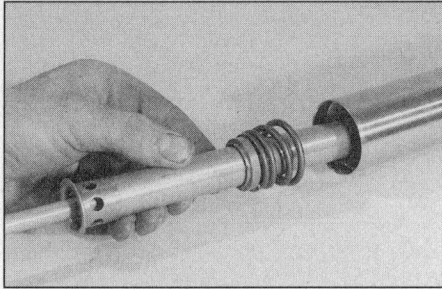

7.19 Slide the damper rod assembly into the fork tube

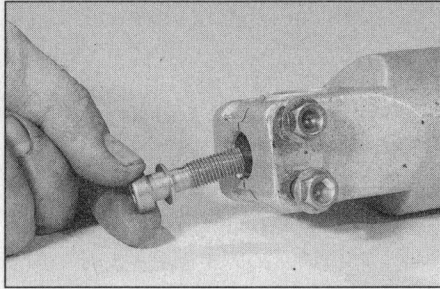

7.20 Fit a new washer to the damper rod Allen bolt; apply thread-lock before installing it in the bottom of the fork slider

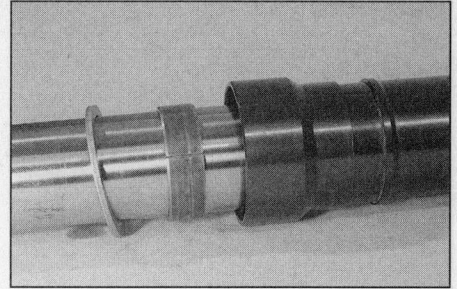

7.21a Lubricate the top bush and washer and slide them over the fork tube

7.21b Use a piece of tubing as a drift to install the top bush in the slider

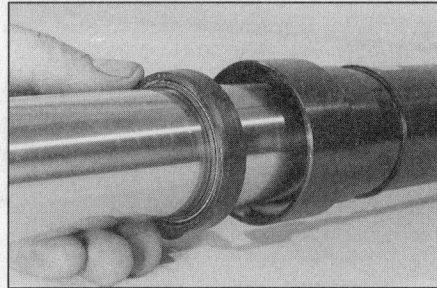

7.22 Install the new oil seal making sure it is the right way up (see text)

place so that it projects fully from the bottom of the tube **(see illustration)**. Install a new O-ring onto the damper rod assembly seat (later models only), then install the seat on the bottom of the damper rod assembly.

20 Oil the fork tube and bottom bush and insert the assembly into position in the fork slider. Fit a new washer to the damper rod Allen bolt, then apply a drop of non-permanent locking compound to its threads and install the bolt into the bottom of the slider **(see illustration)**. Tighten the bolt to the specified torque setting. **Note:** *If the damper rod assembly turns inside the fork tube, preventing tightening of the Allen bolt, wait until the fork has been reassembled so that spring pressure holds the damper in place (see Step 29).*

21 Push the fork tube fully into the slider, then oil the top bush and slide it down over the tube. Press the bush squarely into its recess in the slider as far as possible by hand, then install the washer **(see illustration)**. It will be necessary to use the Honda service tool or to devise an alternative tubular drift to tap the bush fully into place. The best method is to use a length of tubing slightly larger in diameter than the fork tube, that will bear squarely on the bush's washer, then tap the bush home using the tubing as a slide-hammer **(see illustration)**. Take care not to scratch the fork tube during this operation; it is best to make sure that the fork tube is pushed fully into the slider so that any accidental scratching is confined to the area above the oil seal.

22 When the bush is seated fully and squarely in its recess in the slider, so that the washer seats on the oil seal's recess (remove the

washer to check, wipe the recess clean, then reinstall the washer), install the new oil seal. Note that the seal has two springs, one larger than the other. Smear the seal's lips with fork oil and slide it over the tube so that the larger spring faces down toward the slider **(see illustration)**.

23 Place a large plain washer against the oil seal and drive it into place as described in Step 21 above until the snap-ring groove is just visible above the seal. Once the seal is correctly seated, remove the washer and fit the snap-ring, making sure it is correctly located in its groove **(see illustration)**.

24 Lubricate the lips of the dust seal then slide it down the fork tube and press it into position **(see illustration)**.

25 Slowly pour in the specified amount and grade of fork oil whilst pushing the damper rod piston up and down. Once the oil has been added, pump the fork tube in and out at least five times, and pump the damper rod piston at least another 10 times. This will ensure that the fork oil is evenly distributed. Fully insert both the tube and damper rod then check the fork oil level. Add or subtract fork oil until the oil is at the level specified in this Chapter's Specifications **(see illustration)**.

26 Clamp the slider securely in a vice and fully extend the damper rod. Tie a piece of wire around the rod; the wire can then be used to hold the rod in the extended position whilst the slotted spring collar is installed.

27 Insert the fork spring, ensuring that its tapered end is at the bottom, followed by the spring seat, the spacer and the washer (later

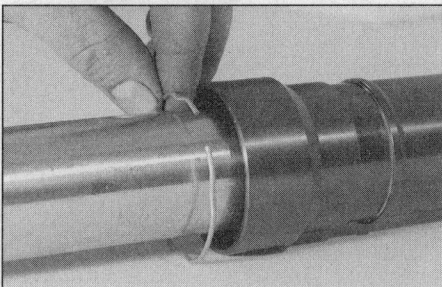

7.23 Secure the seal in place with the snap-ring making sure it is properly located in its groove

7.24 Slide the dust seal into position in the slider

7.25 Fill the fork with the specified type and amount of fork oil then check the oil level with the fork held vertical

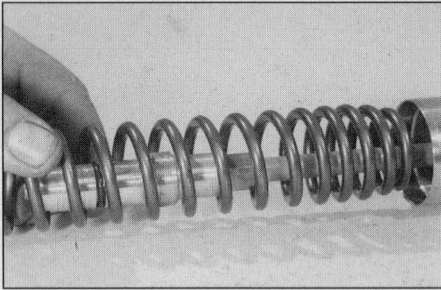

7.27a Install the fork spring making sure its tapered end (and tighter-pitched coils) faces downwards . . .

7.27b . . . then install the spring seat, spacer and washer (later models only)

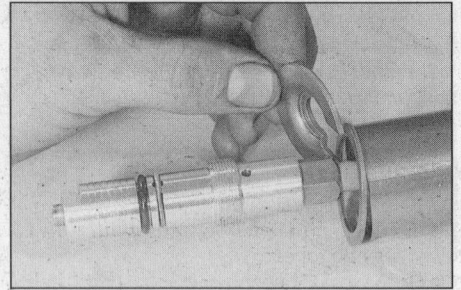

7.28a Slide the slotted collar into position

7.28b Screw the fork top bolt on the preload adjuster

7.29 Carefully screw the top bolt into the fork tube

models only) **(see illustrations)**. Fit a new O-ring to the preload adjuster and the fork top bolt and lubricate them with a little fork oil.

28 With the aid of an assistant push down on the spring seat, compressing the fork spring, and slide the slotted spring collar into position. Screw the top bolt fully on the preload adjuster, counter-holding the locknut, then slowly release the fork spring, making sure the slotted collar is correctly seated against the base of the preload adjuster **(see illustrations)**.

29 Carefully screw the top bolt into the fork tube making sure it is not cross-threaded **(see illustration)**. **Note:** *The top bolt can be tightened to the specified torque setting at this stage if the tube is held between the padded jaws of a vise, but do not risk distorting the tube by doing so. A better method is to tighten the top bolt when the fork leg has been installed and is securely held in the triple clamps. If the damper rod Allen bolt requires tightening (see Step 20), clamp the fork slider between the padded jaws of a vise and have an assistant compress the tube into the slider so that maximum spring pressure is placed on the damper rod head - tighten the damper Allen bolt to the specified torque.*

30 Install the forks as described in Section 6.

8   Steering stem - removal and installation

Caution: *Although not strictly necessary, before removing the steering stem it is recommended that the fuel tank be removed. This will prevent accidental damage to the paintwork.*

Removal

Refer to illustration 8.2

1   Remove the forks as described in Section 6 of this Chapter.

2   Unscrew the bolts securing the brake hose union assembly to the bottom triple clamp **(see illustration)**. Support the brake calipers so that no strain is placed on the hydraulic hose.

3   Disconnect the horn wires, then unscrew the horn mounting bolt and remove the horn.

4   Pry off the cap from the steering stem top nut. Remove the steering stem nut and and lift off the top triple clamp.

5   Straighten the tabs of the steering stem adjuster nut lock washer, then using a suitable C-wrench, slacken and remove the adjuster nut locknut.

6   Remove the lock washer and discard it; a new one must be fitted on reassembly.

7   Support the bottom triple clamp and slacken the adjuster nut.

8   Remove the nut, dust seal, inner race and upper bearing from the top of the steering head.

9   Gently lower the bottom triple clamp and steering stem out of the frame. Remove the lower bearing from the steering stem.

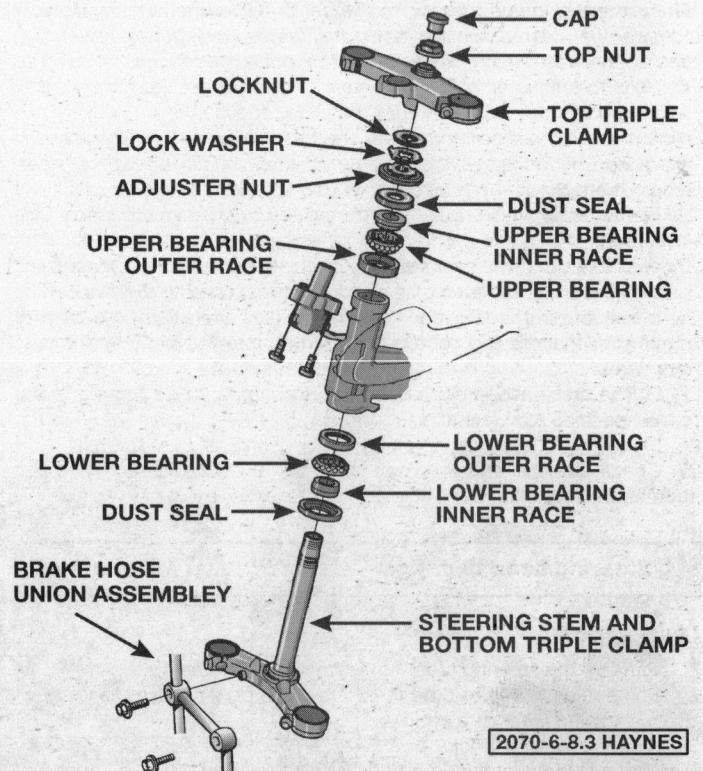

8.2 Exploded view of steering stem and bearings

CAP
TOP NUT
TOP TRIPLE CLAMP
LOCKNUT
LOCK WASHER
ADJUSTER NUT
DUST SEAL
UPPER BEARING INNER RACE
UPPER BEARING
UPPER BEARING OUTER RACE
LOWER BEARING OUTER RACE
LOWER BEARING
LOWER BEARING INNER RACE
DUST SEAL
BRAKE HOSE UNION ASSEMBLEY
STEERING STEM AND BOTTOM TRIPLE CLAMP

2070-6-8.3 HAYNES

10 Remove all traces of old grease from the bearings and races and check them for wear or damage as described in Section 9. **Note:** *Do not attempt to remove the outer races from the frame or the lower inner race and dust seal from the steering stem unless they are to be replaced.*

Installation

11 Smear a liberal quantity of grease on both inner and outer races and the steering stem. Work the grease well into both the upper and lower bearing races.
12 Fit the lower bearing on the steering stem.
13 Carefully lift the steering stem into position and fit the upper bearing and inner race.
14 Apply grease to the underside of the dust seal and fit it to the steering stem.
15 Apply clean engine oil to the threads of the adjuster nut and tighten it using hand pressure only.
16 To preload the bearings to the torque specified by the manufacturer (see Specifications at the beginning of the Chapter) it will be necessary to use the service tool, Pt. No. 07916-3710100 (US models), 3710101 (UK models), which is a socket that fits the adjuster nut. Using the service tool, tighten the adjuster nut to the specified torque setting then turn the steering stem from lock to lock approximately 5 times to settle the bearings and races in position. After preloading the bearings, slacken the adjuster nut one full turn then tighten it again to the specified torque setting. **Note:** *It is important to check the feel of the steering afterwards as described below; if it is too tight readjust the bearings as described below.*
17 If the service tool is not available, tighten the adjuster nut hard using a conventional C-wrench to preload the bearings then adjust as follows.
18 Slacken the adjuster nut slightly until pressure is just released, then turn it slowly clockwise until resistance is just evident. The object is to set the adjuster nut so that the bearings are under a very light loading, just enough to remove any free play. **Caution:** *Take great care not to apply excessive pressure because this will cause premature failure of the bearings.*
19 With the bearings correctly adjusted, fit a new lock washer to the adjuster nut. Bend down two opposite lock washer tabs (the shorter tabs on the washer) into the grooves of the adjuster nut.
20 Hold the adjuster nut to prevent it from moving, then install the locknut and tighten it finger-tight.
21 Again holding the adjuster nut, tighten the locknut approximately 90° more until its slots align with the remaining lock washer tabs. Secure the locknut in position by bending up the tabs into its slots.
22 Fit the top triple clamp to the steering stem and install the stem top nut.
23 Temporarily fit the fork legs, to align the triple clamps, then tighten the steering stem top nut to the specified torque setting. Fit the cap.
24 Install the horn onto its mounting lug on the underside of the bottom triple clamp and tighten the retaining bolt securely, then fit the horn wires.
25 Fit the brake hose union assembly, and tighten the retaining bolts to the specified torque setting.
26 Install the fork legs as described in Section 6 of this Chapter.
27 Check that the steering head bearings are correctly adjusted as soon as the forks and front wheel are installed (see Chapter 1).

9 Steering head bearings - inspection and replacement

Inspection

1 Remove the steering stem as described in Section 8.
2 Remove all traces of old grease from the bearings and races and check them for wear or damage.
3 The ball bearing tracks of the races should be polished and free from indentations. Inspect the ball bearings for signs of wear, damage or discoloration, and examine the bearing retainer cage for signs of cracks or splits. If there are signs of wear on any of the above

components both upper and lower bearing assemblies must be replaced as a set.

Replacement

Refer to illustration 9.6

4 The outer races are an interference fit in the steering head and can be tapped from position with a suitable drift. Tap firmly and evenly around each race to ensure that it is driven out squarely. It may prove advantageous to curve the end of the drift slightly to improve access.
5 Alternatively, the races can be removed using a slide-hammer type bearing extractor; these can often be hired from tool shops.
6 The new races can be pressed into the head using the drawbolt arrangement **(see illustration)**, or by using a large diameter tubular drift which bears only on the outer edge of the race. Ensure that the drawbolt washer or drift (as applicable) bears only on the outer edge of the race and does not contact the race bearing surface.
7 To remove the inner race from the steering stem, use two screwdrivers placed on opposite sides of the race to work it free.
8 With the inner race removed, lift off the dust seal. Inspect the seal for wear or damage and replace it if necessary.
9 Install the dust seal and slide on the new inner race. A length of tubing with an internal diameter slightly larger than the steering stem will be needed to tap the new race into position. Ensure that the drift bears only on the inner edge of the race and does not contact the bearing surface.
10 Install the steering stem as described in Section 8.

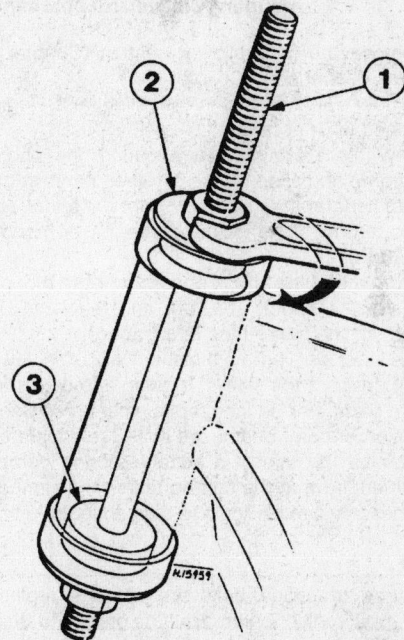

9.6 Drawbolt arrangement for fitting steering stem bearing outer races

1 *Long bolt or threaded bar*   3 *Guide for lower outer race*
2 *Thick washer*

10 Rear shock absorber - removal and installation

Removal

Refer to illustrations 10.3, 10.4, 10.5 and 10.6

1 In order to remove the shock absorber, it is necessary to support the bike securely in an upright position using an auxiliary stand or hoist. The bike must be supported so that the rear wheel is just resting on the ground (ie so that the bike's weight is off the rear suspension), but also so that the rear suspension will not drop, with the possible risk of personal injury, when the shock absorber mounting bolts are removed.

10.3 Unscrew the shock absorber reservoir clamp and release the reservoir

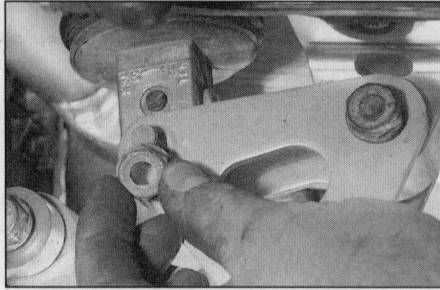

10.4 Unscrew the shock absorber lower mounting nut and remove the bolt

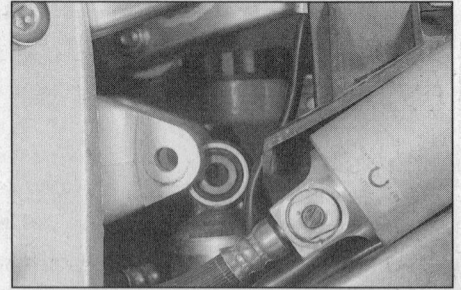

10.5 Unscrew the shock absorber upper mounting nut and remove the bolt

10.6 Examine the shock absorber lower mounting sleeve and bearings for wear

11.3 Remove the collar and rubber (arrow) from the exhaust mounting lug

If necessary, remove the fairing lower section (Chapter 8), and the exhaust system as described in Chapter 4.

2   Remove the rider's seat and open the pillion seat. Remove the side covers as described in Chapter 8.

3   Fully unscrew the retaining clip securing the shock absorber reservoir to the frame and slide off the clip (see illustration). Caution: *Do not attempt to separate the reservoir from the shock absorber.*

4   Unscrew and remove the shock absorber lower mounting nut and bolt (see illustration).

5   Unscrew and remove the shock absorber upper mounting nut and bolt and maneuver the shock absorber and reservoir assembly out from the right side of the frame (see illustration).

6   Withdraw the inner sleeve from the shock absorber lower mounting (see illustration). Inspect the sleeve, bearing and dust seals for wear; replace any worn parts. For bearing replacement, see Section 11.

7   All shock absorber components are available separately. However, disassembly requires the use of a suitable spring compressor. It is therefore recommended that the unit be taken to a Honda dealer who will have the necessary service tools to dismantle the unit.

Installation

8   Check that the mounting bolts are unworn, replacing them if necessary, and apply molybdenum disulfide grease to their shanks.

9   Install the shock absorber so that the rebound damping adjuster is on the left side of the bike.

10   Make sure the reservoir hose is correctly routed and insert both the shock absorber mounting bolts from the right side.

11   Fit the mounting bolt nuts and tighten them to the specified torque.

12   Install the reservoir into its recess and tighten the retaining clip.

13   Install the side covers and rider's seat as described in Chapter 8.

14   Check the operation of the rear suspension and adjust the suspension settings (see Section 12) before taking the bike on the road.

11 Rear suspension linkage - removal, inspection and installation

Removal

Refer to illustrations 11.3, 11.5 and 11.6

1   In order to remove the rear suspension linkage, it is necessary to support the bike securely in an upright position using an auxiliary stand or hoist. The bike must be supported so that the rear wheel is just resting on the ground (ie so that the bike's weight is off the rear suspension) but also so that the rear suspension will not drop, with the possible risk of personal injury, when the rear suspension linkage pivot bolts are removed. If necessary, remove the fairing lower section as described in Chapter 8, and the exhaust system as described in Chapter 4.

2   If not already done, remove the exhaust system as described in Chapter 4. Alternatively, disconnect its rear mountings and support the exhaust system in such a way that no strain is placed on its front mountings.

3   Remove the exhaust mounting collar and rubber from the mounting lug (see illustration).

4   Unscrew and remove the shock absorber lower mounting nut and bolt (see illustration 10.4).

5   Unscrew and remove the nut and pivot bolt securing the shock link arm to the frame (see illustration). Withdraw the bolt through the exhaust mounting lug.

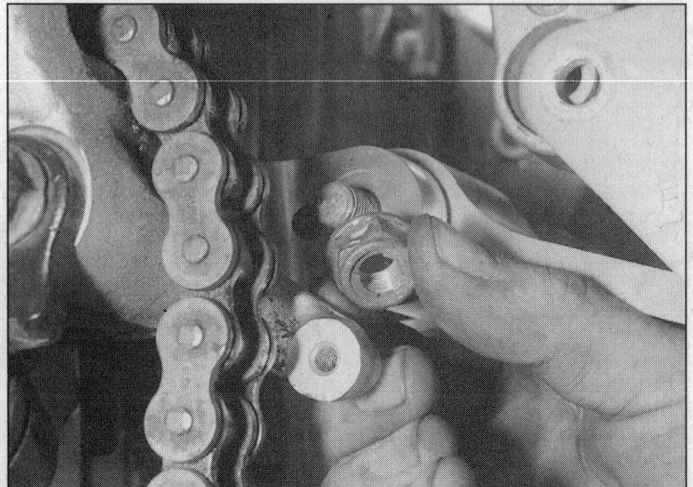

11.5 Unscrew the nut and remove the bolt securing the shock link arm to the frame

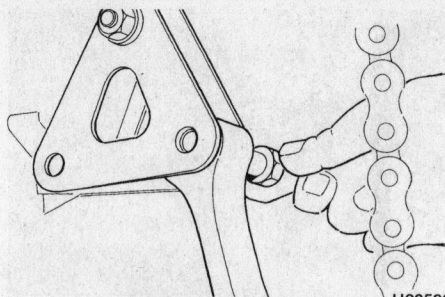

11.6 Unscrew the nut and remove the bolt securing the shock link arm to the shock link plates

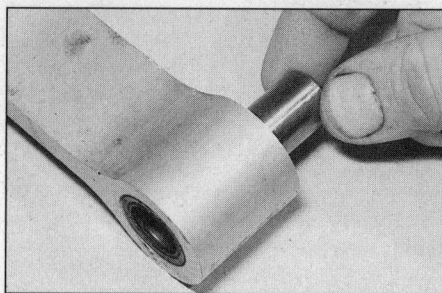

11.8a Remove the inner sleeves from the shock link arm . . .

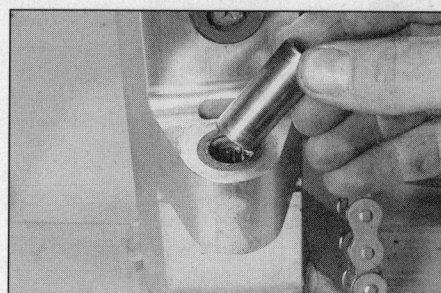

11.8b . . . and the swingarm

6   Unscrew and remove the nut and pivot bolt securing the shock link arm to the shock link plates, and remove the arm **(see illustration)**.

7   Unscrew and remove the nut and pivot bolt securing the shock link plates to the swingarm. Note the "F" and the arrow on each plate indicating the forward facing edge.

Inspection

Refer to illustrations 11.8a and 11.8b

8   Withdraw the inner sleeves from both ends of the shock link arm and from the swingarm **(see illustrations)**.

9   Thoroughly clean all components, removing all traces of dirt, corrosion and grease.

10   Inspect all components closely, looking for obvious signs of wear such as heavy scoring, or for damage such as cracks or distortion.

11   Carefully lever out the dust seals, using a flat-bladed screwdriver, and check them for signs of wear or damage; replace them if necessary.

12   Worn bearings can be drifted out of their bores, but note that removal will destroy them; new bearings should be obtained before work commences. The new bearings should be pressed or drawn into their bores rather than driven into position. In the absence of a press, a suitable drawbolt arrangement can be made up as described below.

13   It will be necessary to obtain a long bolt or a length of threaded rod from a local engineering works or some other supplier. The bolt or rod should be about one inch longer than the combined length of either link, and one bearing. Also required are suitable nuts and two large and robust washers having a larger outside diameter than the bearing housing. In the case of the threaded rod, fit one nut to one end of the rod and stake it in place for convenience.

14   Fit one of the washers over the bolt or rod so that it rests against the head, then pass the assembly through the relevant bore. Over the projecting end place the bearing, which should be greased to ease installation, followed by the remaining washer and nut.

15   Holding the bearing to ensure that it is kept square, slowly tighten the nut so that the bearing is drawn into its bore.

16   Once it is fully home, remove the drawbolt arrangement and, if necessary, repeat the procedure to fit the other bearings. The dust seals can then be pressed into place.

17   Lubricate all the seals, needle roller bearings, inner sleeves and the pivot bolts with molybdenum disulfide grease. Insert the sleeves into the bearings.

Installation

Refer to illustrations 11.20a, 11.20b and 11.20c

18   If not already done, lubricate the seals, needle roller bearings, inner sleeves and the pivot bolts with molybdenum disulfide grease. When installing the linkage components, insert all the pivot bolts from the right side.

19   Install the shock link plates on the swingarm mount, making sure that the "F" mark and the arrow face forward.

20   Connect the shock link arm between the link plates and the frame, passing the frame mounting pivot bolt through the exhaust mounting lug, then connect the shock absorber to the link plates **(see illustrations)**. Tighten all the nuts to the torque setting specified at the beginning of the Chapter **(see illustration)**.

21   Install the exhaust mounting rubber and collar, then install the exhaust system as described in Chapter 4.

22   Install the fairing lower section as described in Chapter 8.

23   Check the operation of the rear suspension before taking the machine on the road.

12 Suspension - adjustments

Front forks

Caution: *Always ensure that both front fork settings are the same. Uneven settings will upset the handling of the machine and could cause it to become unstable.*

Spring preload

Refer to illustration 12.1

1   The front fork spring preload adjuster is located in the center of each fork top bolt and is adjusted using a suitable spanner **(see illustration)**.

2   The preload is indicated by the number of grooves which are

6

11.20a Offer up the shock link arm to the frame and the link plates to the shock absorber

11.20b Note that the shock link arm front mounting bolt must pass through the exhaust mounting lug for access (shown with link arm removed for clarity)

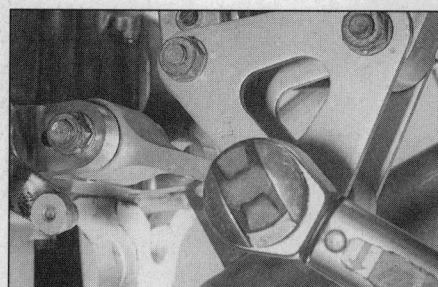

11.20c Tighten all suspension linkage nuts to the specified torque setting

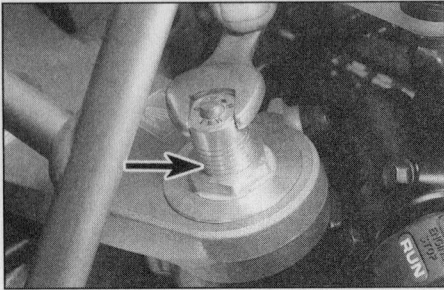

12.1 Adjust the front fork preload using a wrench - note preload setting grooves (arrow)

12.6 Adjust the front fork rebound damping using a flat-bladed screwdriver

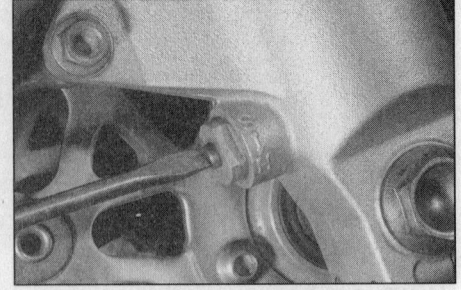

12.13 Adjust the front fork compression damping using a flat-bladed screwdriver (later models only)

visible on the adjuster above the top surface of the top bolt hexagon. To set the preload to the standard amount, turn the adjuster until the third groove from the top aligns with the top surface of the hexagon.

3   To reduce the preload (ie soften the ride), rotate the adjuster counterclockwise (anti-clockwise).

4   To increase the preload (ie stiffen the ride), rotate the adjuster clockwise.

5   Always ensure both adjusters are set to the same position.

Rebound damping

Refer to illustration 12.6

6   The rebound damping adjuster is in the centre of the preload adjuster. Adjust using a flat-bladed screwdriver **(see illustration)**.

7   Damping positions are identified by counting the clicks emitted by the adjuster when it is turned. There are ten damping positions.

8   The standard setting recommended by Honda is seven clicks counterclockwise (anti-clockwise) from the maximum damping setting. In this position, the punch mark on the damping adjuster aligns with the reference mark on the preload adjuster body.

9   To establish the present setting, turn one of the adjusters fully clockwise whilst counting the number of clicks emitted, then rotate it back to its original position. Repeat the procedure on the other adjuster to ensure both are set in the same position.

10   To reduce the rebound damping, turn the adjuster counterclockwise (anti-clockwise) as denoted by the S arrow on the adjuster.

11   To increase the rebound damping, turn the adjuster clockwise as denoted by the H arrow on the adjuster.

12   Always ensure both adjusters are set to the same position.

Compression damping - US 1995 model, UK CBR900RR-R and RR-S only

Refer to illustration 12.13

13   The compression damping adjuster is situated at the bottom of each fork slider and is adjusted using a flat-bladed screwdriver **(see illustration)**

14   Damping positions are indicated by counting the number of clicks emitted by the adjuster when it is turned.

15   The standard setting recommended by Honda is six clicks counterclockwise (anti-clockwise) from the maximum damping setting. In this position, the punch mark on the adjuster aligns with the reference mark on the damper unit.

16   To establish the present setting, rotate one of the adjusters fully clockwise whilst counting the number of clicks emitted, then rotate it back to its original position. Repeat the procedure on the other adjuster to ensure both are set in the same position.

17   To reduce the compression damping, turn the adjuster counterclockwise (anti-clockwise) in the direction of the S arrow.

18   To increase the compression damping, turn the adjuster clockwise in the direction of the H arrow.

19   Always ensure both adjusters are set to the same position.

Rear shock absorber

Spring preload - all models

Refer to illustration 12.20

20   The rear shock absorber spring preload adjuster is in the form of a stepped collar fitted to the top of the shock absorber, and is adjusted by rotating the collar using a suitable C-wrench **(see illustration)**. One is supplied in the bike's toolkit.

21   The preload adjuster has seven numbered positions stamped on it, number 1 being the minimum (ie softest) setting and number seven the maximum (hardest) setting. Honda recommend position two as the standard setting. Align the number for the setting required with the adjuster lug on the shock absorber body.

Rebound damping

Refer to illustration 12.22

22   The rear suspension rebound damping adjuster is situated at the base of the shock absorber on its left side and is adjusted using a flat-bladed screwdriver **(see illustration)**.

12.20 Adjust the rear shock absorber preload using a suitable C-wrench as supplied in the bike's toolkit

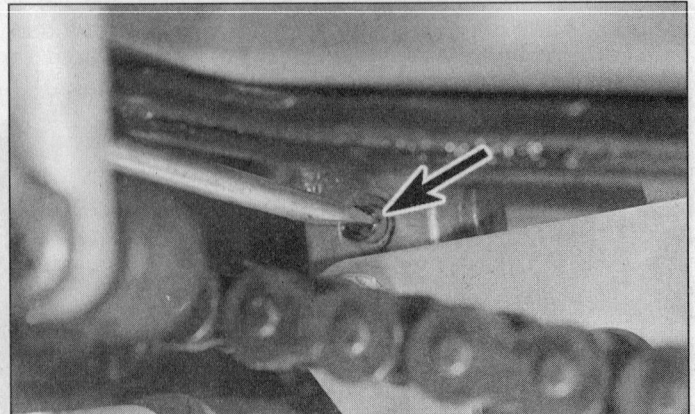

12.22 Adjust the rear shock absorber rebound damping using a flat-bladed screwdriver

12.27 Adjust the rear shock absorber compression damping using a flat-bladed screwdriver

23 The standard setting recommended by Honda is one turn counter-clockwise (anti-clockwise) from the maximum damping setting. In this position, the punch mark on the adjuster aligns with the reference mark on the shock absorber.
24 To establish the current setting, turn the adjuster fully clockwise whilst counting the number of turns until it seats, then rotate it back to its original position. Reposition the adjuster by the required amount.
25 To reduce the rebound damping, turn the adjuster counter-clockwise (anti-clockwise).
26 To increase the rebound damping, turn the adjuster clockwise.

Compression damping

Refer to illustration 12.27

27 The rear suspension compression damping adjuster is situated on the shock absorber reservoir, mounted on the left side of the bike, and is adjusted using a flat-bladed screwdriver **(see illustration)**.
28 The standard setting recommended by Honda is one and a half turns counterclockwise (anti-clockwise) from the maximum damping setting. In this position, the punch mark on the adjuster aligns with the reference mark on the reservoir.
29 To establish the current setting, turn the adjuster fully clockwise whilst counting the number of turns until it seats, then rotate it back to its original position. Reposition the adjuster by the required amount.
30 To reduce the compression damping, turn the adjuster counter-clockwise (anti-clockwise).
31 To increase the compression damping, turn the adjuster clockwise.

13 Swingarm bearings - check

1 Remove the rear wheel as described in Chapter 7, then remove the rear shock absorber as described in Section 10 of this Chapter.
2 Grasp the rear of the swingarm with one hand and place your other hand at the junction of the swingarm and the frame. Try to move the rear of the swingarm from side-to-side. Any wear (play) in the bearings should be felt as movement between the swingarm and the frame at

the front. The swingarm will actually be felt to move forward and backward at the front (not from side-to-side). If any play is noted, the bearings should be replaced (see Section 15).
3 Next, move the swingarm up and down through its full travel. It should move freely, without any binding or rough spots. If it does not move freely, refer to Section 14 for servicing procedures.

14 Swingarm - removal and installation

Removal

Refer to illustrations 14.6, 14.9, 14.10 and 14.11

1 Unscrew the bolt securing the brake hose clamp to the swingarm.
2 Remove the rear brake caliper and its pads as described in Chapter 7. Do not separate the hose from the caliper. Support the caliper so that no strain is placed on the hose.
3 Remove the rear wheel as described in Chapter 7. If the swingarm is to be moved away from the bike for bearing replacement, remove the drive chain as described in Section 16.
4 Remove the rear brake caliper mounting bracket from its slot in the swingarm noting how it fits.
5 Remove the swingarm-mounted rear fender/mudguard as described in Chapter 8.
6 Maneuver the brake caliper out through the front of the swingarm, taking care not to twist the hose **(see illustration)**. Once the caliper is free of the swingarm, support it so that no strain is placed on the hose.
7 Remove the rear shock absorber as described in Section 10.
8 Unscrew and remove the nut and pivot bolt securing the shock link plates to the swingarm.
9 Working on the left side of the bike, unscrew the swingarm pivot nut **(see illustration)**.
10 Working on the right side of the bike, unscrew the swingarm pivot locknut. This requires the use of a Honda service tool, Pt. No. 07908-4690002, which is a special wrench that fits the locknut **(see illustration)**. There is no alternative to the use of this tool; if you do not have access to it, the swingarm pivot locknut must be unscrewed and later tightened by a Honda dealer.
11 The swingarm pivot bolt fits into the head of the swingarm pivot adjusting bolt (actually a threaded sleeve). Using an Allen key, unscrew the swingarm pivot bolt which will unscrew the adjusting bolt **(see illustration)**.
12 When the adjusting bolt is fully unscrewed, have an assistant support the swingarm, then withdraw the pivot bolt and the adjusting bolt and remove the swingarm.
13 Remove the front and rear drive chain guards and the chain slider from the swingarm if necessary.
14 Remove the collar from the left side of the swingarm. Inspect all components for wear or damage as described in Section 15.

Installation

Refer to illustrations 14.17, 14.18a thru 14.18c, 14.20a, 14.20b and 14.21

15 If removed, install the front and rear drive chain guards and the chain slider. Apply a drop of non-permanent locking compound to the chain slider mounting screws.

6

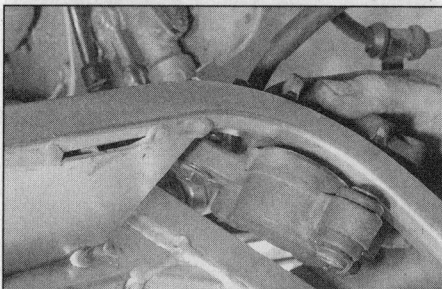

14.6 Maneuver the rear brake caliper out through the front of the swingarm

14.9 Unscrew the swingarm pivot nut

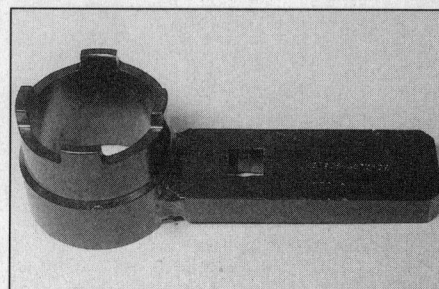

14.10 This Honda service tool is essential for removing and installing the swingarm

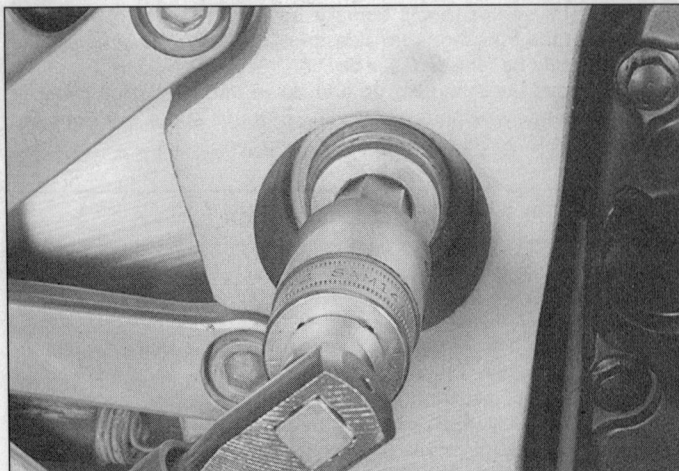

14.11 Use an Allen key in the head of the swingarm pivot bolt to unscrew the adjusting bolt

14.17 Install the collar into the left side of the swingarm

16 Lubricate the dust seals, bearings, collar, center spacer and the pivot bolt with grease.
17 Fit the collar to the left side of the swingarm **(see illustration)**.
18 Offer up the swingarm, and have an assistant hold it in place. Install the swingarm pivot adjusting bolt and tighten it as much as possible by hand. Then install the swingarm pivot bolt through the swingarm, and engage its flats with those of the adjusting bolt and tighten the adjusting bolt further by turning the pivot bolt using a suitable Allen key. Tighten the adjusting bolt in this way to the torque setting specified at the beginning of the Chapter **(see illustrations)**.
19 Check the movement of the swingarm.
20 Install the swingarm pivot locknut and tighten it as much as possible by hand. Tighten the locknut further using the service tool as described in Step 10, using an Allen key applied through its middle to prevent the swingarm pivot bolt and adjusting bolt from turning. Then, using a torque wrench applied to the hole in the arm of the service tool, tighten the locknut to the specified torque setting **(see illustrations)**.
Note: The specified torque setting takes into account the extra

leverage given by the service tool and cannot be duplicated without it.
21 Install the swingarm pivot nut on the pivot bolt left end, and whilst preventing the pivot bolt from turning using the Allen key in its right end, tighten the nut to the specified torque setting **(see illustration)**. Recheck the movement of the swingarm as described in Section 13.
22 Install the shock link plates on the swingarm and tighten the nut to the specified torque setting.
23 Install the rear shock absorber as described in Section 10.
24 Maneuver the rear brake caliper through the front of the swingarm, install its pads and mount on the bracket as described in Chapter 7. Install the bracket into the slot in the swingarm.
25 Install the rear fender/mudguard as described in Chapter 8.
26 Install the rear wheel as described in Chapter 7. If removed, install the drive chain as described in Section 16.
27 Install the brake hose clamp on the swingarm and tighten the bolt securely.
28 Check the operation of the rear suspension before taking the machine on the road.

14.18a Install the adjusting bolt into the swingarm and tighten as much as possible by hand . . .

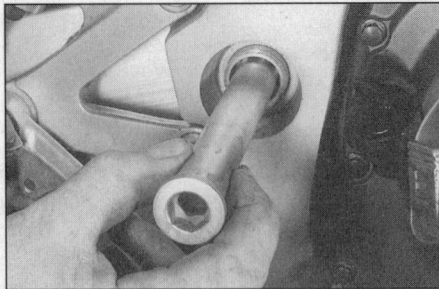

14.18b . . . then insert the swingarm pivot bolt and engage it with the adjusting bolt . . .

14.18c . . . and tighten the adjusting bolt to the specified torque setting using an Allen key in the head of the pivot bolt

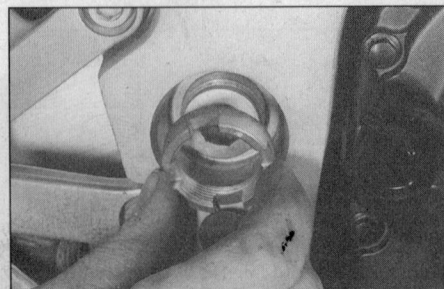

14.20a Install the locknut onto the adjusting bolt and tighten as much as possible by hand . . .

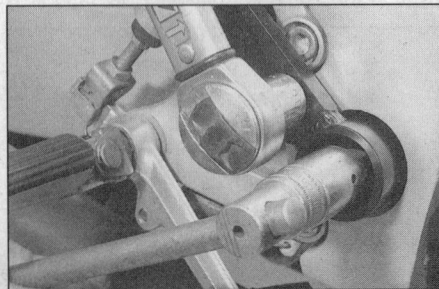

14.20b . . . then use the service tool as shown to tighten the locknut to the specified torque setting

14.21 Tighten the swingarm pivot nut to the specified torque setting

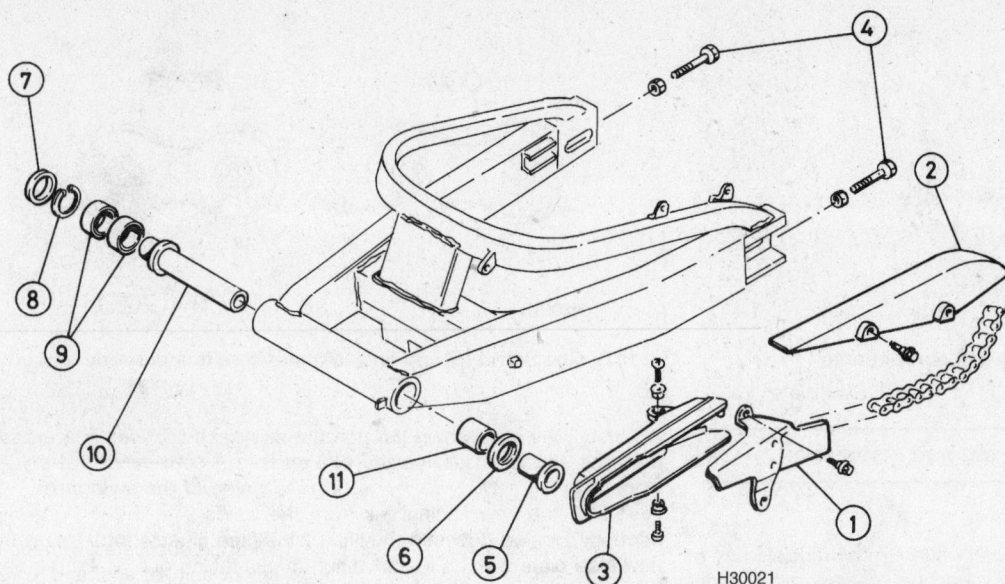

15.1 Exploded view of swingarm components

1 Chain front guard
2 Chain rear guard
3 Drive chain slider
4 Drive chain adjusting bolt
5 Collar
6 Dust seal
7 Dust seal
8 Snap-ring
9 Ball bearings
10 Center spacer
11 Needle roller bearing

H30021

15 Swingarm - inspection and bearing replacement

Inspection

Refer to illustration 15.1

1 Thoroughly clean all components, removing all traces of dirt, corrosion and grease **(see illustration)**.

2 Inspect all components closely, looking for obvious signs of wear such as heavy scoring, and cracks or distortion due to accident damage. Inspect the wear indicator line on the front of the drive chain slider. If it is worn to the line it must be replaced. Any damaged or worn component must be replaced.

Bearing replacement

Refer to illustrations 15.4 and 15.12

3 Lever out the dust seals, using a flat-bladed screwdriver, and inspect them for signs of wear or damage; replace them if necessary. Note that the left and right seals are a different size.

4 Using snap ring pliers, remove the snap ring from the right side of the swingarm **(see illustration)**.

5 The two right side bearings can be driven out of position simultaneously, using a hammer and suitable drift inserted from the left side of the swingarm. Move the drift around the face of the bearing whilst driving it out of position, so that the bearings leave the swingarm

squarely. If difficulty is experienced, a special tool is available from Honda, Pt. No. 07936-3710001. An alternative is to use an internally-expanding bearing puller which are available commercially or can be hired. Otherwise the bearings must be replaced by a dealer service agent or bike repair specialist. Once the bearings have been removed, tip out the center spacer, noting which way round it fits.

6 Wash the bearings thoroughly in a high flash-point solvent to remove all traces of the old grease.

7 Check the bearing tracks and balls for wear, pitting or damage to hardened surfaces. A small amount of side movement in the bearing is normal, but no radial movement should be detectable.

8 Check the bearings for play and roughness when they are spun by hand. All bearings will emit a small amount of noise when spun but they should not chatter or sound rough. If there is any doubt about the condition of the bearings they should be replaced.

9 Install the center spacer, then pack the bearings with grease and drift them separately into position using a suitable tubular drift which bears only on the outer race of the bearing.

10 Secure the bearings in position with the snap ring, ensuring that it is correctly seated in its groove.

11 The needle roller bearing fitted to the left side of swingarm can, if necessary, be replaced as described in Section 11 of this Chapter.

12 Press the dust seals into position using a suitable tubular spacer which bears only on the hard outer edge of the seal **(see illustration)**. Make sure the correct size seal is fitted to each side.

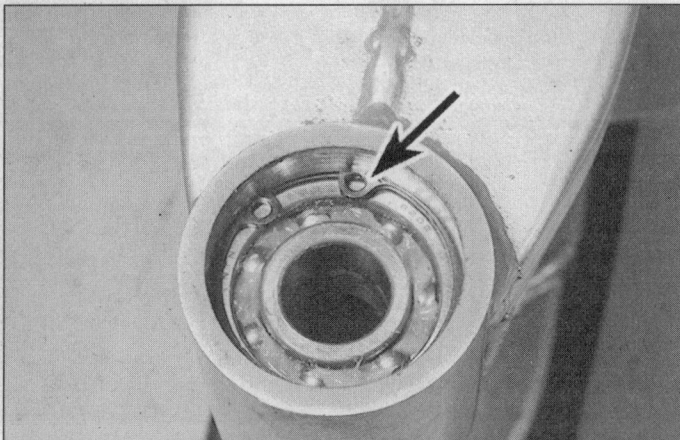

6

15.4 Swingarm right side bearings are retained by a snap-ring (arrow)

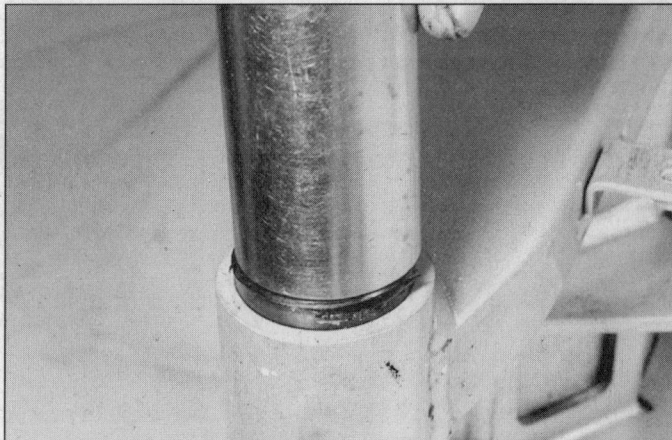

15.12 Press the dust seals into position using a suitable tubular spacer

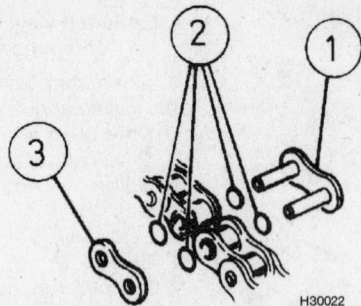

16.7 Drive chain joining link components

1 Joining link   2 O-rings   3 Side plate

16 Drive chain - removal, cleaning and installation

Removal

Note: *The original equipment drive chain fitted to all models has a staked-type master (joining) link which can be disassembled using either Honda service tool, Pt. No. 07HMH-MR10102, or one of several commercially-available drive chain cutting/staking tools. Such chains can be recognised by the master link side plate's identification marks (and usually its different color), as well as by the staked ends of the link's two pins which look as if they have been deeply center-punched, instead of peened over as with all the other pins.*

Warning: *NEVER install a drive chain which uses a clip-type master (split) link. Use ONLY the correct service tools to secure the staked-type of master link - if you do not have access to such tools, have the chain replaced by a dealer service department or bike repair shop to be sure of having it securely installed.*

1   Locate the joining link in a suitable position to work on by rotating the back wheel.

2   Slacken the drive chain as described in Chapter 1.

3   Split the chain at the joining link using the chain cutter, following carefully the manufacturer's operating instructions. Remove the chain from the bike, noting its routing through the swingarm.

Cleaning

4   Soak the chain in kerosene (paraffin) for five or six minutes. **Caution:** *Don't use gasoline (petrol), solvent or other cleaning fluids which might damage its internal sealing properties. Don't use high-pressure water. Remove the chain, wipe it off, then blow dry it with compressed air immediately. The entire process shouldn't take longer than ten minutes - if it does, the O-rings in the chain rollers could be damaged.*

Installation

Refer to illustrations 16.7 and 16.8

5   Remove the front sprocket cover as described in Section 17, Steps 1 through 3.

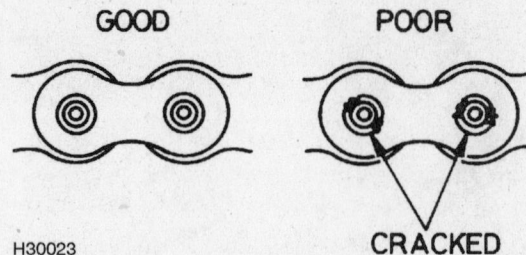

16.8 Check the joining link to make sure it is properly staked

6   Install the drive chain through the swingarm sections and around the front sprocket, leaving the two ends in a convenient position to work on.

7   Install the new joining link from the inside with the four O-rings correctly located between the link plates **(see illustration)**. Install the new side plate with its identification marks facing out. Stake the new link using the drive chain cutting/staking tool, following carefully the instructions of both the chain manufacturer and the tool manufacturer. DO NOT re-use old joining link components.

8   After staking, check the joining link and staking for any signs of cracking. If there is any evidence of cracking, the joining link, O-rings and side plate must be replaced **(see illustration)**.

9   Install the front sprocket cover as described in Section 17, Steps 11 and 12.

10 On completion, adjust and lubricate the chain following the procedures described in Chapter 1. **Caution:** *Use only the recommended lubricant.*

17 Sprockets - check and replacement

Check

Refer to illustrations 17.1, 17.2 and 17.3

1   Unscrew the two speedometer sensor mounting bolts and remove the sensor (later models only) **(see illustration)**.

2   Unscrew the two bolts securing the front sprocket cover to the engine unit, noting the position of the wiring clamp on later models **(see illustration)**.

3   Remove the sprocket cover and the drive chain guide plate **(see illustration)**.

4   Check the wear pattern on both sprockets (see Chapter 1). If the sprocket teeth are worn excessively, replace the chain and both sprockets as a set. Whenever the drive chain is inspected, the sprockets should be inspected also. If you are replacing the chain, replace the sprockets as well.

5   Fit the drive chain guide and sprocket cover. Fit the retaining

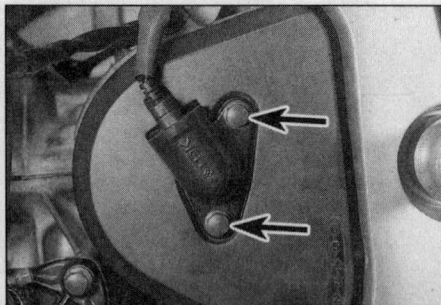

17.1 Unscrew the speedometer sensor mounting bolts (arrows) and remove the sensor (later models only)

17.2 Unscrew the sprocket cover retaining bolts (arrows) . . .

17.3 . . . and remove the cover and the drive chain guide plate

17.9 Unscrew the front sprocket retaining bolt (arrow) . . .

17.10 . . . and remove the front sprocket from countershaft and the drive chain

screws, not forgetting the wiring clamp on the upper screw (if fitted), and tighten them securely.

6 Install the speedometer sensor, making sure it fits properly over the bolt head, and tighten the bolts securely (later models only).

7 Adjust and lubricate the chain following the procedures described in Chapter 1. **Caution:** *Use only the recommended lubricant.*

Replacement

Front sprocket

Refer to illustrations 17.9 and 17.10

8 Remove the sprocket cover (refer to Steps 1 through 3 above).

9 Shift the transmission into gear while an assistant sits on the seat and applies the rear brake hard. This will lock the transmission and allow you to slacken and remove the engine sprocket bolt **(see illustration)**.

10 Pull the engine sprocket and chain off the shaft, then separate the sprocket from the chain **(see illustration)**.

11 Engage the new sprocket with the chain and slide it on the shaft. Install the washer and retaining bolt and tighten it to the specified torque setting whilst locking the transmission as described above

12 Install the engine sprocket cover (refer to Steps 5 through 7).

Rear sprocket

Refer to illustration 17.14

13 To remove the rear sprocket, remove the rear wheel as described

in Chapter 7. Unscrew the nuts holding the sprocket to the wheel coupling and lift it off. Check the condition of the rubber damper under the rear wheel coupling (see Section 18).

14 Fit the new sprocket. Apply a smear of clean oil to the threads of the retaining nuts prior to fitting and tighten them to the torque setting specified at the beginning of the Chapter **(see illustration)**.

18 Rear wheel coupling/rubber damper - check and replacement

Refer to illustration 18.4

1 Remove the rear wheel as described in Chapter 7.

2 Remove the spacer from the center of the sprocket coupling.

3 Lift the sprocket coupling away from the wheel leaving the rubber dampers in position in the wheel. Take care not to lose the spacer from the inside of the coupling bearing.

4 Lift the rubber damper segments from the wheel and check them for cracks, hardening and general deterioration **(see illustration)**. Replace the rubber dampers as a set if necessary.

5 Checking and replacement procedures for the sprocket coupling bearing are described in Section 16 of Chapter 7.

6 Installation is the reverse of the removal procedure, making sure that the sprocket coupling spacers are correctly positioned.

7 Install the rear wheel as described in Chapter 7.

17.14 Tighten the rear sprocket retaining nuts to the specified torque setting

18.4 Remove the rubber dampers from the rear wheel and inspect them for wear and damage

6

Chapter 7 Brakes, wheels and tires

Contents

Specifications

Brakes

Brake fluid type	See Chapter 1
Disc thickness	
Front	
New	4.5 mm (0.18 in)
Service limit	3.5 mm (0.14 in)
Rear	
New	5.0 mm (0.20 in)
Service limit	4.0 mm (0.16 in)
Disc maximum runout (front and rear)	0.3 mm (0.012 in)
Caliper bore ID	
Front	
Cylinder A	
New	30.230 to 30.280 mm (1.1902 to 1.1921 in)
Service limit	30.29 mm (1.193 in)
Cylinder B	
New	27.000 to 27.050 mm (1.0630 to 1.0650 in)
Service limit	27.06 mm (1.065 in)
Rear	
New	38.180 to 38.230 mm (1.5031 to 1.5051 in)
Service limit	38.240 mm (1.506 in)
Caliper piston OD	
Front	
Piston A	
New	30.148 to 30.198 mm (1.1869 to 1.1889 in)
Service limit	30.14 mm (1.1866 in)
Piston B	
New	26.935 to 26.968 mm (1.0604 to 1.0617 in)
Service limit	26.927 mm (1.0601 in)
Rear	
New	38.115 to 38.148 mm (1.5006 to 1.5019 in)
Service limit	38.107 mm (1.5003 in)

7

Master cylinder bore ID
  Front
    New.. 14.000 to 14.043 mm (0.5512 to 0.5529 in)
    Service limit .... 14.055 mm (0.5533 in)
  Rear
    New.. 15.870 to 15.913 mm (0.6248 to 0.6265 in)
    Service limit .... 15.925 mm (0.627 in)
Master cylinder piston OD
  Front
    New.. 13.957 to 13.984 mm (0.5495 to 0.5506 in)
    Service limit .... 13.945 mm (0.5490 in)
  Rear
    New.. 15.827 to 15.854 mm (0.6231 to 0.6242 in)
    Service limit .... 15.815 mm (0.6226 in)

Wheels

Maximum wheel runout (front and rear)
  Axial (side-to-side) .... 2.0 mm (0.08 in)
  Radial (out-of-round)... 2.0 mm (0.08 in)
Maximum axle runout (front and rear)....................................... 0.2 mm (0.01 in)

Tires

Tire pressures .... See Chapter 1
Tire sizes*
  Front .... 130/70 ZR 16
  Rear .... 180/55 ZR 17
*Refer to the machine's handbook or the tire information label on the drive chain guard for approved tire brands

Torque settings

	Nm	ft-lbs
Front brake caliper		
Mounting bolts		
US 1993 and 1994 models, UK CBR900RR-N and RR-P	27	20
US 1995 model, UK CBR900RR-R and RR-S	31	22
Joining Torx bolts..	33	24
Pad retaining pin..	18	13
Pad pin plug..	2.5	1.8
Front brake disc retaining bolts	20	14
Front brake master cylinder bracket bolt....................................	6	4.3
Front brake lever pivot bolt...	1	0.7
Front brake lever pivot bolt locknut	6	4.3
Rear brake caliper		
Mounting bolt...	23	17
Pad retaining pin..	18	13
Pad pin plug..	2.5	1.8
Slider pin...	28	21
Rear brake disc retaining bolts	43	31
Brake hose banjo fitting bolts	35	25
Brake pipe flare nuts	17	12
Brake caliper bleed valves	6	4.3
Front axle bolt	60	43
Front axle clamp bolts...	22	16
Rear axle nut	95	69

1 General information

The models covered in this manual have cast aluminum wheels designed to accept tubeless tires. Both front and rear brakes are hydraulically operated disc brakes, the front using a twin disc set up and the rear a single disc. The front brake calipers are of the dual opposed piston type; the rear is a sliding caliper with a single piston.
Caution: *Disc brake components rarely require disassembly. Do not disassemble components unless absolutely necessary. If any hydraulic brake line is loosened, the entire system must be disassembled, drained, cleaned and then properly filled and bled upon reassembly. Do not use solvents on internal brake components. Solvents will cause the seals to swell and distort. Use only clean brake fluid or alcohol for cleaning. Use care when working with brake fluid as it can injure your eyes and it will damage painted surfaces and plastic parts.*

2 Front brake pads - replacement

Refer to illustrations 2.2, 2.3, 2.4, 2.5, 2.10 and 2.11
Warning: *When replacing the front brake pads always replace the pads in BOTH calipers - never just on one side. The dust created by the brake system may contain asbestos, which is harmful to your health. Never blow it out with compressed air and don't inhale any of it. An approved filtering mask should be worn when working on the brakes.*
1 Remove the brake pad inspection cover noting its correct fitted position.
2 Unscrew the pad retaining pin plug from the caliper body to reveal the pad retaining pin **(see illustration)**.
3 Unscrew and remove the pad retaining pin from the caliper **(see illustration)**.
4 Remove the pad spring from the caliper noting its correct fitted

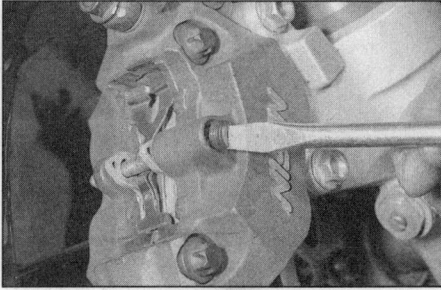

2.2 Unscrew the plug from the caliper body

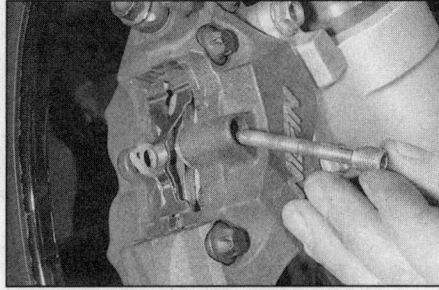

2.3 Unscrew the pad pin and withdraw it from the caliper

2.4 Note the position of the pad spring before removing it from the caliper

position, and withdraw the brake pads from the caliper body **(see illustration)**.

5   Inspect the surface of each pad for contamination and check that the friction material has not worn beyond its service limit groove **(see illustration)**. If either pad is worn to or beyond the service limit groove (ie the grooves are no longer visible), fouled with oil or grease, or heavily scored or damaged by dirt and debris, both pads must be replaced as a set. Note that it is not possible to degrease the friction material; if the pads are contaminated in any way they must be replaced.

6   If the pads are in good condition clean them carefully, using a fine wire brush which is completely free of oil and grease to remove all traces of road dirt and corrosion. Using a pointed instrument, clean out the grooves in the friction material and dig out any embedded particles of foreign matter. Any areas of glazing may be removed using emery cloth.

7   Check the condition of the brake discs (see Section 4).

8   Remove all traces of corrosion from the pad pin. Inspect the pin for signs of damage and replace if necessary.

9   Push the pistons as far back into the caliper as possible using hand pressure only. Due to the increased friction material thickness of new pads, it may be necessary to remove the master cylinder reservoir cover, plate and diaphragm and syphon out some fluid.

10 Smear the backs of the pads with copper-based grease, making sure that none gets on the front or sides of the pads. Insert the pads into the caliper so that the friction material of each pad is facing the disc **(see illustration)**. Insert the pad spring making sure that it is correctly fitted.

11 Insert the pad retaining pin making sure it passes through the holes in the pads and sits correctly on the pad spring, then tighten it to the torque setting specified at the beginning of the Chapter **(see illustration)**.

12 Fit the pad pin plug and tighten to its specified torque setting, then install the pad inspection cover.

13 Top up the master cylinder reservoir if necessary (see Chapter 1), and replace the reservoir cover, plate and diaphragm if removed.

14 Operate the brake lever several times to bring the pads into contact with the disc. Check the master cylinder fluid level (see Chapter 1) and the operation of the brake before riding the motorcycle.

3   Front brake calipers - removal, overhaul and installation

Warning: *If a caliper indicates the need for an overhaul (usually due to leaking fluid or sticky operation), all old brake fluid should be flushed from the system. Also, the dust created by the brake system may contain asbestos, which is harmful to your health. Never blow it out with compressed air and don't inhale any of it. An approved filtering mask should be worn when working on the brakes. Do not, under any circumstances, use petroleum-based solvents to clean brake parts. Use clean brake fluid, brake cleaner or denatured alcohol only.*

Removal

1   Unscrew the reflector mounting bolt and remove the reflector (US models only).

2   Remove the brake hose banjo fitting bolt, noting its correct alignment with the boss on the caliper, and separate the hose from the caliper. Plug the hose end or wrap a plastic bag tightly around it to minimise fluid loss and prevent dirt entering the system. Discard the sealing washers; new ones must be used on installation. **Note:** *If you are planning to overhaul the caliper and don't have a source of compressed air to blow out the pistons, just loosen the banjo bolt at this stage and retighten it lightly. The bike's hydraulic system can then be used to force the pistons out of the body with the pads removed. Disconnect the hose once the pistons have been sufficiently displaced.*

3   Unscrew the caliper mounting bolts and slide the caliper away from the disc **(see illustration 3.20a)**. Remove the brake pads (Section 2).

Overhaul

Refer to illustration 3.4

4   Clean the exterior of the caliper with denatured alcohol or brake system cleaner **(see illustration)**.

5   Remove the pistons from the caliper body, either by pumping them out by operating the front brake lever until the pistons are displaced, or by forcing them out using compressed air. Note that two sizes of piston are used (see sizes A and B in the Specifications at the beginning of this Chapter), and that different size seals are fitted accordingly. If the compressed air method is used, place a wad of rag between the

7

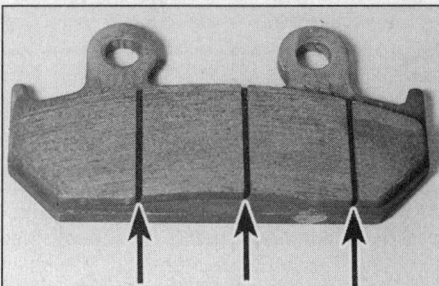

2.5 Brake pads must be replaced when the wear grooves (arrows) are no longer visible (typical pad design shown)

2.10 Insert the pads into the caliper so that their friction material faces the disc

2.11 Make sure the pad pin fits correctly through the brake pad holes and in the groove on the pad spring (arrow) before tightening to its specified torque setting

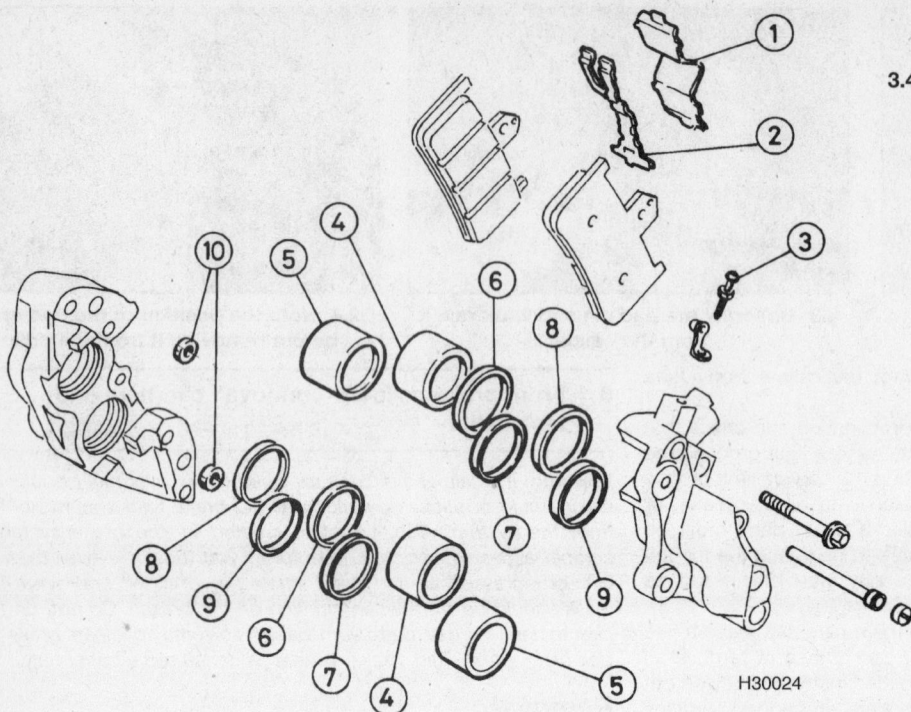

3.4 Exploded view of a front brake caliper

1 *Pad inspection cover*
2 *Pad spring*
3 *Bleed valve*
4 *Piston A*
5 *Piston B*
6 *Dust seal A*
7 *Dust seal B*
8 *Piston seal A*
9 *Piston seal B*
10 *Caliper seals*

H30024

pistons to act as a cushion, then use compressed air directed into the fluid inlet to force the pistons out of the body. Use only low pressure to ease the pistons out and make sure both pistons are displaced at the same time. If the air pressure is too high and the pistons are forced out, the caliper and/or pistons may be damaged. **Warning:** *Never place your fingers in front of the pistons in an attempt to catch or protect them when applying compressed air, as serious injury could result.*

6   Unscrew the four caliper assembly Torx bolts and split the caliper body into its two halves. Extract and discard the two caliper seals as they must be replaced with new ones.

7   Using a wooden or plastic tool, remove the dust seals from the caliper bores and discard them. New seals must be used on installation. If a metal tool is being used, don't damage the caliper bores.

8   Remove and discard the piston seals in the same way.

9   Clean the pistons and bores with denatured alcohol, clean brake fluid or brake system cleaner. **Caution:** *Do not, under any circumstances, use a petroleum-based solvent to clean brake parts.* If compressed air is available, use it to dry the parts thoroughly (make sure it's filtered and unlubricated).

10 Inspect the caliper bores and pistons for signs of corrosion, nicks and burrs and loss of plating. If surface defects are present, the caliper assembly must be replaced. If the caliper is in bad shape the master cylinder should also be checked.

11 If the necessary measuring equipment is available, compare the dimensions of the caliper bores and pistons to those given in the Specifications Section of this Chapter, replacing any component that is worn beyond the service limit.

12 Lubricate the new piston seals with clean brake fluid and install them in their grooves in the caliper bores. Note that two sizes of bore and piston are used and care must therefore be taken when installing the new seals to ensure that the correct size seals are fitted to the correct bores. The same care must be taken when fitting the new dust seals and the pistons.

13 Lubricate the new dust seals with clean brake fluid and install them in their grooves in the caliper bores.

14 Lubricate the pistons with clean brake fluid and install them closed-end first into the caliper bores. Using your thumbs, push the pistons all the way in, making sure they enter the bore squarely.

15 Lubricate the new caliper seals and install them into one half of the caliper body.

16 Join the two halves of the caliper body together, making sure that the caliper seals are correctly seated in their recesses.

17 Apply a drop of non-permanent locking compound to the threads of the four caliper assembly Torx bolts, then install them in the caliper body and tighten to the specified torque setting.

Installation

Refer to illustrations 3.20a, 3.20b and 3.21

18 Install the brake pads as described in Section 2.

19 Install the caliper on the brake disc making sure the pads sit squarely either side of the disc.

20 Retain the caliper with two new mounting bolts and tighten them to the torque setting specified at the beginning of this Chapter **(see illustrations)**. **Note:** *The mounting bolt threads contain a locking*

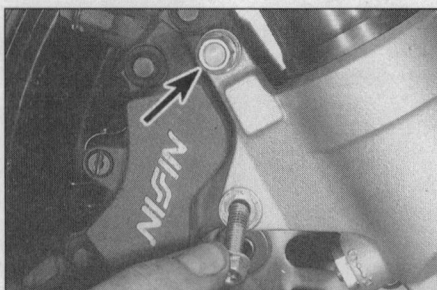

3.20a Install the new caliper mounting bolts (arrow) . . .

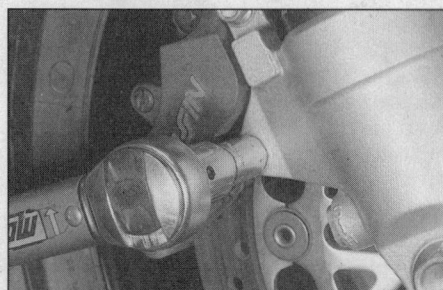

3.20b . . . tightening them to the specified torque setting

3.21 Align the brake hose so that it butts against the boss on the caliper (arrow)

4.2 Using a dial indicator to measure disc runout

4.3a The minimum thickness of the disc is stamped on the disc itself (arrow)

4.3b Using a micrometer to measure disc thickness

4.5 Unscrew the disc bolts a little at a time and in a criss-cross pattern to avoid distorting the disc

agent, necessitating that they be renewed if disturbed.

21 Connect the brake hose to the caliper, using new sealing washers on each side of the fitting. Align the brake hose so that it butts against the boss on the caliper **(see illustration)**. Tighten the banjo fitting bolt to the torque setting specified at the beginning of the Chapter.

22 Fit the reflector and tighten the reflector mounting bolt (US models).

23 Fill the master cylinder with the recommended brake fluid (see Chapter 1) and bleed the hydraulic system as described in Section 11.

24 Check for leaks and thoroughly test the operation of the brake before riding the motorcycle.

4 Front brake discs - inspection, removal and installation

Inspection

Refer to illustrations 4.2, 4.3a and 4.3b

1 Visually inspect the surface of the discs for score marks and other damage. Light scratches are normal after use and won't affect brake operation, but deep grooves and heavy score marks will reduce braking efficiency and accelerate pad wear. If the discs are badly grooved they must be machined or replaced.

2 To check disc runout, the front wheel must be raised off the ground using an auxiliary stand or hoist. Always make sure that the bike is properly supported. It may be necessary to use a jack and a block of wood under the engine, in which case the fairing lower section (see Chapter 8) and the exhaust system (see Chapter 4) must be removed. Mount a dial indicator to a fork leg, with the plunger on the indicator touching the surface of the disc about 10 mm (1/2 inch) from the outer edge **(see illustration)**. Rotate the wheel and watch the indicator needle, comparing your reading with the limit listed in this Chapter's Specifications. If the runout is greater than the service limit, check the hub bearings for play. If the bearings are worn, replace them and repeat this check. If the disc runout is still excessive, it will have to be replaced, although machining by an engineering shop may be possible.

3 The disc must not be machined or allowed to wear down to a thickness less than the service limit as listed in this Chapter's Specifications and stamped on the disc itself. The thickness of the disc can be checked with a micrometer **(see illustrations)**. If the thickness of the disc is less than the service limit, it must be replaced.

Removal

Refer to illustration 4.5

4 Remove the wheel as described in Section 14. **Caution:** *Do not lay the wheel down and allow it to rest on one of the discs - the disc could become warped. Set the wheel on wood blocks so the disc doesn't support the weight of the wheel.*

5 Mark the relationship of the disc to the wheel, so it can be installed in the same position. Unscrew the disc retaining bolts and remove the disc from the wheel **(see illustration)**. Loosen the bolts a little at a time, in a criss-cross pattern, to avoid distorting the disc.

6 Remove the disc. If both discs are to be removed mark them LEFT and RIGHT to ensure they are correctly positioned on installation.

Installation

Refer to illustration 4.7

7 Install the disc on the wheel, aligning the previously applied matchmarks (if you're reinstalling the original disc). Make sure the arrow stamped on the disc marking the normal direction of rotation is pointing in the direction of wheel rotation and is on the outer face of the disc **(see illustration)**.

4.7 Make sure the directional arrow on the disc points in the direction of normal wheel rotation (arrow)

7

8 Install the bolts and tighten them in a criss-cross pattern evenly and progressively to the torque setting specified at the beginning of the Chapter **(see illustration 4.5)**. Clean off all grease from the brake disc(s) using acetone or brake system cleaner. If new brake discs have been installed, remove any protective coating from their working surfaces.

9 Install the wheel as described in Section 14.

10 Operate the brake lever several times to bring the pads into contact with the disc. Check the operation of the brakes carefully before riding the bike.

5 Front brake master cylinder - removal, overhaul and installation

1 If the master cylinder is leaking fluid, or if the lever does not produce a firm feel when the brake is applied, and bleeding the brakes does not help (see Section 11), and the hydraulic hoses are all in good condition, then master cylinder overhaul is recommended. Before disassembling the master cylinder, read through the entire procedure and make sure that you have the correct rebuild kit. Also, you will need some new, clean brake fluid of the recommended type, some clean rags and internal snap-ring pliers. **Note:** *To prevent damage to the paint from spilled brake fluid, always cover the fuel tank when working on the master cylinder.*

2 **Caution:** *Disassembly, overhaul and reassembly of the brake master cylinder must be done in a spotlessly clean work area to avoid contamination and possible failure of the brake hydraulic system components.*

Removal

3 Loosen, but do not remove, the screws holding the reservoir cover in place.

4 Disconnect the electrical connectors from the brake light switch.

5 Unscrew the banjo fitting bolt and separate the brake hose from the master cylinder. Note the alignment of the hose with the boss on the master cylinder. Discard the two sealing washers as these must be replaced with new ones. Wrap the end of the hose in a clean rag and suspend the hose in an upright position or bend it down carefully and place the open end in a clean container. The objective is to prevent

excessive loss of brake fluid, fluid spills and system contamination.

6 Remove the locknut from the underside of the brake lever pivot bolt, then unscrew the bolt and remove the brake lever.

7 Remove the master cylinder mounting bolts to free the clamp, then lift the master cylinder and reservoir away from the handlebar. Note the "UP" mark on the clamp indicating its correct position for instalment, and the punch mark on the handlebar which must align with the mating surfaces of the clamp. **Caution:** *Do not tip the master cylinder upside down or brake fluid will run out.*

Overhaul

Refer to illustration 5.8

8 Unscrew the reservoir cover screws and remove the reservoir cover, the diaphragm plate and the rubber diaphragm **(see illustration)**. Drain the brake fluid from the reservoir into a suitable container. Wipe any remaining fluid out of the reservoir with a clean rag.

9 Remove the reservoir bracket bolt from the master cylinder, and release the reservoir hose clamp at the master cylinder end. Carefully pull the hose from the elbow and separate the fluid reservoir from the master cylinder. Check the hose for cracks and replace if necessary.

10 Using a pair of snap-ring pliers, remove the snap-ring from under the dust cover on the reservoir hose elbow and withdraw the elbow from the master cylinder. Remove and discard the O-ring as a new one must be fitted on installation.

11 Undo the brake light switch retaining screw and remove the switch.

12 Carefully remove the boot from the end of the piston.

13 Using snap-ring pliers, remove the snap-ring and slide out the piston assembly and the spring. Lay the parts out in the proper order to prevent confusion during reassembly.

14 Clean all of the parts with brake system cleaner (available at auto parts stores), denatured alcohol or clean brake fluid. **Caution:** *Do not, under any circumstances, use a petroleum-based solvent to clean brake parts.* If compressed air is available, use it to dry the parts thoroughly (make sure it's filtered and unlubricated).

15 Check the master cylinder bore for corrosion, scratches, nicks and score marks. If damage is evident, the master cylinder must be replaced with a new one. If the master cylinder is in poor condition, then the calipers should be checked as well.

16 If the necessary measuring equipment is available, compare the dimensions of the master cylinder bore and piston to those given in the

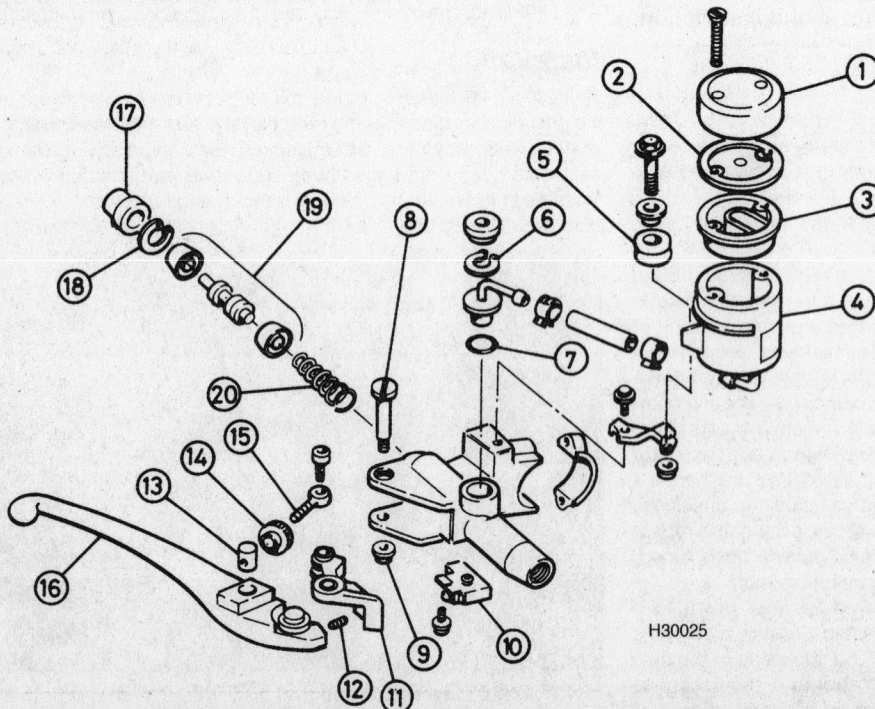

5.8 Exploded view of the front brake master cylinder

1 Reservoir cover
2 Diaphragm plate
3 Rubber diaphragm
4 Fluid reservoir
5 Damping rubber
6 Snap-ring
7 O-ring
8 Lever pivot bolt
9 Locknut
10 Brake light switch
11 Span adjuster arm
12 Spring
13 Joint pin
14 Span adjuster wheel
15 Span adjuster rod
16 Brake lever
17 Dust boot
18 Snap-ring
19 Piston assembly
20 Spring

H30025

5.24a Install the master cylinder clamp so that the "UP" mark is the right way up (arrow)

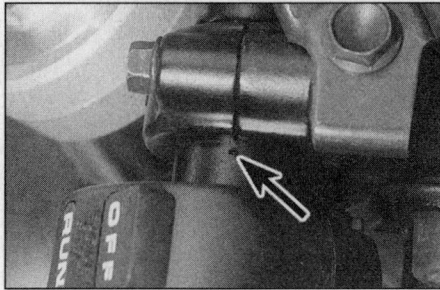

5.24b Align the mating surfaces of the clamp with the punchmark on the handlebar (arrow)

5.25 Align the brake hose so that it butts against the boss on the master cylinder (arrow)

6.1a Unscrew the plug from the rear brake caliper body . . .

6.1b . . . then slacken the pad pin with an Allen wrench . . .

6.1c . . . and withdraw the pin from the caliper

Specifications Section at the beginning of this Chapter, replacing any component that it is worn beyond the service limit.

17 The boot, piston assembly and spring are included in the rebuild kit. Use all of the new parts, regardless of the apparent condition of the old ones.

18 Before reassembling the master cylinder, soak the piston and the rubber cup seals in clean brake fluid for ten minutes. Lubricate the master cylinder bore with clean brake fluid, then carefully insert the piston and related parts in the reverse order of disassembly. Make sure the lips on the cup seals do not turn inside out when they are slipped into the bore and ensure the spring is fitted the correct way around.

19 Depress the piston, then install the snap-ring making sure it is properly seated in the groove. Install the rubber dust boot and make sure the lip is seated properly in the piston groove.

20 Install the brake light switch and securely tighten the screw.

21 Install a new O-ring on the reservoir hose elbow and insert the elbow into the master cylinder. Install the snap-ring, making sure it is properly seated in its groove, and replace the dust cover.

22 Inspect the damping rubber at the fluid reservoir mounting and replace if deteriorated.

23 Connect the reservoir hose and hose clamp and install the fluid reservoir on the master cylinder, tightening the reservoir bracket bolt securely.

Installation

Refer to illustrations 5.24a, 5.24b and 5.25

24 Attach the master cylinder to the handlebar and fit the clamp making sure the "UP" mark is facing upwards. Align the upper mating surfaces of the clamp with the punch mark on the handlebar, then tighten the upper clamp bolt first followed by the lower one **(see illustrations)**.

25 Connect the brake hose to the master cylinder, using new sealing washers on each side of the fitting. Align the hose so that it butts against the boss on the master cylinder **(see illustration)**. Tighten the banjo union bolt to the specified torque setting.

26 Install the brake lever and pivot bolt and tighten the bolt to the specified torque setting. Install the pivot bolt locknut and tighten it to the specified torque setting. The front brake lever has an adjuster mechanism which alters the span of the lever from the handlebar according to the rider's requirements. Adjust this as necessary by

turning the adjuster either clockwise or anti-clockwise until the desired position is achieved (see Chapter 1).

27 Connect the brake light switch wiring.

28 Fill the fluid reservoir with the specified brake fluid as described in Chapter 1. Refer to Section 11 of this Chapter and bleed the air from the system.

6 Rear brake pads - replacement

Refer to illustrations 6.1a thru 6.1c, 6.2, 6.3a, 6.3b, 6.4, 6.7a and 6.7b
Warning: *The dust created by the brake system may contain asbestos, which is harmful to your health. Never blow it out with compressed air and don't inhale any of it. An approved filtering mask should be worn when working on the brakes.*

1 Unscrew the plug from the caliper body to reveal the pad retaining pin, then slacken and remove the pad pin **(see illustrations)**.

2 Unscrew the caliper mounting bolt **(see illustration)**, then release the brake hose from its retaining clips and pivot the caliper upwards. Do not place any excess strain on the hydraulic hose.

6.2 Unscrew the caliper mounting bolt

7

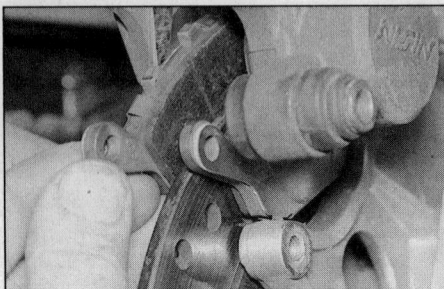

6.3a Pivot the caliper away from the disc, and recover the inner . . .

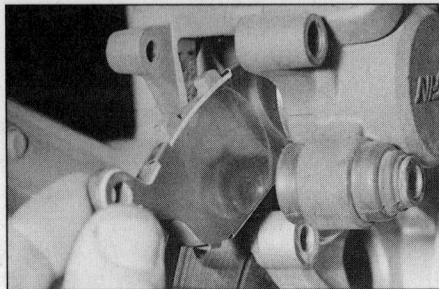

6.3b . . . and outer pads noting the correct fitted locations of the insulating pads and shims

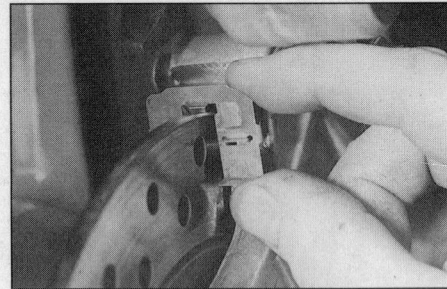

6.4 Remove the small anti-rattle spring from the caliper bracket

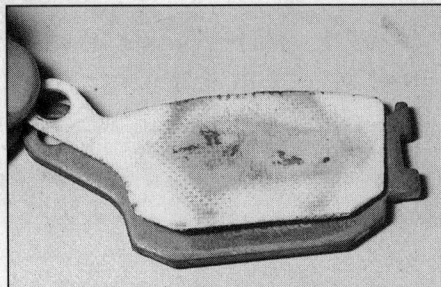

6.7a Prior to installing, fit the insulating pad . . .

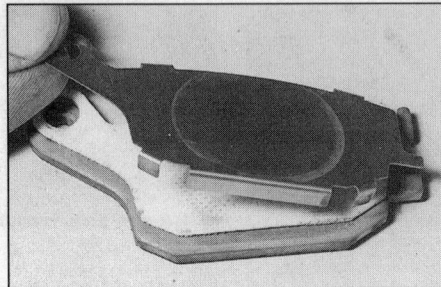

6.7b . . . and shim to the rear of each rear brake pad

3   Slide out the pads along with the insulating pad and shim which are fitted to the rear of each pad (see illustrations).

4   Remove the anti-rattle springs from the caliper bracket and body, noting their correct fitted positions (see illustration).

5   Inspect the brake pads and associated components as described in Steps 5 through 7 of Section 2.

6   Fit the small anti-rattle spring to the caliper bracket and clip the large spring into the caliper body. Make sure both springs are correctly fitted and clipped securely in position.

7   Install the insulating pad and shim to the rear of each pad backing plate (see illustrations).

8   Insert the pads into the caliper, ensuring the anti-rattle springs remain correctly positioned, so that the friction material of each pad is facing the disc. Insert the pad retaining pin making sure that it passes through the hole in both pads as well as the insulating pads and shims.

9   Pivot the caliper assembly down onto the disc and fit the caliper mounting bolt, tightening it to the torque setting specified at the beginning of the chapter.

10 Check that the brake hose is located in the retaining clamp on the swingarm.

11 Tighten the pad retaining pin to the specified torque setting, then install the pad pin plug and tighten it to the specified torque setting.

12 Refill the master cylinder reservoir (see Chapter 1) and fit the diaphragm and cap.

13 Operate the brake pedal several times to bring the pads into contact with the disc. Check the master cylinder fluid level (see Chapter 1) and the operation of the brake before riding the bike.

7   Rear brake caliper - removal, overhaul and installation

Warning: *If the caliper indicates the need for an overhaul (usually due to leaking fluid or sticky operation), all old brake fluid should be flushed from the system. Also, the dust created by the brake system may contain asbestos, which is harmful to your health. Never blow it out with compressed air and don't inhale any of it. An approved filtering mask should be worn when working on the brakes. Do not, under any circumstances, use petroleum-based solvents to clean brake parts. Use clean brake fluid, brake cleaner or denatured alcohol only.*

Removal

1   Remove the brake hose banjo fitting bolt and separate the hose from the caliper. Plug the hose end or wrap a plastic bag tightly around it to minimise fluid loss and prevent dirt entering the system. Discard the sealing washers; new ones must be used on installation. **Note:** *If you are planning to overhaul the caliper and don't have a source of compressed air to blow out the pistons, just loosen the banjo bolt at this stage and retighten it lightly. The bike's hydraulic system can then be used to force the piston out of the body once the pads are removed. Disconnect the hose once the piston has been sufficiently displaced.*

2   Remove the brake pads and anti-rattle springs (see Section 6).

3   Slide the caliper away from its mounting bracket and recover the slider pin rubber boot.

Overhaul

Refer to illustration 7.4

4   Clean the exterior of the caliper with denatured alcohol or brake system cleaner (see illustration).

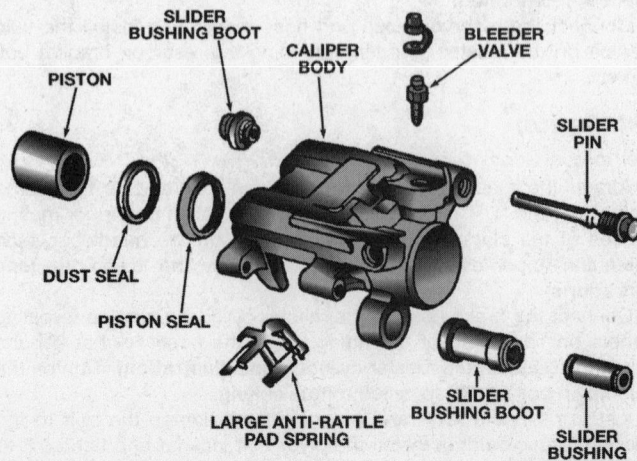

7.4 Exploded view of the rear brake caliper

5   If the piston wasn't forced out using the bike's hydraulic system, place a wad of rag between the piston and caliper frame to act as a cushion, then use compressed air directed into the fluid inlet to force the piston out of the body. Use only low pressure to ease the piston out. If the air pressure is too high and the piston is forced out, the caliper and/or piston may be damaged. **Warning:** *Never place your fingers in front of the piston in an attempt to catch or protect it when applying compressed air, as serious injury could result.*

6   Using a wooden or plastic tool, remove the dust seal from the caliper bore. If a metal tool is being used, take great care not to damage the caliper bore.

7   Remove the piston seal in the same way.

8   Clean the piston and bore with denatured alcohol, clean brake fluid or brake system cleaner. **Caution:** *Do not, under any circumstances, use a petroleum-based solvent to clean brake parts.* If compressed air is available, use it to dry the parts thoroughly (make sure it's filtered and unlubricated).

9   Inspect the caliper bore and piston for signs of corrosion, nicks and burrs and loss of plating. If surface defects are present, the caliper assembly must be replaced. If the caliper is in bad shape the master cylinder should also be checked.

10   If the necessary measuring equipment is available, compare the dimensions of the caliper bore and piston to those given in the Specifications Section of this Chapter, replacing any component that is worn beyond the service limit.

11   Temporarily install the caliper to its mounting bracket. Make sure that it slides smoothly in-and-out and check that the slider pin and the slider bush are a snug fit in their bores. If not check the slider pin and bush and their bores for burrs or excessive wear, replacing worn components as necessary. The slider pin and bush boots should be replaced as a matter of course.

12   Lubricate the new piston seal with clean brake fluid and install it in the groove in the caliper bore.

13   Lubricate the new dust seal with clean brake fluid and install it in the groove in the caliper bore.

14   Lubricate the piston with clean brake fluid and install it closed-end first in the caliper bore. Using your thumbs, push the piston all the way in, making sure it enters the bore squarely.

15   If the slider pin has been removed, apply a few drops of non-permanent locking compound to its threads. Fit the pin to the caliper body and tighten it to the specified torque setting.

16   Install the new slider pin and bush boots.

17   Apply a thin coat of PBC (poly butyl cuprysil) grease, or silicone grease designed for high-temperature brake applications, to the slider pin and bush. Slide the slider bush into position in the caliper body ensuring the boot is correctly seated in the groove on each end of the bush.

Installation

18   Slide the caliper into position on the mounting bracket. Make sure the slider pin boot is correctly located in its grooves on the body and bracket.

19   Install the anti-rattle springs and brake pads as described in Section 6.

20   Connect the brake hose to the caliper, using new sealing washers on each side of the fitting. Align the hose with the boss on the caliper. Tighten the banjo fitting bolt to the specified torque setting. Check that the brake hose is located in the retaining clamp on the swingarm.

21   Fill the master cylinder with the recommended brake fluid (see Chapter 1) and bleed the hydraulic system as described in Section 11.

22   Check for leaks and thoroughly test the operation of the brake before riding the motorcycle on the road.

8   Rear brake disc - inspection, removal and installation

Inspection

1   Refer to Section 4 of this Chapter, noting that the dial indicator should be attached to the swingarm.

8.3 Rear brake disc is retained by four bolts (arrows)

Removal

Refer to illustration 8.3

2   Remove the wheel as described in Section 15. **Caution:** *Don't lay the wheel down and allow it to rest on the disc or the sprocket - they could become warped. Set the wheel on wood blocks so the disc doesn't support the weight of the wheel.*

3   Mark the relationship of the disc to the wheel, so it can be installed in the same position. Remove the bolts that retain the disc to the wheel. Loosen the bolts a little at a time, in a criss-cross pattern, to avoid distorting the disc, then remove the disc **(see illustration)**.

Installation

4   Position the disc on the wheel, aligning the previously applied matchmarks (if you're reinstalling the original disc). Make sure the arrow (stamped on the disc) marking the direction of rotation is pointing in the proper direction and is on the outer face of the disc.

5   Install the bolts and tighten them in a criss-cross pattern evenly and progressively to the torque setting specified at the beginning of this Chapter. Clean off all grease from the brake disc using acetone or brake system cleaner. If a new brake disc has been installed, remove any protective coating from its working surfaces.

6   Install the wheel as described in Section 15.

7   Operate the brake pedal several times to bring the pads into contact with the disc. Check the operation of the brake carefully before riding the bike.

9   Rear brake master cylinder - removal, overhaul and installation

1   If the master cylinder is leaking fluid, or if the pedal does not produce a firm feel when the brake is applied, and bleeding the brakes does not help (see Section 11), and the hydraulic hoses are all in good condition, then master cylinder overhaul is recommended. Before disassembling the master cylinder, read through the entire procedure and make sure that you have the correct rebuild kit. Also, you will need some new, clean brake fluid of the recommended type, some clean shop towels and internal snap-ring pliers.

2   **Caution:** *Disassembly, overhaul and reassembly of the brake master cylinder must be done in a spotlessly clean work area to avoid contamination and possible failure of the brake hydraulic system components.*

Removal

3   Unscrew the banjo union bolt from the top of the master cylinder, noting the alignment of the hose with the boss on the cylinder. Discard

7

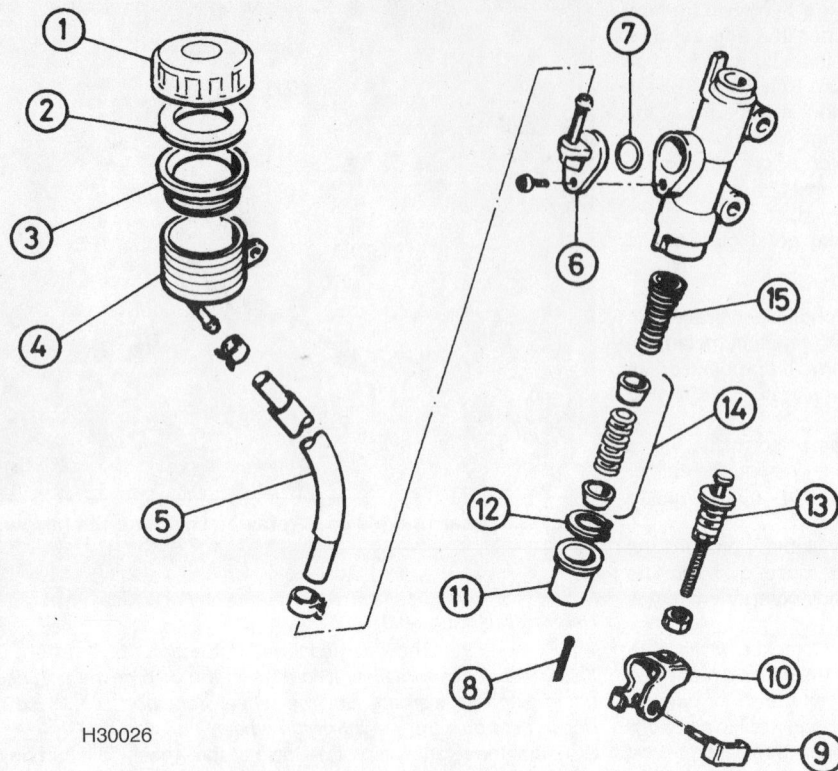

1 Reservoir cap
2 Diaphragm plate
3 Rubber diaphragm
4 Fluid reservoir
5 Connecting hose
6 Elbow
7 O-ring
8 Cotter (split) pin
9 Clevis pin
10 Clevis
11 Dust boot
12 Snap-ring
13 Pushrod
14 Piston assembly
15 Spring

H30026

the sealing washers as new ones must be fitted. Wrap the end of the hose in a clean rag and suspend the hose in an upright position or bend it down carefully and place the open end in a clean container. The objective is to prevent excessive loss of brake fluid, fluid spills and system contamination.

4   Remove the cotter pin (split pin) and withdraw the clevis pin which secures the master cylinder pushrod to the brake pedal.

5   Remove the bolt securing the master cylinder fluid reservoir to the frame. Unscrew the reservoir cap and pour the contents into a container.

6   Unscrew the master cylinder mounting bolts which pass through the step guard, and remove the step guard and master cylinder assembly.

7   Separate the fluid reservoir hose from the elbow on the master cylinder by releasing the hose clamp.

Overhaul

Refer to illustration 9.8

8   Unscrew the fluid reservoir hose elbow retaining screw from the master cylinder and remove the elbow. Discard the O-ring as a new one must be fitted on installation **(see illustration)**. Inspect the reservoir hose for cracks or splits and replace if necessary.

9   Before removing the clevis from the end of the pushrod, it is advised that note is made of its original position, either by measuring the distance from the center of the clevis pin to the base of the master cylinder body, or by a spot of paint or other identifying mark on the pushrod threads. Hold the clevis with a pair of pliers and loosen the locknut. Unscrew the clevis and locknut from the pushrod and carefully remove the rubber dust boot from the pushrod.

10   Depress the pushrod and, using snap-ring pliers, remove the snap-ring. Slide out the piston assembly and spring. Lay the parts out in the proper order to prevent confusion during reassembly.

11   Clean all of the parts with brake system cleaner (available at auto parts stores), denatured alcohol or clean brake fluid. **Caution:** *Do not, under any circumstances, use a petroleum-based solvent to clean brake parts.* If compressed air is available, use it to dry the parts thoroughly (make sure it's filtered and unlubricated).

12   Check the master cylinder bore for corrosion, scratches, nicks and

score marks. If damage is evident, the master cylinder must be replaced with a new one. If the master cylinder is in poor condition, then the caliper should be checked as well.

13   If the necessary measuring equipment is available, compare the dimensions of the master cylinder bore and piston to those given in the Specifications Section of this Chapter, replacing any component that is worn beyond the service limit.

14   A new piston and spring are included in the rebuild kit. Use them regardless of the condition of the old ones.

15   Before reassembling the master cylinder, soak the piston and the rubber cup seals in clean brake fluid for ten or fifteen minutes. Lubricate the master cylinder bore with clean brake fluid, then carefully insert the parts in the reverse order of disassembly, ensuring the tapered end of the spring is facing the piston. Make sure the lips on the cup seals do not turn inside out when they are slipped into the bore.

16   Depress the pushrod, then install the snap-ring, making sure it is properly seated in the groove. Note that the snap-ring must be fitted with its chamfered edge towards the master cylinder piston.

17   Install the rubber dust boot, making sure the lip is seated properly in the groove.

18   Install the locknut and clevis to the end of the pushrod and position them as noted on removal (see Step 9). Hold the clevis and tighten the locknut against it.

19   Fit a new O-ring to the fluid reservoir hose elbow and fit the elbow to the master cylinder. Install the retaining screw and tighten it securely. Reconnect the fluid reservoir hose and secure with its clip.

Installation

Refer to illustration 9.20, 9.22 and 9.23

20   Install the mounting bolts into the step guard and master cylinder and attach the assembly to the frame, tightening the bolts securely **(see illustration)**.

21   Secure the master cylinder reservoir to the frame with its mounting bolt.

22   Align the brake pedal with the master cylinder clevis and slide in the clevis pin. Check the pedal height at this stage. If it requires adjusting, withdraw the clevis pin, slacken off the locknut and rotate the clevis on the pushrod threads. Tighten the locknut when the pedal

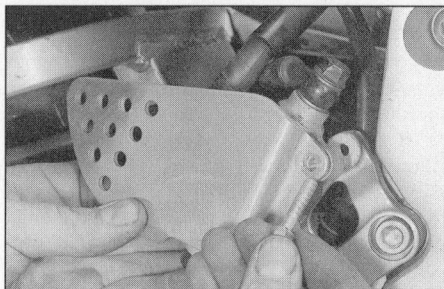

9.20 Install the mounting bolts through the step guard and the rear master cylinder and tighten them securely

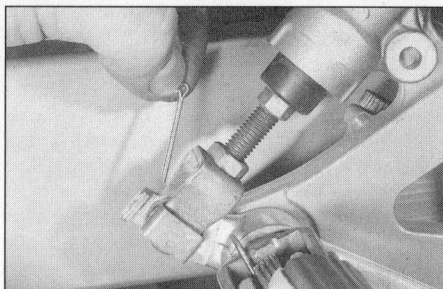

9.22 Fit a new cotter (split) pin to the clevis pin

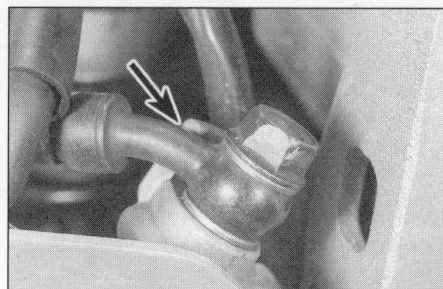

9.23 Align the hose so that it butts against the boss on the master cylinder

height is as required. Install the clevis pin and secure it in position with a new cotter pin (split pin) **(see illustration)**.

23 Connect the brake hose banjo fitting to the top of the master cylinder, using a new sealing washer on each side of the fitting. Align the hose so that it butts against the boss on the master cylinder **(see illustration)**, and tighten the banjo union bolt to the torque setting specified at the beginning of this Chapter.

24 Fill the fluid reservoir with the specified fluid (see Chapter 1) and bleed the system following the procedure in Section 11.

25 Check the position of the brake pedal and adjust the brake light switch (see Chapter 1). Check the operation of the brake carefully before riding the bike.

10 Brake pipe and hoses - inspection and replacement

Inspection

Refer to illustration 10.4

1 Once a week, or if the motorcycle is used less frequently, before every ride, check the condition of the brake hoses.

2 Twist and flex the rubber hoses while looking for cracks, bulges and seeping fluid. Check extra carefully around the areas where the hoses connect with the banjo fittings, as these are common areas for hose failure.

3 Inspect the metal banjo union fittings connected to the brake hoses. If the fittings are rusted, scratched or cracked, replace them.

4 Inspect the metal pipe attached to the front of the lower triple clamp which links the two front brake hose unions **(see illustration)**. If the plating on the metal pipe is chipped or scratched, the lines may rust. If the fittings are rusted, scratched or cracked, replace the pipe assembly.

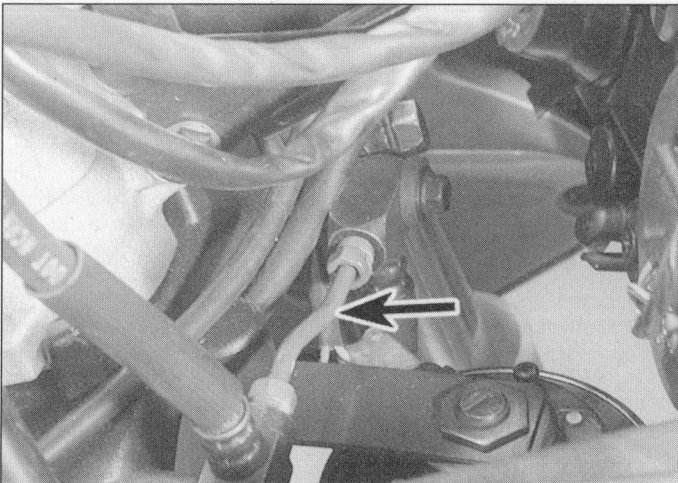

10.4 The metal linking pipe is mounted on the lower triple clamp (arrow)

Replacement

5 The brake hoses have banjo union fittings on each end of the hose. Cover the surrounding area with plenty of rags and unscrew the banjo bolt on each end of the hose. Detach the hose from any clips that may be present and remove the hose. Discard the sealing washers.

6 Position the new hose, making sure it isn't twisted or otherwise strained, between the two components. Make sure the metal tube portion of the banjo fitting is located adjacent to the boss on the component it connects to. Install the banjo bolts, using new sealing washers on both sides of the fittings, and tighten them to the torque setting specified at the beginning of this Chapter.

7 The metal pipe has a flare nut at each end, and can be removed once both nuts have been slackened. Apply a few drops of non-permanent locking compound to each of the flare nut threads then install the new pipe and tighten the flare nuts to the specified torque.

8 Flush the old brake fluid from the system, refill with the recommended fluid (see Chapter 1) and bleed the air from the system (see Section 11). Check the operation of the brakes carefully before riding the bike.

11 Brake system bleeding

1 Bleeding the brakes is simply the process of removing all the air bubbles from the brake fluid reservoirs, the hoses and the brake calipers. Bleeding is necessary whenever a brake system hydraulic connection is loosened, when a component or hose is replaced, or when the master cylinder or caliper is overhauled. Leaks in the system may also allow air to enter, but leaking brake fluid will reveal their presence and warn you of the need for repair.

2 To bleed the brakes, you will need some new, clean brake fluid of the recommended type (see Chapter 1), a length of clear vinyl or plastic tubing, a small container partially filled with clean brake fluid, some rags and a wrench to fit the brake caliper bleed valves.

3 Cover the fuel tank and other painted components to prevent damage in the event that brake fluid is spilled.

4 Remove the reservoir cap or cover and slowly pump the brake lever or pedal a few times, until no air bubbles can be seen floating up from the holes at the bottom of the reservoir. Doing this bleeds the air from the master cylinder end of the line. Reinstall the reservoir cap or cover.

5 Attach one end of the clear vinyl or plastic tubing to the bleed valve and submerge the other end in the brake fluid in the container.

6 Remove the reservoir cap or cover and check the fluid level. Do not allow the fluid level to drop below the lower mark during the bleeding process. To top up the rear master cylinder reservoir, unscrew the retaining bolt and move the reservoir away from the frame for better access, taking care not to spill any fluid.

7 Carefully pump the brake lever or pedal three or four times and hold it in (front) or down (rear) while opening the caliper bleed valve. When the valve is opened, brake fluid will flow out of the caliper into the clear tubing and the lever will move toward the handlebar or the pedal will move down.

7

8   Retighten the bleed valve (note the torque setting in the Specifications of this Chapter), then release the brake lever or pedal gradually. Repeat the process until no air bubbles are visible in the brake fluid leaving the caliper and the lever or pedal action is firm.

9   Replace the rubber diaphragm, diaphragm plate and reservoir cap or cover, wipe up any spilled brake fluid and check the entire system for leaks. **Note:** *If bleeding is difficult, it may be necessary to let the brake fluid in the system stabilize for a few hours (it may be aerated). Repeat the bleeding procedure when the tiny bubbles in the system have settled out.*

12   Wheels - inspection and repair

Refer to illustration 12.2

1   In order to carry out a proper inspection of the wheels, it is necessary to support the bike with an auxiliary stand or hoist so that either wheel can be raised from the ground. Clean the wheels thoroughly to remove mud and dirt that may interfere with the inspection procedure or mask defects. Make a general check of the wheels and tires as described in Chapter 1.

2   With the bike on the auxiliary stand or hoist and the wheel to be inspected in the air, attach a dial indicator to the fork slider or the swingarm and position its stem against the side of the rim **(see illustration)**. Spin the wheel slowly and check the side-to-side (axial) runout of the rim, then compare your readings with the value listed in this Chapter's Specifications. In order to accurately check radial runout with the dial indicator, the wheel would have to be removed from the machine. With the axle clamped in a vice, the wheel can be rotated to check the runout.

3   An easier, though slightly less accurate, method is to attach a stiff wire pointer to the fork slider or the swingarm and position the end a fraction of an inch from the wheel (where the wheel and tire join). If the wheel is true, the distance from the pointer to the rim will be constant as the wheel is rotated. **Note:** *If wheel runout is excessive, check the wheel bearings very carefully before replacing the wheel.*

4   The wheels should also be visually inspected for cracks, flat spots on the rim and other damage. Since tubeless tires are fitted, look very closely for dents in the area where the tire bead contacts the rim. Dents in this area may prevent complete sealing of the tire against the rim, which leads to deflation of the tire over a period of time.

5   If damage is evident, or if runout in either direction is excessive, the wheel will have to be replaced with a new one. Never attempt to repair a damaged cast aluminum wheel.

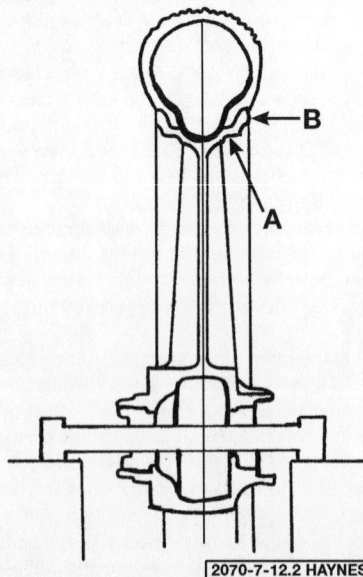

2070-7-12.2 HAYNES

12.2 Use a dial indicator to measure wheel runout

A Radial runout   *B Axial runout*

13   Wheels - alignment check

1   Misalignment of the wheels, which may be due to a cocked rear wheel or a bent frame or triple clamps, can cause strange and possibly serious handling problems. If the frame or triple clamps are at fault, repair by a frame specialist or replacement with new parts are the only alternatives.

2   To check the alignment you will need an assistant, a length of string or a perfectly straight piece of wood and a ruler graduated in 1/64 inch increments. A plumb bob or other suitable weight will also be required.

3   In order to make a proper check of the wheels it is necessary to support the bike in an upright position with the aid of an auxiliary stand. Measure the width of both tires at their widest points. Subtract the smaller measurement from the larger measurement, then divide the difference by two. The result is the amount of offset that should exist between the front and rear tires on both sides.

4   If a string is used, have your assistant hold one end of it about half way between the floor and the rear axle, touching the rear sidewall of the tire.

5   Run the other end of the string forward and pull it tight so that it is roughly parallel to the floor. Slowly bring the string into contact with the front sidewall of the rear tire, then turn the front wheel until it is parallel with the string. Measure the distance from the front tire sidewall to the string.

6   Repeat the procedure on the other side of the motorcycle. The distance from the front tire sidewall to the string should be equal on both sides.

7   As was previously pointed out, a perfectly straight length of wood may be substituted for the string. The procedure is the same.

8   If the distance between the string and tire is greater on one side, or if the rear wheel appears to be cocked, refer to Chapter 6, *"Swingarm bearings - check"*, and make sure the swingarm is tight.

9   If the front-to-back alignment is correct, the wheels still may be out of alignment vertically.

10 Using the plumb bob, or other suitable weight, and a length of string, check the rear wheel to make sure it is vertical. To do this, hold the string against the tire upper sidewall and allow the weight to settle just off the floor. When the string touches both the upper and lower tire sidewalls and is perfectly straight, the wheel is vertical. If it is not, place thin spacers under one leg of the auxiliary stand.

11 Once the rear wheel is vertical, check the front wheel in the same manner. If both wheels are not perfectly vertical, the frame and/or major suspension components are bent.

14   Front wheel - removal and installation

Removal

Refer to illustrations 14.4a, 14.4b, 14.5 and 14.6

1   In order to raise the front wheel off the ground it is necessary to use an auxiliary stand or hoist. Depending on the type of stand used it may also be necessary to raise the wheel by tying down the rear of the machine. If this is not possible, remove the fairing lower section (see Chapter 8) and the exhaust system (see Chapter 4) and place a floor jack, with a wood block on the jack head, under the engine and raise the jack to lift the wheel off the ground. Always make sure the bike is properly supported.

2   Unscrew both left and right reflector mounting bolts and remove the reflectors (US models only).

3   Unscrew both left and right brake caliper mounting bolts and slide the calipers off the discs. Support each caliper with a piece of wire or a bungee cord so that no strain is placed on the hydraulic hoses. Do not disconnect the brake hoses from the calipers.

4   Unscrew the speedometer cable retaining screw and detach the cable from its drive unit in the front wheel **(see illustrations)** (early models only).

5   Unscrew the front axle bolt then loosen the four axle clamp bolts **(see illustration)**.

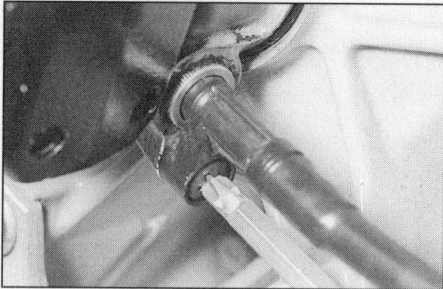

14.4a Unscrew the retaining screw . . .

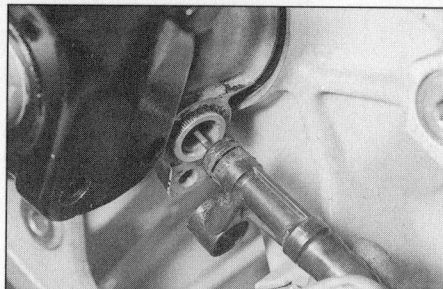

14.4b . . . and detach the speedometer cable from its drive unit

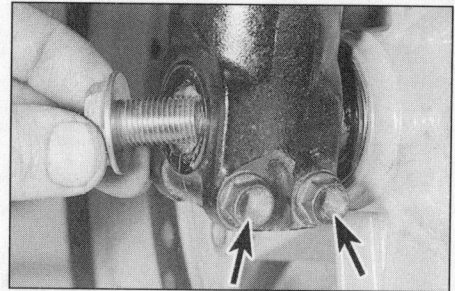

14.5 Remove the axle bolt and loosen the axle clamp bolts on both sides (arrows)

14.6 Withdraw the axle from the left side of the bike

14.10 Align speedometer drive gear slots with driveplate tabs (arrows) when fitting speedometer drive - early models

14.11 Do not omit the spacer from the right side of the wheel

6 Support the wheel, then withdraw the axle and carefully lower the wheel **(see illustration)**.

7 Remove the spacer from the right side of the wheel. Remove the speedometer drive (early models) or spacer (later models) from the left side. **Caution:** *Don't lay the wheel down and allow it to rest on one of the discs - the disc could become warped. Set the wheel on wood blocks so the disc doesn't support the weight of the wheel.* **Note:** *Do not operate the front brake lever with the wheel removed.*

8 Check the axle for straightness by rolling it on a flat surface such as a piece of plate glass (if the axle is corroded first remove the corrosion using fine emery cloth). If the axle is at all bent, replace it.

9 Check the condition of the wheel bearings (see Section 16).

Installation

Refer to illustrations 14.10, 14.11, 14.12, 14.13 and 14.19

10 On early models, fit the speedometer drive to the wheel's left side, aligning its drive gear slots with the driveplate tabs **(see illustration)**. On later models install the spacer in the left side of the wheel.

11 Install the spacer in the right side of the wheel **(see illustration)**.

12 Maneuver the wheel into position. Note the arrow cast into the wheel showing its correct direction of rotation **(see illustration)**. Apply a thin coat of grease to the axle.

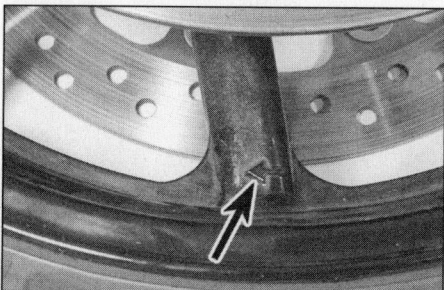

13 Lift the wheel into position making sure the spacer(s) remains in place. On early models, position the speedometer drive lug against the rear of the lug on the fork slider **(see illustration)**.

14 Slide the axle into position from the left side. Note that the groove in the axle's head should align with the outside edge of the left fork. Tighten the left axle clamp bolts to the specified torque setting.

15 Install the axle bolt and tighten it to the specified torque setting.

16 Install the brake calipers making sure the pads sit squarely on either side of the disc. Retain each caliper with two new mounting bolts and tighten them to the torque setting specified at the beginning of this Chapter. **Note:** *The mounting bolt threads contain a locking agent, necessitating that they be renewed if disturbed.*

17 On early models, connect the speedometer cable to the drive, aligning the inner cable slot with the drive dog, and securely tighten its retaining screw.

18 Rest the front wheel on the ground and remove the auxiliary stand or hoist. Pump the front forks a few times to settle all components in position and tighten the right axle clamp bolts to the specified torque setting.

19 Using feeler gauges, check the clearance between each brake disc and the caliper body **(see illustration)**. There should be at least 0.7 mm (0.028 in) clearance present on each side of the disc. If not, slacken the axle clamp bolts and push or pull the fork slider inwards or outwards (as

14.12 Note the arrow on the wheel which indicates the correct direction of normal rotation (arrow)

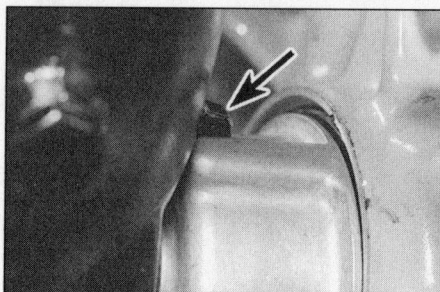

14.13 On early models, rotate the speedometer drive unit so that its lug abuts the rear of the fork slider lug (arrow)

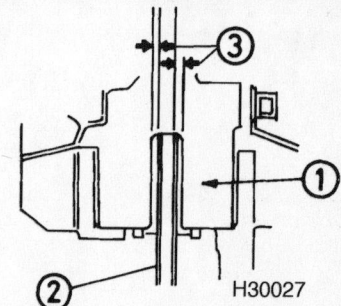

14.19 Use a feeler gauge to check the caliper body-to-disc clearance

1 Caliper body
2 Disc
3 0.7 mm (0.028 in) clearance

H30027

**15.8 Install the collared spacer into the
left side of the rear wheel**

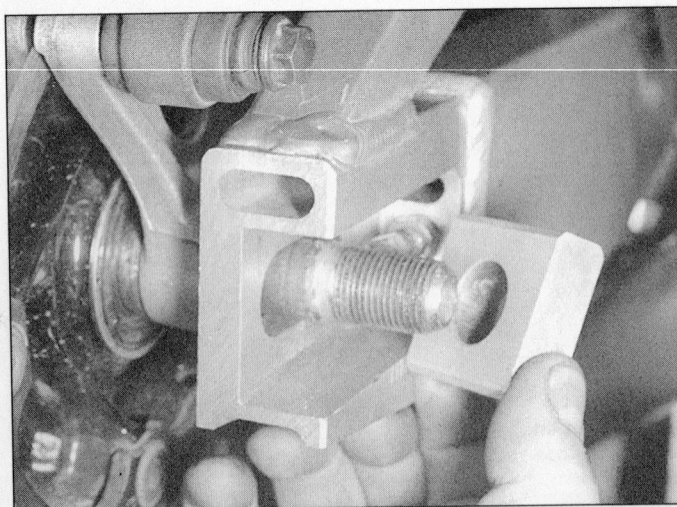

**15.9 Position the axle in the chain adjuster block
in the left side of the swingarm**

applicable) until the required clearance is present. With the slider correctly positioned, tighten the clamp bolts to the specified torque.
20 Apply the front brake, pump the forks up and down several times and check for proper brake operation.
21 Install both left and right side reflectors and tighten the mounting bolts securely (US models only).

15 Rear wheel - removal and installation

Removal

1 Use an auxiliary stand or hoist to hold the bike upright and the rear wheel off the ground. Always make sure the bike is properly supported.
2 Back off their locknuts and rotate the drive chain adjuster bolts so that their heads release pressure on the chain adjuster blocks.
3 Unscrew and remove the rear axle nut.
4 Support the wheel then slide out the axle from the left and lower the wheel to the ground.
5 Disengage the chain from the sprocket and remove the wheel from the swingarm. Note which spacer fits on which side of the wheel, and remove them from the wheel. **Caution:** *Do not lay the wheel down and allow it to rest on the disc or the sprocket - they could become warped. Set the wheel on wood blocks so the disc or the sprocket doesn't support the weight of the wheel. Do not operate the brake pedal with the wheel removed.*

6 Check the axle for straightness by rolling it on a flat surface such as a piece of plate glass (if the axle is corroded, first remove the corrosion with fine emery cloth). If the axle is bent at all, replace it.
7 Check the condition of the wheel bearings (see Section 16).

Installation

Refer to illustrations 15.8, 15.9, 15.11a and 15.11b

8 Apply a thin coat of grease to the seal lips, then slide the spacers into their proper positions on both sides of the hub, noting that the left spacer is shouldered **(see illustration)**.
9 Apply a thin coat of grease to the axle and slide the axle into position in the left chain adjuster block **(see illustration)**.
10 Engage the drive chain with the sprocket and lift the wheel into position. Make sure both spacers remain in the wheel and the disc fits correctly in the caliper, with the brake pads sitting squarely on each side of the disc.
11 Slide the axle into position and fit the right side chain adjuster block and axle nut, but do not tighten it yet **(see illustrations)**.
12 Adjust the chain slack as described in Chapter 1.
13 Tighten the axle nut to the specified torque setting.
14 Remove the auxiliary stand or hoist.
15 Operate the brake pedal several times to bring the pads into contact with the disc. Check the operation of the rear brake carefully before riding the bike.

15.11a Fit the right side chain adjuster block . . .

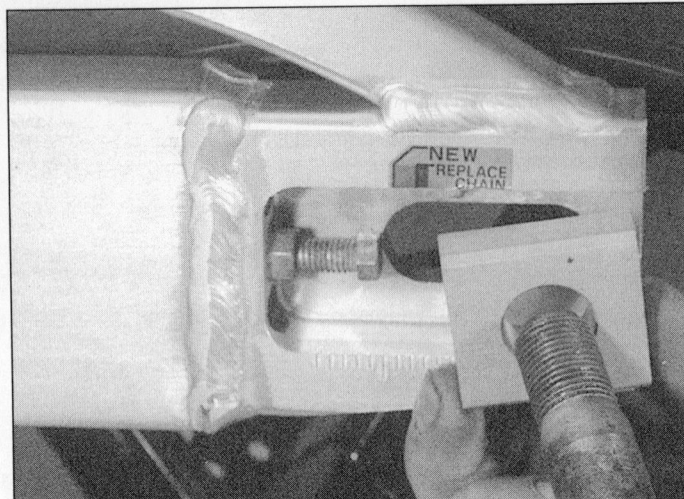

15.11b . . . followed by the axle nut

TIRE CHANGING SEQUENCE - TUBELESS TIRES

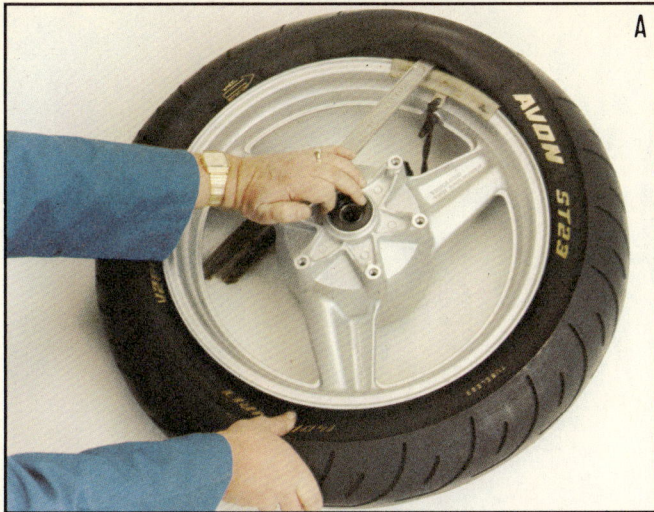

Deflate tire. After releasing beads, push tire bead into well of rim at point opposite valve. Insert lever next to valve and work bead over edge of rim.

Use two levers to work bead over edge of rim. Note use of rim protectors.

When first bead is clear, remove tire as shown.

Before installing, ensure that tire is suitable for wheel. Take note of any sidewall markings such as direction of rotation arrows.

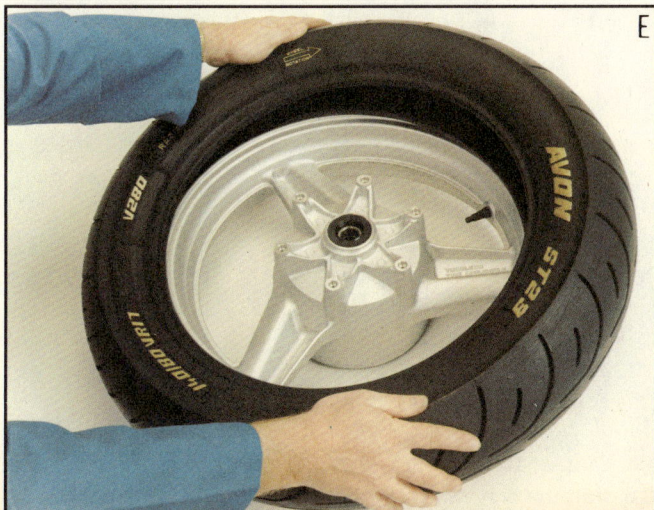

Work first bead over the rim flange.

Use a tire lever to work the second bead over rim flange.

BRAKE DISC

SPACER

RIGHT WHEEL
BEARING

LEFT WHEEL
BEARING

GREASE
SEAL

BRAKE DISC

SPEEDOMETER
DRIVE GEAR

THRUST
WASHERS

DISC RETAINING
BOLT

SHIM

GREASE
SEAL

SPACER

SPEEDOMETER
DRIVEPLATE

SHIM

DISC RETAINING
BOLT

SPEEDOMETER
DRIVE

SPACER
-LATER
MODELS

FRONT WHEEL

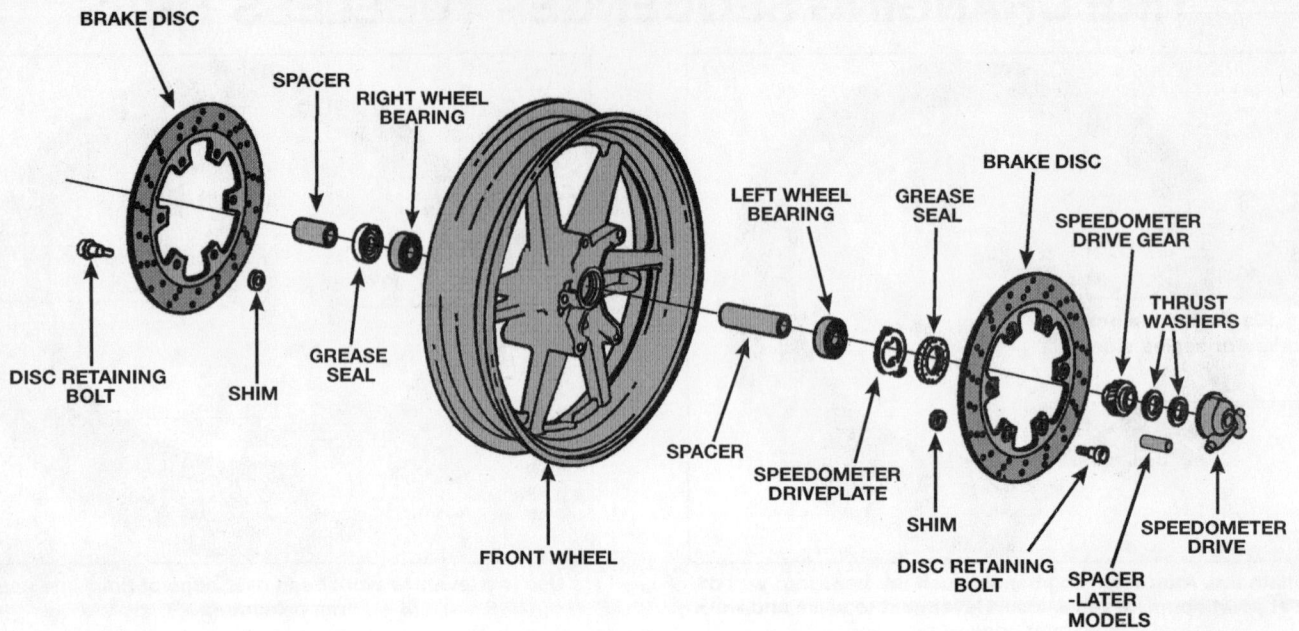

2070-7-16.3 HAYNES

16.3a Exploded view of the front wheel and associated components

16 Wheel bearings - removal, inspection and installation

Front wheel bearings

Refer to illustrations 16.3a, 16.3b, 16.5, 16.10a, 16.10b, 16.11, 16.12, 16.13a and 16.13b

Note: *Always replace the wheel bearings in pairs. Never replace the bearings individually. Avoid using a high pressure cleaner on the wheel bearing area.*

1   Remove the wheel as described in Section 14.

2   Set the wheel on blocks so as not to allow the weight of the wheel to rest on the brake discs.

3   Using a flat-bladed screwdriver, pry out the grease seal first from the right side of the wheel then from the left side **(see illustrations)**.

4   Withdraw the speedometer driveplate from the left side of the wheel (early models only).

5   Using a metal rod (preferably a brass drift punch) inserted through the center of the hub bearing, tap evenly around the inner race of the left side bearing to drive it from the hub **(see illustration)**. The bearing spacer will also come out.

6   Lay the wheel on its other side and remove the right side bearing using the same technique.

7   If the bearings are of the unsealed type or are only sealed on one side, clean them with a high flash-point solvent (one which won't leave any residue) and blow them dry with compressed air (don't let the bearings spin as you dry them). Apply a few drops of oil to the bearing. **Note:** *If the bearing is sealed on both sides don't attempt to clean it.*

8   Hold the outer race of the bearing and rotate the inner race - if the bearing doesn't turn smoothly, has rough spots or is noisy, replace it with a new one.

9   If the bearing is good and can be re-used, wash it in solvent once again and dry it, then pack the bearing with high-quality wheel bearing grease.

10 Thoroughly clean the hub area of the wheel. Install the right side bearing into the recess in the hub, with the marked or sealed side

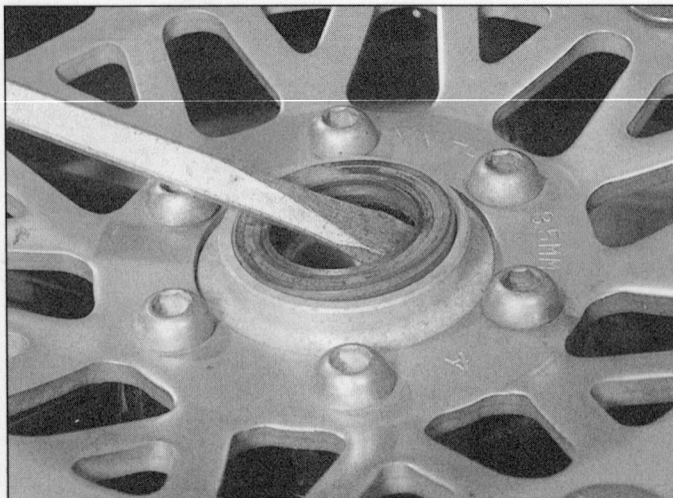

16.3b Prise out the grease seal using a flat-bladed screwdriver

16.5 Drift the first wheel bearing out using a metal rod passed through from the opposite side of the wheel

16.10a Install the bearing with the marked or sealed side facing out . . .

16.10b . . . then drive the bearing squarely into the hub using a drift or socket which bears only on the outer race

16.11 Install the spacer into the wheel before fitting the other bearing

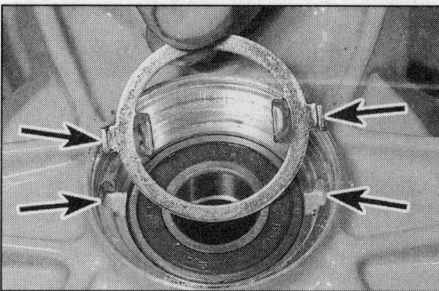

16.12 Fit speedometer driveplate to left side of wheel, making sure its tangs are correctly seated in hub slots (arrows) . . .

16.13a . . . then press in a new grease seal . . .

16.13b . . . and drift squarely into position using a suitably sized drift or socket

facing out. Using a bearing driver or a socket large enough to contact the outer race of the bearing, drive it in until it's completely seated **(see illustrations)**.

11 Turn the wheel over and install the bearing spacer **(see illustration)**. Unless the bearings are sealed on both sides, pack the remaining space no more than 2/3 full of high-melting point wheel bearing grease. Once the grease is packed in, drive the left side bearing into place as described above.

12 Fit the speedometer driveplate to the left side of the wheel on early models; ensure its locating tangs are correctly located in the hub slots **(see illustration)**.

13 Install new grease seals, using a seal driver, large socket or a flat piece of wood to drive them into place **(see illustrations)**.

14 Clean off all grease from the brake discs using acetone or brake system cleaner then install the wheel as described in Section 14.

Sprocket coupling bearing

Refer to illustrations 16.15, 16.16, 16.17a, 16.17b, 16.18, 16.20 and 16.22

15 Remove the rear wheel as described in Section 15 and remove the shouldered spacer from the left side of the wheel **(see illustration)**.

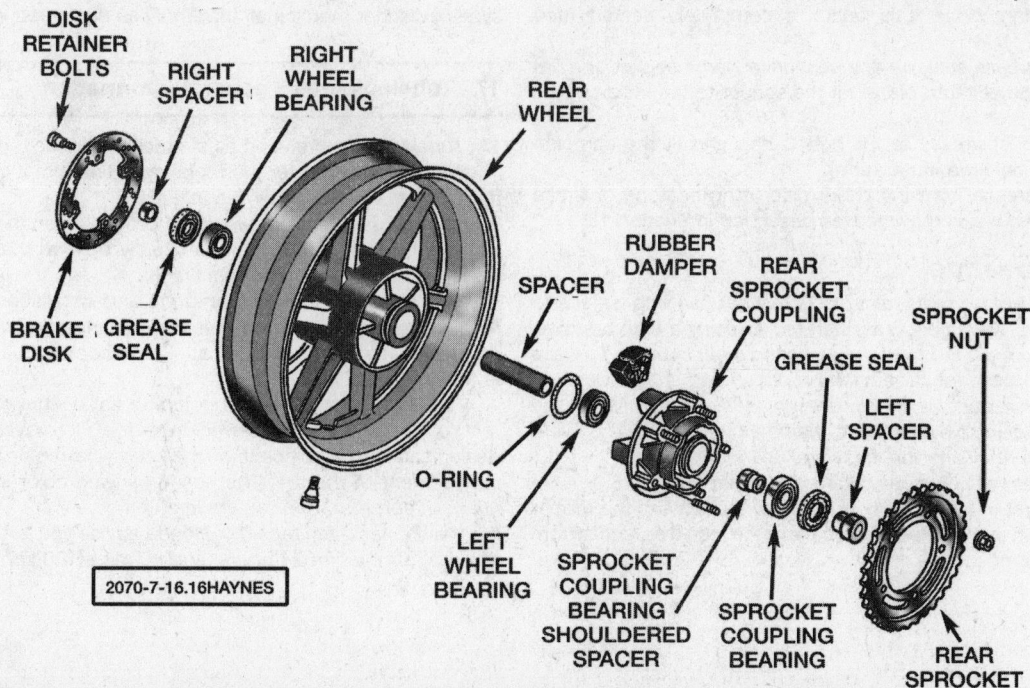

16.15 Exploded view of the rear wheel and associated components

16.16 Remove the sprocket coupling ensuring the rubber dampers remain in position in the wheel

16.17a Remove the spacer from inside the coupling bearing . . .

16.17b . . . then pry out the grease seal from the coupling

16.18 Support the coupling and drive the bearing out from the inside using a hammer and bearing driver or socket

16.20 Drive the new coupling bearing into position using a large tubular drift which bears only on the bearing's outer race

16.22 Ensure the hub O-ring is in position on the wheel and apply a smear of grease to it to aid installation

16 Lift the sprocket coupling away from the wheel leaving the rubber dampers in position in the wheel **(see illustration)**.
17 Remove the shouldered spacer from the inside of the coupling bearing and pry out the grease seal from the outside of the coupling **(see illustrations)**.
18 Support the coupling on blocks of wood and drive out the bearing with a bearing driver or socket **(see illustration)**.
19 Inspect the bearing as described above in Steps 7 through 9.
20 Thoroughly clean the bearing recess then install the bearing into the recess in the coupling, with the marked or sealed side facing out. Using a bearing driver or a socket large enough to contact the outer race of the bearing, drive it in until it's completely seated **(see illustration)**.
21 Install a new grease seal, using a seal driver, large socket or a flat piece of wood to drive it into place. Fit the spacer to the inside of the bearing.
22 Apply a smear of grease to the hub O-ring and fit the sprocket coupling to the wheel **(see illustration)**.
23 Clean off all grease from the brake disc using acetone or brake system cleaner then install the wheel as described in Section 15.

Rear wheel bearings

Note: *Early model grease seals were prone to wear, allowing water into the wheel bearings. This effect is accelerated by using a high pressure water cleaner directly on the rear wheel bearing area. Improved grease seals, spacers and bearings have been produced and should be used instead of the original parts when they are replaced. Always replace the wheel bearings in pairs. Never replace the bearings individually. Avoid using a high pressure cleaner on the wheel bearing area.*

24 Remove the rear wheel as described in Section 15.
25 Lift the sprocket coupling away from the wheel leaving the rubber dampers in position in the wheel. Take care not to lose the spacer from inside the sprocket coupling.

26 Pry out the grease seal from the right side of the wheel.
27 Set the wheel on blocks so as not to allow the weight of the wheel to rest on the brake disc or sprocket.
28 Remove, inspect and install the bearings as described above in Steps 5 through 11.
29 Install a new grease seal to the right side of the wheel, using a seal driver, large socket or a flat piece of wood to drive it into place.
30 Install a new hub O-ring and smear it with grease **(see illustration 16.22)**. Fit the sprocket coupling to the wheel, making sure both the spacers are in position.
31 Clean off all grease from the brake disc using acetone or brake system cleaner then install the wheel as described in Section 15.

17 Tubeless tires - general information

1 Tubeless tires are used as standard equipment on this motorcycle. They are generally safer than tube-type tires but if problems do occur they require special repair techniques.
2 The force required to break the seal between the rim and the bead of the tire is substantial, and is usually beyond the capabilities of an individual working with normal tire irons.
3 Also, repair of the punctured tire and installation on the wheel rim requires special tools, skills and experience that the average do-it-yourselfer lacks. Ensure that the wheel is balanced after tire replacement or repair.
4 For these reasons, if a puncture or flat occurs with a tubeless tire, the wheel should be removed from the motorcycle and taken to a dealer service department or a motorcycle repair shop for repair or replacement of the tire. The accompanying color illustrations can be used to replace a tubeless tire in an emergency.
5 If a tire has been repaired, Honda advise you not to exceed 50 mph (80 kph) for the first 24 hours, and 80 mph (130 kph) thereafter.

Chapter 8 Fairing and bodywork

Contents

1 General information

This Chapter covers the procedures necessary to remove and install the fairing and other body parts. Since many service and repair operations on these motorcycles require the removal of the fairing and/or other body parts, the procedures are grouped here and referred to from other Chapters.

In the case of damage to the fairing or other body parts, it is usually necessary to remove the broken component and replace it with a new (or used) one. The material that the fairing and other body parts are composed of doesn't lend itself to conventional repair techniques. There are however some shops that specialise in "plastic welding", so it may be worthwhile seeking the advice of one of these specialists before consigning an expensive component to the bin.

When attempting to remove any fairing section, first study it closely, noting any fasteners and associated fittings, to be sure of returning everything to its correct place on installation. In most cases the aid of an assistant will be required when removing sections, to help avoid the risk of damage to paintwork. Once the evident fasteners have been removed, try to withdraw the section as described but DO NOT FORCE it - if it will not release, check that all fasteners have been removed and try again. Where a section engages another by means of tabs, be careful not to break the tab or its mating slot or to damage the paint work. Remember that a few moments of patience at this stage will save you a lot of money in replacing broken fairing sections!

Several types of fastener are used on the models covered here. Usually bolts or screws of varying sizes and head form are used, requiring flat-bladed or cross-head screwdrivers, or Allen wrenches. Most screw into nuts, trim nuts or threads in mounting brackets (as applicable) and are released by unscrewing them counterclockwise

(anti-clockwise) in the usual way. Some small (usually cross-head) screws are threaded into plastic inserts, spreading the insert apart behind the section(s) to secure them; unscrew the screw and withdraw the insert to release it. Some types of clip, known as "quick screws" are used, which are released by turning them (either with a cross-head screwdriver or an Allen key, as appropriate) counterclockwise (anti-clockwise) through 90°; where their locations are known, this type are noted in the text, but otherwise their use will be self-evident on unscrewing them. Where trim clips are used, these are released by pulling out the center pin, and fastened by pushing it in.

When installing a fairing section, first study it closely, noting any fasteners and associated fittings removed with it, to be sure of returning everything to its correct place. Check that all fasteners are in good condition, including all trim nuts or clips and damping/rubber mounts; any of these must be replaced if faulty before the section is reassembled. Check also that all mounting brackets are straight and repair or replace them if necessary before attempting to install the section. Where assistance was required to remove a section, make sure your assistant is on hand to install it.

Note that a small amount of lubricant (liquid soap or similar) applied to the mounting rubbers will assist the section retaining pegs to engage without the need for undue pressure. Carefully settle the section in place, following the specific instructions provided, and check that it engages correctly with its partners (where applicable) before tightening any of the fasteners. Where a section engages another by means of tabs, be careful not to break the tab or its mating slot or to damage the paint work.

Tighten the fasteners securely, but be careful not to overtighten any of them or the section may break (not always immediately) due to the uneven stress.

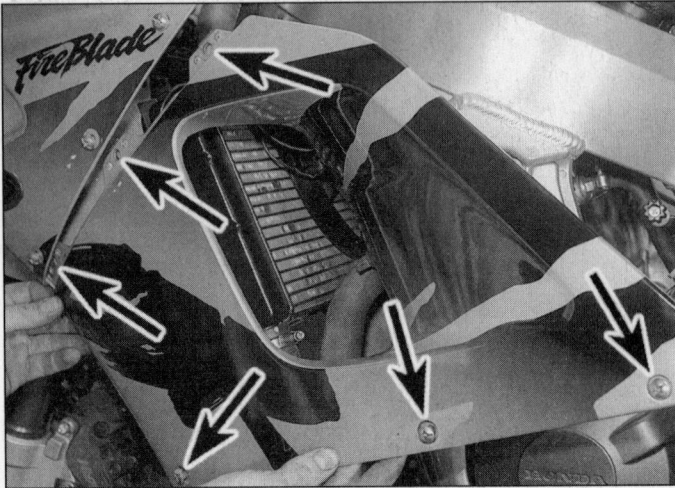

2.2a Undo six "quick screws" to release each fairing middle section (arrows)

2 Fairing sections - removal and installation

Fairing middle sections

Removal

Refer to illustration 2.2a and 2.2b

1 Remove the lower trim clip from the inner middle fairing panel.
2 Unscrew the six "quick screws" which secure the middle section to both the upper section and the lower section, noting the locations of the longer screws, then carefully remove the middle section, noting how it engages with both other panels **(see illustrations)**.

Installation

3 Installation is the reverse of removal. Make sure the longer "quick screws" are installed in their correct locations.

Fairing lower section

Removal

Refer to illustrations 2.6, 2.7a and 2.7b

4 Remove the middle sections as described above.
5 The lower section can be removed as a single unit (see Step 8) or as independent halves (see Steps 6 and 7).
6 Unscrew the Allen bolt at the base of the assembly which secures both halves of the lower section to each other and to the inner panel **(see illustration)**.
7 Remove the fastener at the top left of the inner panel which secures it to the lower section half **(see illustration 2.9)**. Supporting the lower section half, unscrew the three Allen bolts which secure it to the bike, and withdraw the section **(see illustrations)**.
8 To remove the entire lower fairing as an assembly, support the fairing from underneath then remove the three Allen bolts from each side which secure the assembly to the bike **(see illustrations 2.7a and 2.7b)** and lower it gently to the ground.

Installation

Refer to illustration 2.9

9 Installation is the reverse of removal. If the lower section components were separated, make sure the tabs of the inner panel engage correctly **(see illustration)**.

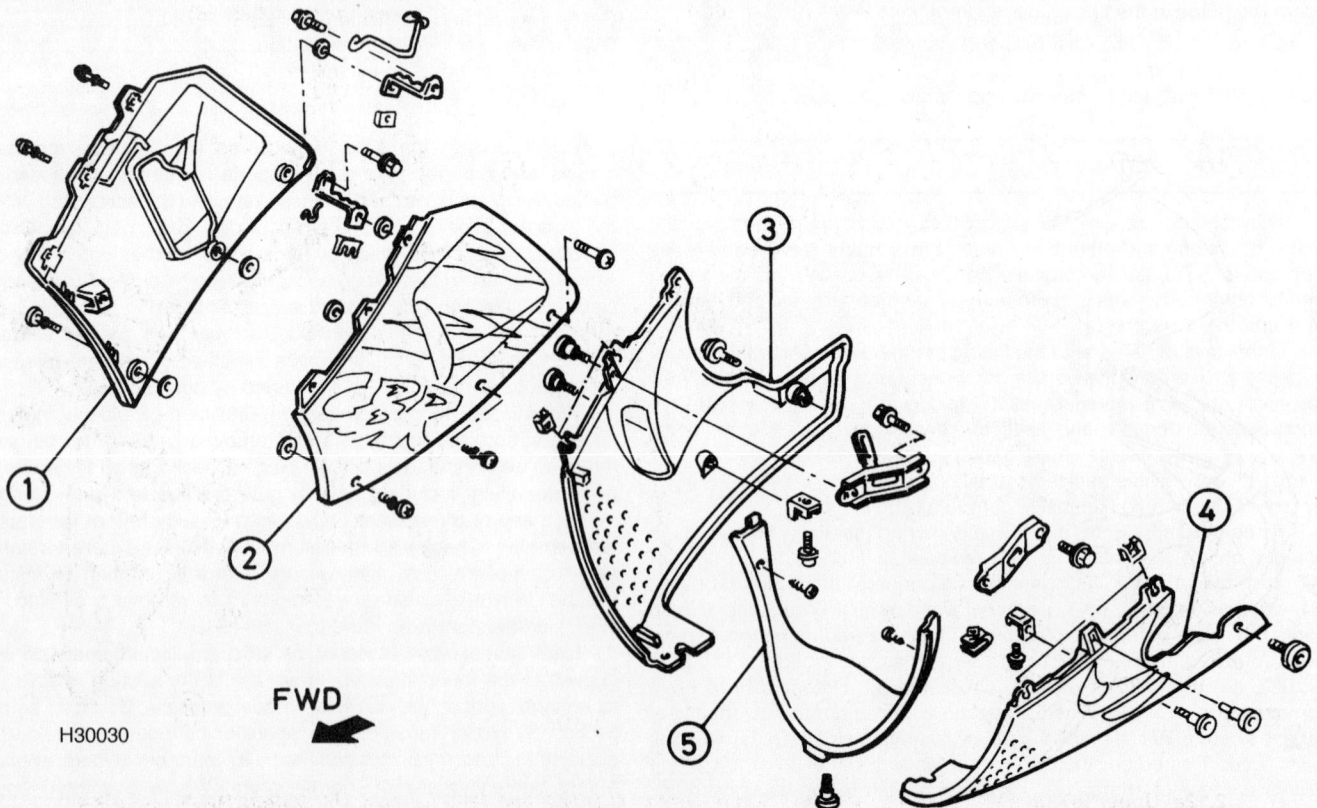

2.2b Middle and lower fairing section assemblies

1 *Middle fairing panel - right*
2 *Middle fairing panel - left*
3 *Lower fairing panel - right*
4 *Lower fairing panel - left*
5 *Inner lower fairing panel*

2.6 Remove the Allen bolt joining the two halves of the lower section together

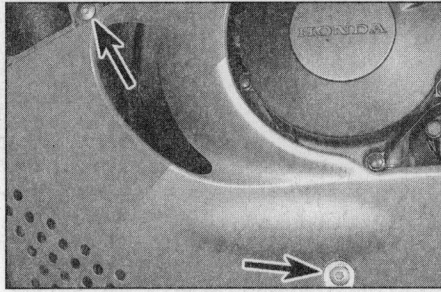

2.7a Unscrew the two front bolts (arrows) . . .

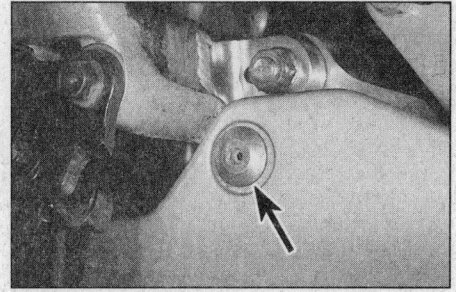

2.7b . . . and single rear bolt (arrow) to free each side of the lower section

Fairing upper section

Removal

Refer to illustrations 2.12a, 2.12b, 2.12c, 2.13, 2.16 and 2.18

10 Remove both middle sections as described above.

11 Remove the rear view mirrors as described in Section 4.

12 Remove the fasteners from the left side upper fairing inner cover, carefully noting the positions of the various fasteners and collars (later models), and remove the cover **(see illustrations)**. On later models, also remove the trim clip to release the relay holder bracket from the fairing, noting how it fits **(see illustration)**.

13 Remove the fasteners from the right side upper fairing inner cover. Note the positions of the various fasteners and collars (later models), and remove the cover. Remove the trim clip to release the connector block holder and fusebox from the fairing, noting how they fit **(see illustration)**.

14 Trace the wiring back from each turn signal assembly and disconnect at the wiring connectors. Note the routing of the wires.

15 Disconnect the speedometer cable from the speedometer; release it from the guide in the fairing, noting its routing (early models only).

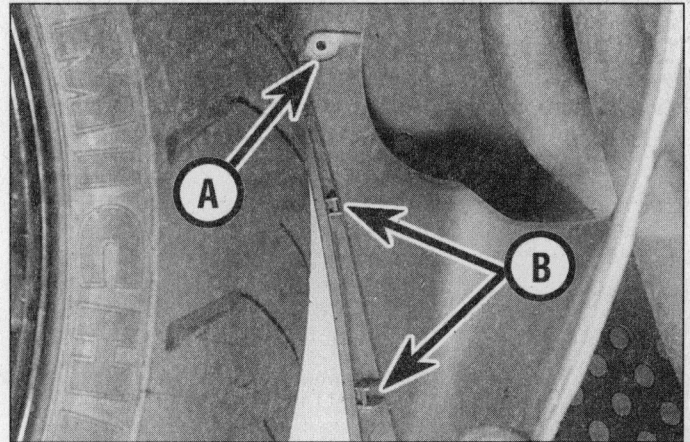

2.9 Lower section inner panel retaining screw location (A) and locating tabs (B)

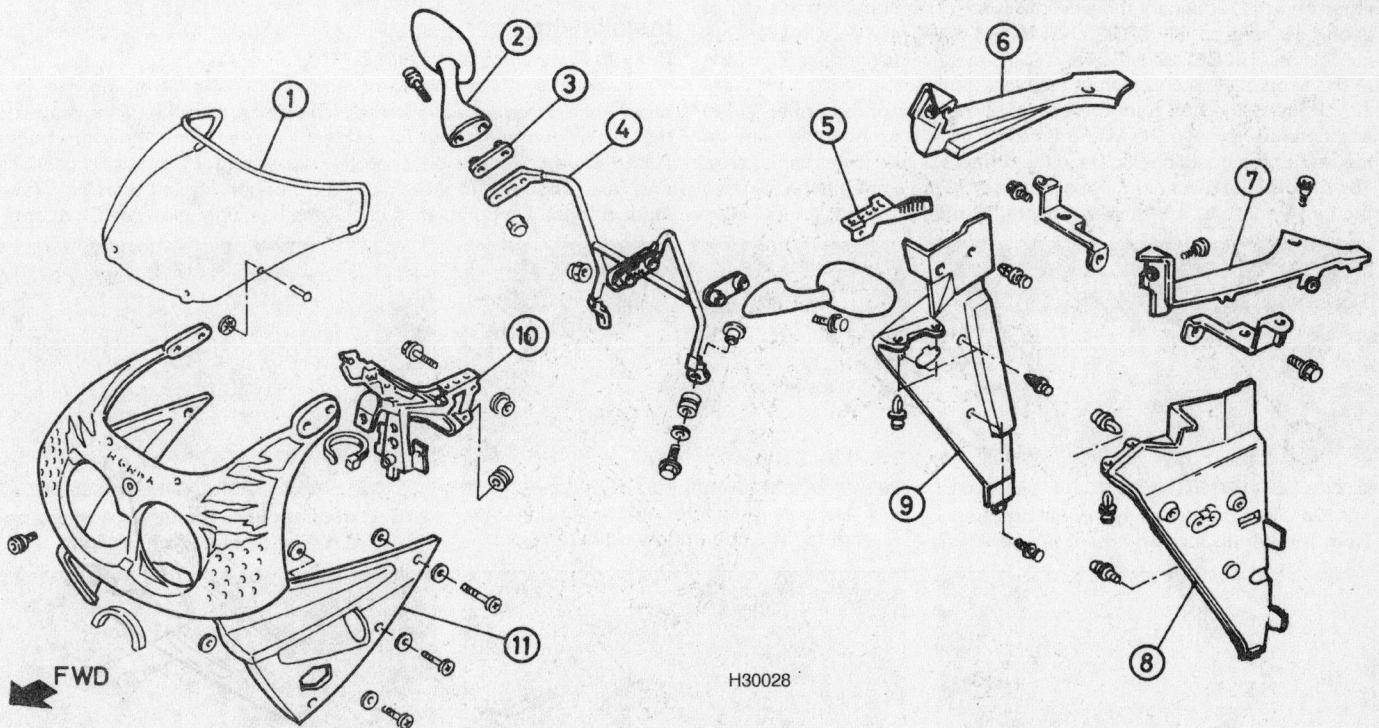

2.12a Upper fairing section and associated components - US 1993 and 1994 models, UK CBR900RR-N and RR-P

1 Windshield
2 Rear view mirror
3 Rubber mounting
4 Upper stay
5 Wiring connector bracket
6 Inner cover - right
7 Inner cover - left
8 Inner middle panel - left
9 Inner middle panel - right
10 Lower stay
11 Upper fairing section

8

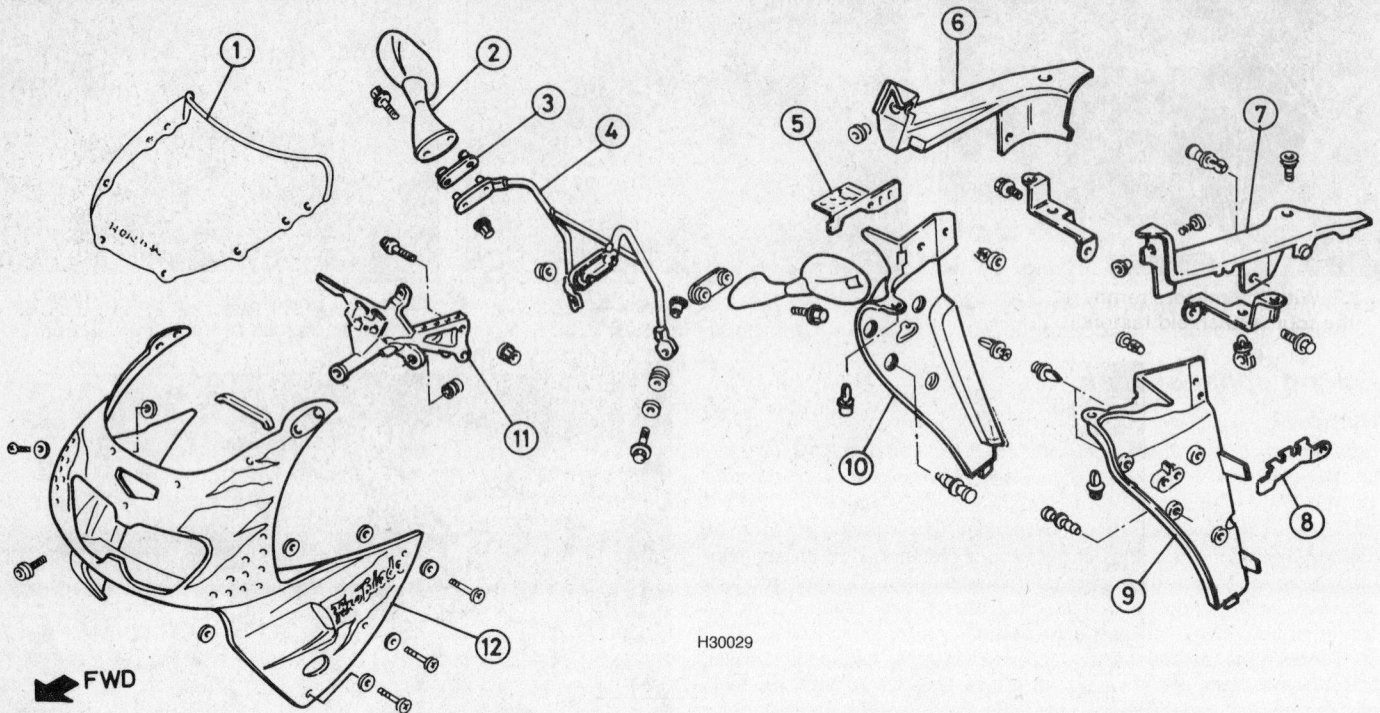

2.12b Upper fairing section and associated components - US 1995 model, UK CBR900RR-R and RR-S

1 Windshield	4 Upper stay	7 Inner cover - left	10 Inner middle panel - right
2 Rear view mirror	5 Wiring connector bracket	8 Relay bracket	11 Lower stay
3 Rubber mounting	6 Inner cover - right	9 Inner middle panel - left	12 Upper fairing section

16 Disconnect the wiring plug from the rear of each headlight **(see illustration)**. Disconnect the sidelight wire connectors from the rear of each headlight on UK CBR900RR-N and RR-P models. On US 1995 models and UK CBR900RR-R and RR-S models, disconnect the wiring to the separate sidelight unit at the block connector.

17 Unscrew the bolt from the center of the front of the upper fairing and carefully remove the fairing from the bike. Note how the pegs on the rear of the headlight locate in the rubbers on the lower fairing stay.

18 To separate the inner middle panels from the upper fairing, remove the turn signal assemblies (see Chapter 9), then withdraw the trim clips which fasten the panels to the fairing **(see illustration)**.

Installation

Refer to illustrations 2.19a and 2.19b

19 Installation is the reverse of removal. Make sure the pegs on the headlight are correctly aligned with the rubbers on the fairing stay and that all cables and wires are correctly routed. Note how the upper fairing inner covers fit between the upper and lower fairing sections and are secured together by the longer "quick screws" **(see illustrations)**. On completion check the headlight aim (see Chapter 9).

2.12c Release the relay holder bracket from the fairing, noting how it fits (arrows)

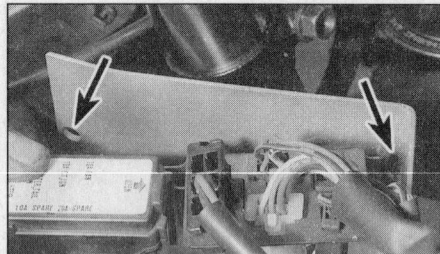

2.13 Release connector and fusebox holder from fairing, noting how it fits (arrows)

2.16 Disconnect the headlight connectors at the back of each headlight

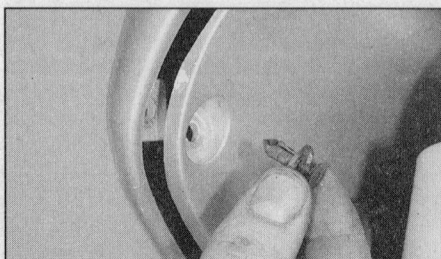

2.18 Pull out trim clips to detach inner middle panels from upper fairing

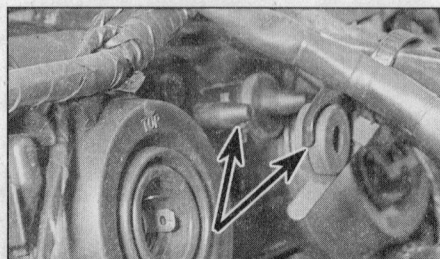

2.19a Align pegs on headlight with rubber mounts on lower fairing stay

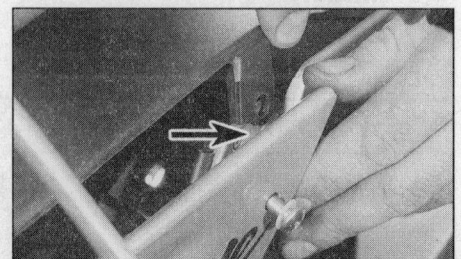

2.19b The inner covers mountings fit between the upper and middle sections

3.2 With the mirrors removed, unscrew the four windshield fasteners (arrows)

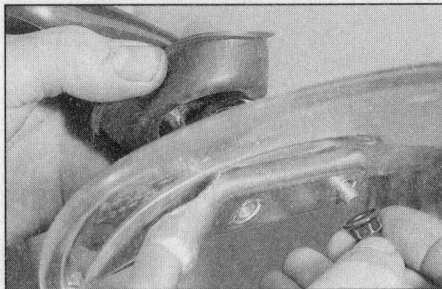

4.1a Pull back the rubber boot to reveal the bolts, then undo the nuts

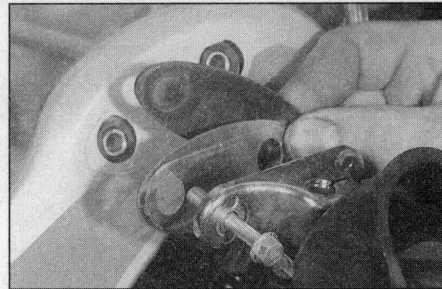

4.1b Note the order of the damping components as the mirror is removed

5.2a The upper stay is mounted to the frame either side of the steering head

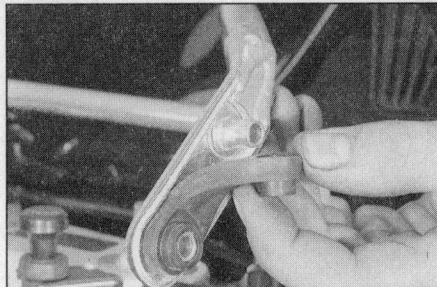

5.2b Note the rubber dampers fitted to each mounting point of the stay

3 Windshield - removal and installation

Removal

Refer to illustration 3.2

1 Remove both rear view mirrors as described in Section 4.
2 Remove the four fasteners securing the windshield to the upper fairing assembly, noting how they fit **(see illustration)**. Withdraw the windshield from between the fairing and fairing stay at each mirror mount, then lift the windshield away from the bike. Don't lose the rubber mounting dampers fitted to each side of the fairing stay.

Installation

3 Installation is the reverse of removal. Make sure the fasteners are correctly and securely fitted.

4 Rear view mirrors - removal and installation

Removal

Refer to illustrations 4.1a and 4.1b

1 Peel back the rubber cover and undo the nuts from the two rear view mirror retaining bolts **(see illustration)**. Remove the rear view mirror along with its mounting plate and rubber, noting the order in which they fit **(see illustration)**.

Installation

2 Installation is the reverse of removal. Make sure the components are fitted in the correct order.

5 Fairing stay - removal and installation

Upper stay

Removal

Refer to illustrations 5.2a and 5.2b

1 Remove the upper fairing assembly as described in Section 2.

2 Unscrew the two bolts which secure the upper fairing stay to the frame either side of the steering head, and carefully lift the stay away from the bike. Note the exact position of all damping rubbers and collars so that they can be returned to their original locations on installation **(see illustrations)**.

Installation

3 Installation is the reverse of removal.

Lower stay

Removal

Refer to illustration 5.7

4 Remove the upper fairing assembly as described in Section 2.
5 Remove the instrument cluster as described in Chapter 9.
6 Carefully note the correct routing of the wiring around the lower fairing stay, and release the wiring from any clips or ties.
7 Unscrew the lower stay mounting nuts and bolts, and remove the stay from the bike **(see illustration)**.

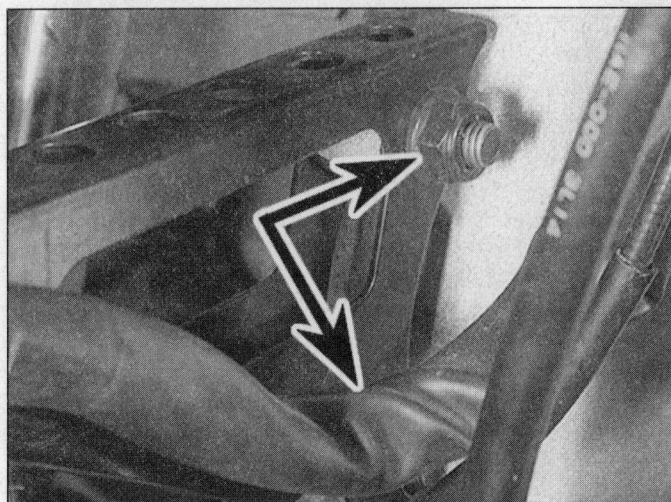

5.7 The lower stay is secured to the frame by two bolts (arrows)

8

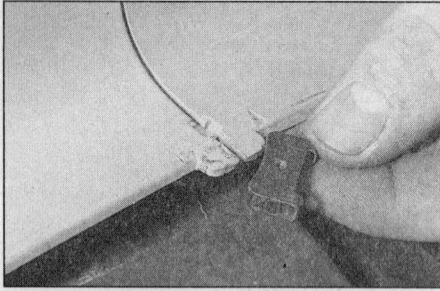

6.2a The side covers are joined by two clips, one under the rider's seat . . .

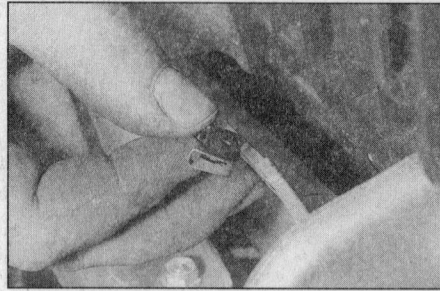

6.2b . . . and one under the pillion seat

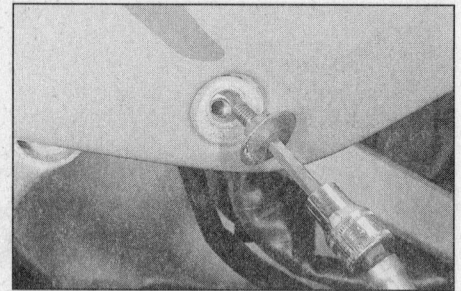

6.3 Remove the side cover front mounting bolt . . .

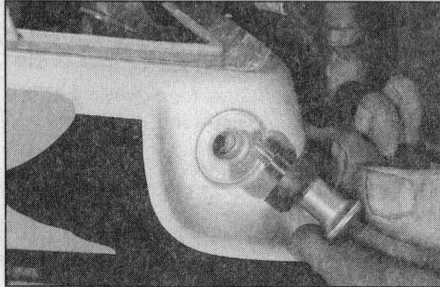

6.4a . . . then the middle one . . .

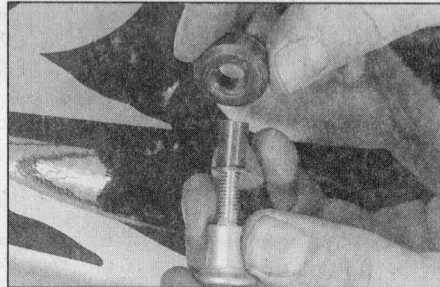

6.4b . . . noting the arrangement of the collars

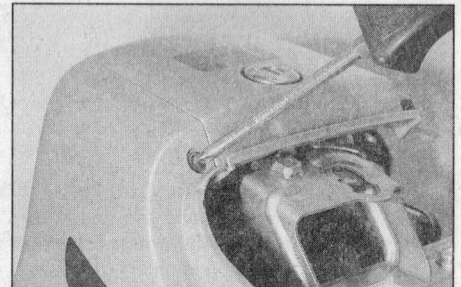

6.5a The rear of the side cover is secured by an upper screw . . .

6.5b . . . and a lower screw . . .

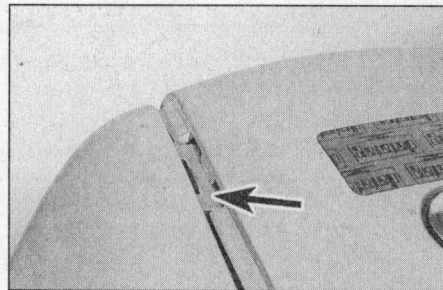

6.5c . . . and slots into a tab on the top cover (arrow)

Installation

8   Installation is the reverse of removal. Make sure all the wiring is correctly routed and secured.

6   Side covers - removal and installation

Removal

Refer to illustrations 6.2a, 6.2b, 6.3, 6.4a, 6.4b, 6.5a, 6.5b and 6.5c

1   Remove the rider's seat as described in Section 9, then open the passenger seat with the ignition key and swing it upright.

2   Remove the two joint clips holding the two side covers together between the rider's seat and the pillion seat **(see illustrations)**.

3   Unscrew the side cover front mounting bolt **(see illustration)**.

4   Unscrew the side cover middle mounting bolt and collars, noting how they fit together **(see illustrations)**.

5   Support the side cover and unscrew the rear mounting screws. Remove the side cover from the machine, noting how the slot in the rear of the cover fits into the tab on the top cover **(see illustrations)**.

Installation

6   Installation is the reverse of removal. Make sure the tab on the top cover fits correctly into the slot in the side cover, and take care not to break either.

7   Front fender/mudguard - removal and installation

Removal

Refer to illustrations 7.1, 7.2a and 7.2b

1   Unscrew the four front fender/mudguard bolts. Note the positions of the brake hose retainers on the rear mounts **(see illustration)**.

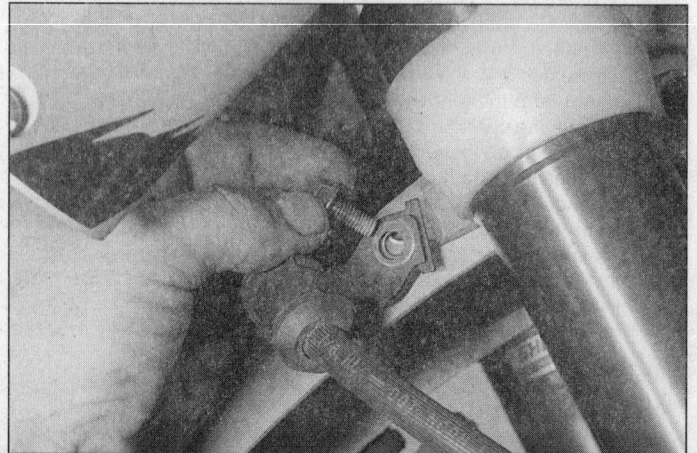

7.1 Note the position of the brake hose retainers

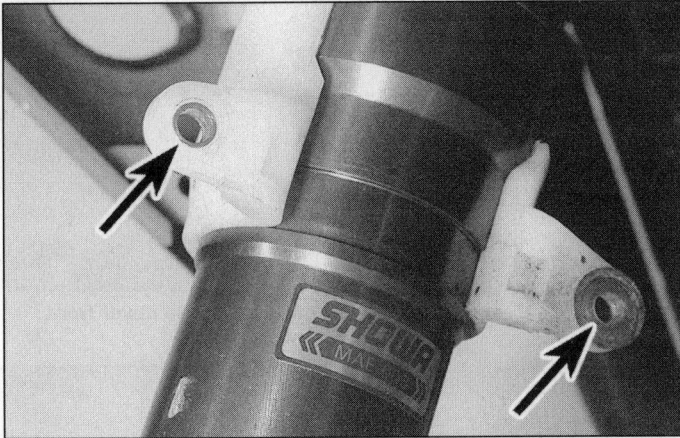

7.2a Take care not to lose the collars in the fender/mudguard mounting brackets (arrows) and note how they fit

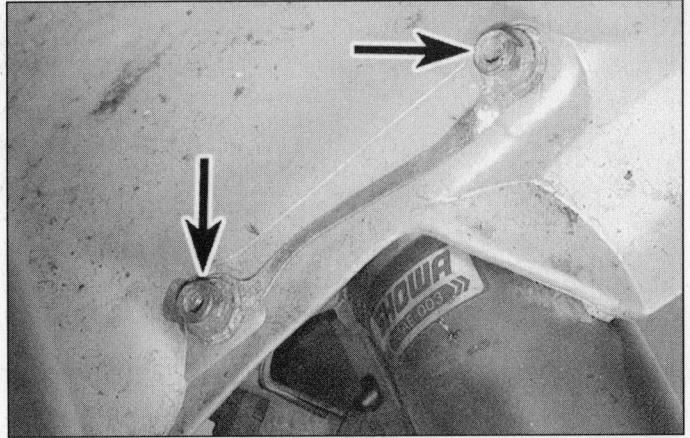

7.2b The brackets on the underside of the fender/mudguard are likely to bend and break if the mounting bolts are overtightened (arrows)

2   Remove the fender/mudguard by withdrawing it forward from the bike. Take care not to lose the fender/mudguard mounting bracket collars, noting how they fit. Note the two brackets on the underside of the fender/mudguard which join the mounting nuts, and take care not to lose them **(see illustrations)**.

Installation

3   Installation is the reverse of removal. Do not forget to position the brake hose retainers on the rear mounts, and take care not to overtighten the mounting bolts as this could cause the mounting nuts to rotate, breaking the bracket that joins them.

---

8   Rear fender/mudguard - removal and installation

Frame-mounted fender/mudguard

Removal

1   Remove the battery as described in Chapter 9.
2   Remove both side covers as described in Section 6.
3   Disconnect the wiring from the rear light assembly at the connector block, then unscrew the two rear light assembly mounting bolts and remove the assembly with the top cover attached.
4   Disconnect the wiring connectors from the rear turn signal assemblies.
5   Remove the rear shock absorber reservoir from its retaining clip on the right side of the frame and support it on the frame tube (see Chapter 6, if necessary). Do not disconnect the reservoir from its hose.
6   Unscrew the rear brake master cylinder reservoir mounting bolt and tie or support the reservoir in an upright position (see Chapter 7, if necessary). Do not disconnect the reservoir from its hose.
7   Disconnect the side stand switch wire connector, the fuel pump wire connector and the alternator wire connector, then unscrew the wire clamp from the fender/mudguard.
8   Remove the spark unit, fuel pump relay and starter relay from their mounting points on the rear fender/mudguard, leaving them connected to the harness and noting their locations and wiring routes. Disconnect the wiring connector from the regulator/rectifier mounted on the right side of the sub-frame, then remove the harness with the spark unit, fuel pump relay and starter relay attached.
9   Unscrew the two rear fender/mudguard mounting bolts located underneath the passenger seat rail at the rear of the sub-frame. Lower the fender/mudguard and carefully draw it back away from the machine, taking care not to snag any wiring.

Installation

10 Installation is the reverse of removal. Make sure all wires are correctly reconnected and routed, and test the rear turn signals, rear light and brake light before riding the bike.

Swingarm-mounted fender/mudguard

Removal

11 Release the rear brake hose from its retaining clamp on the fender/mudguard.
12 Unscrew the drive chain guard screws and remove the guard.
13 Unscrew the fender/mudguard retaining screws and carefully remove the fender/mudguard from the swingarm.

Installation

14 Installation is the reverse of removal. Make sure the brake hose is securely clamped in position.

---

9   Seat - removal and installation

Rider's seat

Removal

Refer to illustrations 9.1a and 9.1b

1   Unscrew the two mounting bolts located under the rear of the seat, then slide the seat back and up to release its locating hook, noting the location of the hook in its bracket **(see illustrations)**.

Installation

2   Installation is the reverse of removal.

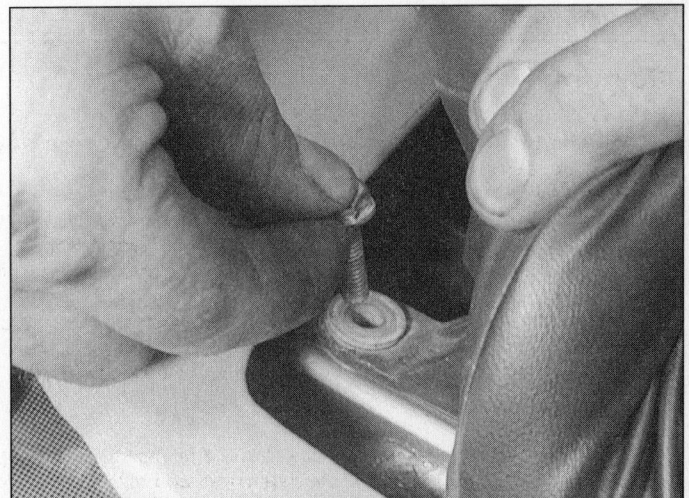

9.1a Pull back the rear corners of the rider's seat to access the mounting bolts

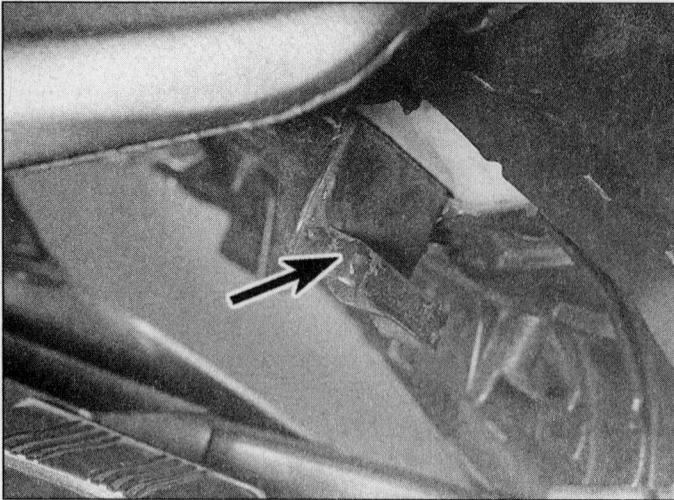

9.1b Note how the hook on the rider's seat (arrow) fits into the frame

9.4 Unscrew the two bolts (arrows) to remove the passenger seat

Passenger seat

Removal

Refer to illustration 9.4

3   Open the passenger seat with the ignition key and swing it upright.

4   Unhook the seat spring, noting carefully how it fits, then unscrew the two seat mounting throughbolts and nuts and remove the seat from the bike **(see illustration)**.

5   To separate the seat from its mounting bracket, unscrew the two bolts which form the hinge between the seat and the bracket.

Installation

6   Installation is the reverse of removal. Make sure the spring is correctly located and that the seat moves freely on its hinges.

Chapter 9 Electrical system

Contents

Specifications

Battery
Capacity ... 12V, 8Ah

Alternator
Type ... Three-phase
Output ... 445 watts at 5000 rpm
Charging voltage ... 13.0 to 15.5 volts @ 5000 rpm
Stator coil resistance ... 0.1 to 0.3 ohms @ 20°C (68°F)

Starter motor
Brush length
New ... 12 to 13 mm (0.47 to 0.51 in)
Service limit ... 4.5 mm (0.18 in)

Fuses
Main fuse (on starter relay) ... 30A
All other fuses (in fusebox) ... 4 x 10A and 1 x 20A*
*See fusebox lid for circuits protected

Bulbs
	US models	UK models
Headlight		
US 1993 and 1994 models, UK CBR900RR-N and RR-P	12V 60/55W	12V 60/55W
US 1995 model, UK CBR900RR-R and RR-S	12V 45/45W	12V 60/55W
Sidelight		
UK CBR900RR-N and RR-P	Not fitted	12V 4W
US 1995 model, UK CBR900RR-R and RR-S	12V 5W	12V 5W
Stop/taillight	12V 32/3CP	12V 21/5W
Front turn signal lights/running lights (US models)	12V 32/3CP	12V 21W
Rear turn signal lights	12V 32CP	12V 21W
Instrument illuminating lights	12V 1.7W	12V 1.7W
Warning lamps	12V 1.7W	12V 1.7W
License plate light	12V 4CP	Not fitted

Torque settings
	Nm	ft-lbs
Ignition (main) switch bolts	24	17
Neutral switch	12	9
Oil pressure switch	12	9
Sidestand switch bolt	10	7
Alternator stator coil bolts	12	9

9

1   General information

Refer to illustration 1.1

The machines covered by this manual are equipped with a 12-volt electrical system. The components include a three-phase alternator and a regulator/rectifier unit.

The regulator/rectifier unit maintains the charging system output within the specified range to prevent overcharging and converts the ac (alternating current) output of the alternator to dc (direct current) to power the lights and other components and to charge the battery.

The starter motor is mounted to the crankcase behind the cylinders. The starting system includes the motor, the battery, the relay and the various wires and switches. If the engine stop switch and the main key switch are both in the "Run" or "On" position, the circuit relay allows the starter motor to operate only if the transmission is in neutral (neutral switch on) or the clutch lever is pulled to the handlebar (clutch switch on) and the sidestand is up (sidestand switch on) **(see illustration)**.

Note: *Keep in mind that electrical parts, once purchased, can't be returned. To avoid unnecessary expense, make sure the faulty part has been positively identified before buying a replacement.*

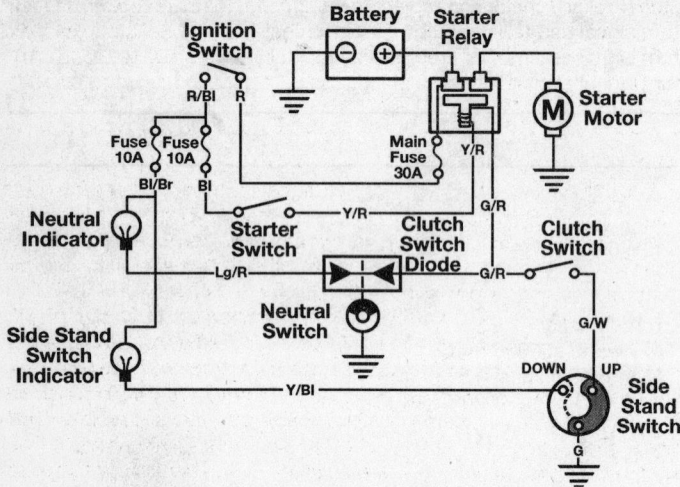

1.1 Starter circuit diagram

Bl Black	R Red
Bl/Br Black and brown	R/Bl Red and black
G/W Green and white	Y/Bl Yellow and black
G/R Green and red	Y/R Yellow and red
Lg/R Light green and red	

2   Electrical troubleshooting

Warning: *To prevent the risk of short circuits, the ignition (main) switch must always be "OFF" and the battery negative (-) cable should be disconnected before any of the bike's other electrical components are disturbed. Don't forget to reconnect the cable securely once work is finished or if battery power is needed for circuit testing.*

Refer to illustration 2.1

A typical electrical circuit consists of an electrical component, the switches, relays, etc. related to that component and the wiring and connectors that hook the component to both the battery and the frame. To aid in locating a problem in any electrical circuit, refer to the wiring diagrams at the end of this Chapter.

Before tackling any troublesome electrical circuit, first study the appropriate diagrams thoroughly to get a complete picture of what makes up that individual circuit. Trouble spots, for instance, can often be narrowed down by noting if other components related to that circuit are operating properly or not. If several components or circuits fail at

one time, chances are the fault lies in the fuse or ground (earth) connection, as several circuits often are routed through the same fuse and ground (earth) connections.

Electrical problems often stem from simple causes, such as loose or corroded connections or a blown fuse. Prior to any electrical troubleshooting, always visually check the condition of the fuse, wires and connections in the problem circuit. Intermittent failures can be especially frustrating, since you can't always duplicate the failure when it's convenient to test. In such situations, a good practice is to clean all connections in the affected circuit, whether or not they appear to be good. All of the connections and wires should also be wiggled to check for looseness which can cause intermittent failure.

If testing instruments are going to be utilized, use the diagrams to plan where you will make the necessary connections in order to accurately pinpoint the trouble spot.

The basic tools needed for electrical troubleshooting include a test light or voltmeter, a continuity tester (which includes a bulb, battery and set of test leads) and a jumper wire, preferably with a circuit breaker incorporated, which can be used to bypass electrical components **(see illustration)**. Specific checks described later in this Chapter may also require an ohmmeter. Ideally a multimeter with resistance, current and voltage measuring facilities should be available.

Voltage checks should be performed if a circuit is not functioning properly. Connect one lead of a test light or voltmeter to either the negative battery terminal or a known good ground (earth). Connect the other lead to a connector in the circuit being tested, preferably nearest to the battery or fuse. If the bulb lights, voltage is reaching that point, which means the part of the circuit between that connector and the battery is problem-free. Continue checking the remainder of the circuit in the same manner. When you reach a point where no voltage is present, the problem lies between there and the last good test point. Most of the time the problem is due to a loose connection. Keep in mind that some circuits only receive voltage when the ignition key is in the "ON" position.

One method of finding short circuits is to remove the fuse and connect a test light or voltmeter in its place to the fuse terminals. There should be no load in the circuit (it should be switched off). Move the wiring harness from side-to-side while watching the test light. If the bulb lights, there is a short to ground (earth) somewhere in that area, probably where insulation has rubbed off a wire. The same test can be performed on other components in the circuit, including the switch.

A ground (earth) check should be done to see if a component is grounded (earthed) properly. Disconnect the battery and connect one lead of a self-powered test light (continuity tester) to a known good ground (earth). Connect the other lead to the wire or ground (earth) connection being tested. If the bulb lights, the ground (earth) is good. If the bulb does not light, the ground (earth) is not good.

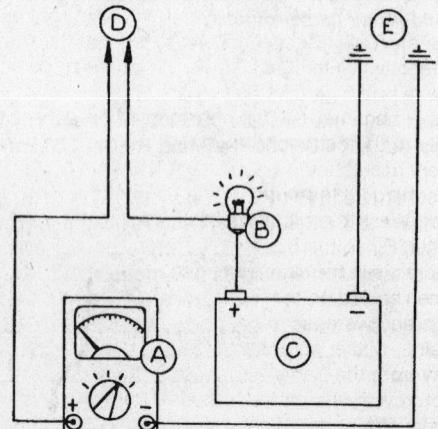

2.1 Simple testing equipment for checking the wiring

A Multimeter	D Positive probe (+)
B Bulb	E Negative probe (-)
C Battery	

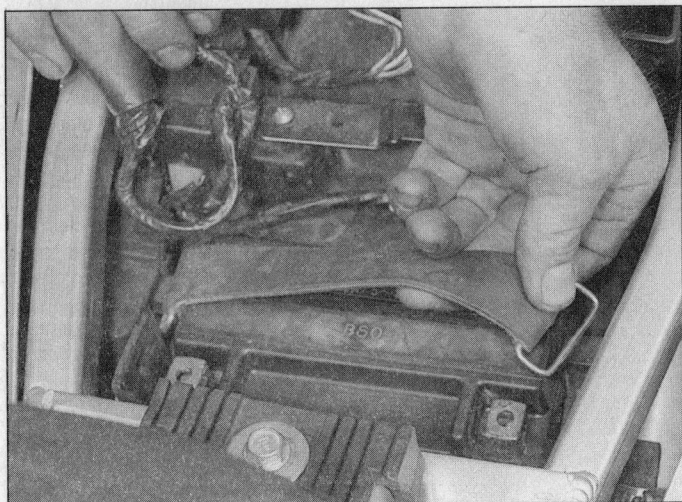

3.2a Unhook the battery retaining strap to release the battery . . .

3.2b . . . and lift the battery out of its holder

A continuity check is performed to see if a circuit, section of circuit or individual component is capable of passing electricity through it. Disconnect the battery and connect one lead of a self-powered test light (continuity tester) to one end of the circuit being tested and the other lead to the other end of the circuit. If the bulb lights, there is continuity, which means the circuit is passing electricity through it properly. Switches can be checked in the same way.

Remember that all electrical circuits are designed to conduct electricity from the battery, through the wires, switches, relays, etc. to the electrical component (light bulb, motor, etc.). From there it is directed to the frame (ground (earth) where it is passed back to the battery. Electrical problems are basically an interruption in the flow of electricity from the battery or back to it.

3 Battery - inspection and maintenance

Refer to illustration 3.2a and 3.2b

1 The battery fitted to the models covered in this manual is of the maintenance free (sealed) type and therefore requires no maintenance as such. However, the following checks should still be regularly performed.

2 Remove the rider's seat (see Chapter 8) to gain access to the battery. Disconnect the battery leads as described below and unhook the battery retaining strap **(see illustration)**. Lift the battery out of its holder **(see illustration)**.

3 Check the battery terminals and leads for tightness and corrosion. If corrosion is evident, disconnect the leads from the battery, disconnecting the negative (-) terminal first, and clean the terminals and lead ends with a wire brush or knife and emery paper. Reconnect the leads, connecting the negative (-) terminal last, and apply a thin coat of petroleum jelly to the connections to slow further corrosion.

4 The battery case should be kept clean to prevent current leakage, which can discharge the battery over a period of time (especially when it sits unused). Wash the outside of the case with a solution of baking soda and water. Rinse the battery thoroughly, then dry it.

5 Look for cracks in the case and replace the battery if any are found. If acid has been spilled on the frame or battery holder, neutralize it with a baking soda and water solution, dry it thoroughly, then touch up any damaged paint. Make sure the battery vent tube (if equipped) is directed away from the frame and is not kinked or pinched.

6 If the motorcycle sits unused for long periods of time, disconnect the cables from the battery terminals. Refer to Section 4 and charge the battery approximately once every month.

7 The condition of the battery can be assessed by measuring the voltage present at the battery terminals. Connect the voltmeter positive (+) probe to the battery positive (+) terminal and the negative (-) probe to the battery negative (-) terminal. When fully charged there should be

approximately 13 volts present. If the voltage falls below 12.3 volts the battery must be removed, disconnecting the negative (-) terminal first, and recharged as described below in Section 4.

4 Battery - charging

Refer to illustration 4.2

1 To charge the battery it is first necessary to remove it from the bike. To do this remove the rider's seat (see Chapter 8) then unhook the battery retaining strap **(see illustration 3.2a)**. Disconnect the leads from the battery, disconnecting the negative (-) terminal first, and lift the battery out of its holder **(see illustration 3.2b)**.

2 Honda recommend that the battery is charged at a rate of 0.9 amps for between 5 and 10 hours. Exceeding this figure can cause the battery to overheat, buckling the plates and rendering it useless. Few owners will have access to an expensive current controlled charger, so if a normal domestic charger is used check that after a possible initial peak, the charge rate falls to a safe level **(see illustration)**. If the battery becomes hot during charging **stop**. Further charging will cause damage. **Note:** *In emergencies Honda state that the battery can be charged at a rate of 4 amps for a period of 1 hour. However, this is not recommended and the low amp charge is by far the safer method of charging the battery.*

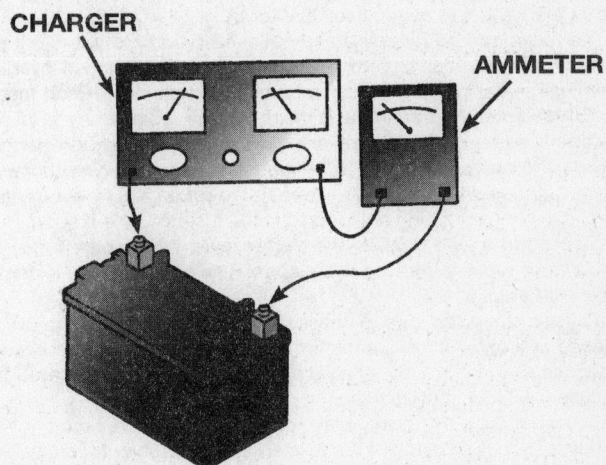

CHARGER

AMMETER

4.2 If the charger doesn't have an ammeter built in, connect one in series as shown; DO NOT connect the ammeter between the battery terminals or it will be ruined

9

5.2a The fusebox is located under the upper fairing right inner cover. Fuse locations and ratings are marked on the top of the fusebox lid

5.2b Unclip the fusebox lid to reveal the fuses

3   If the recharged battery discharges rapidly if left disconnected it is likely that an internal short caused by physical damage or sulphation has occurred. A new battery will be required. A sound item will tend to lose its charge at about 1% per day.

4   On installation, clean the battery terminals and lead ends with a wire brush or knife and emery paper. Reconnect the leads, connecting the negative (-) terminal last, and apply a thin coat of petroleum jelly to the connections to slow further corrosion. Fit the battery retaining strap and install the rider's seat as described in Chapter 8.

5   Fuses - check and replacement

Refer to illustrations 5.2a and 5.2b

1   Most circuits are protected by fuses of different ratings. All fuses except the main fuse are located in the fusebox which is situated under the inner cover in the right side of the upper fairing. The main fuse is fitted to the top of the starter relay which is behind the right side cover **(see illustration 26.7)**. All fusebox fuses are labelled for easy identification.

2   To gain access to the fusebox, remove the fasteners securing the upper fairing right inner cover (see Chapter 8) then unclip and remove the fusebox lid **(see illustrations)**. To gain access to the main fuse remove the right side cover (see Chapter 8).

3   The fuses can be removed and checked visually. If you can't pull the fuse out with your fingertips, use a pair of needle-nose pliers. A blown fuse is easily identified by a break in the element. Each fuse is clearly marked with its rating and must only be replaced by a fuse of the correct rating. **Caution:** *Never put in a fuse of a higher rating or bridge the terminals with any other substitute, however temporary it may be. Serious damage may be done to the circuit, or a fire may start.* Spare fuses of each rating are located in the fusebox, and a spare main fuse is clipped to the base of the starter relay. If the spare fuses are used, always replace them so that a spare fuse of each rating is carried on the bike at all times.

4   If a fuse blows, be sure to check the wiring circuit very carefully for evidence of a short-circuit. Look for bare wires and chafed, melted or burned insulation. If a fuse is replaced before the cause is located, the new fuse will blow immediately.

5   Occasionally a fuse will blow or cause an open-circuit for no obvious reason. Corrosion of the fuse ends and fusebox terminals may occur and cause poor fuse contact. If this happens, remove the corrosion with a wire brush or emery paper, then spray the fuse end and terminals with electrical contact cleaner.

6   Lighting system - check

1   The battery provides power for operation of the headlight, taillight, brake light, license plate light and instrument cluster lights. If none of the lights operate, always check battery voltage before proceeding. Low battery voltage indicates either a faulty battery or a defective charging system. Refer to Section 3 for battery checks and Sections 29 and 30 for charging system tests. Also, check the condition of the fuses and replace any blown fuses with new ones.

Headlight

2   If the headlight fails to work, check the fuse first with the key "ON" (see Section 5), then unplug the electrical connector for the headlight and use jumper wires to connect the bulb directly to the battery terminals. If the light comes on, the problem lies in the wiring or one of the switches in the circuit or the headlight relay. Refer to Section 18 for the switch testing procedures, and also the wiring diagrams at the end of this Chapter.

Taillight/license plate light

3   If the taillight fails to work, check the bulbs and the bulb terminals first, then the fuses, then check for battery voltage at the taillight electrical connector. If voltage is present, check the ground (earth) circuit for an open or poor connection.

4   If no voltage is indicated, check the wiring between the taillight and the ignition switch, then check the switch. On UK models, check the lighting switch as well.

Brake light

5   See Section 12 for the brake light switch checking procedure.

Neutral indicator light

6   If the neutral light fails to operate when the transmission is in neutral, check the fuses and the bulb (see Sections 5 and 15). If the bulb and fuses are in good condition, check for battery voltage at the connector attached to the neutral switch on the left side of the engine. If battery voltage is present, refer to Section 20 for the neutral switch check and replacement procedures.

7   If no voltage is indicated, check the wiring between the switch and the bulb for open-circuits and poor connections.

Oil pressure warning light

8   See Section 16 for the oil pressure switch check.

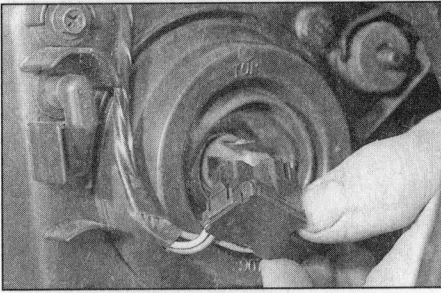

7.1 Disconnect the wiring from the headlight bulb and pull off rubber cover

7.2a Press on the headlight bulb retaining clip to release it . . .

7.2b . . . then swing the clip away and remove the bulb

7.4 Install the dust cover with the "TOP" mark facing up (arrow)

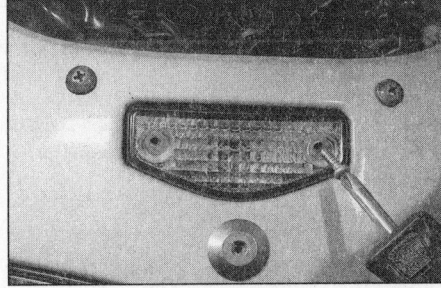

7.9 Unscrew two lens retaining screws to gain access to the sidelight bulb

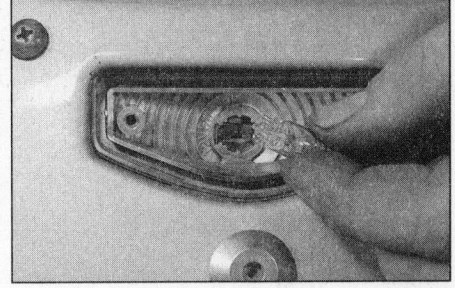

7.10 Gently pull the sidelight bulb out of its holder

Sidestand switch warning light

9   If the sidestand light fails to operate when the stand is extended down, check the fuses and the bulb (see Sections 5 and 15). If the bulb and fuses are in good condition, refer to Section 21 for the sidestand switch check and replacement procedures.

10   If the switch is functioning correctly check for voltage at the switch wiring connector. If no voltage is indicated, check the wiring between the switch and the bulb for open-circuits and poor connections.

7   Headlight bulb and sidelight bulb - replacement

Note: *The headlight bulb is of the quartz-halogen type. Do not touch the bulb glass as skin acids will shorten the bulb's service life. If the bulb is accidentally touched, it should be wiped carefully when cold with a rag soaked in stoddard solvent (methylated spirit) and dried before fitting. Allow the bulb time to cool before removing it if the headlight has been extinguished.*

Warning: *Let the bulb cool first if the headlight has just been used.*

Headlight

Refer to illustrations 7.1, 7.2a, 7.2b and 7.4

1   Disconnect the wiring connector from the headlight bulb and remove the rubber dust cover **(see illustration)**.

2   Release the bulb retaining clip and swing it away, then remove the bulb **(see illustrations)**.

3   Fit the new bulb, bearing in mind the note and warning at the start of this Section and secure it in position with the retaining clip.

4   Install the dust cover, making sure it is correctly seated and with the "TOP" mark facing up **(see illustration)**.

5   Reconnect the wiring and check the operation of the headlight.

Sidelight
UK CBR900RR-N and RR-P

6   Pull the bulbholder out from the base of the headlight. Push the bulb inwards and twist it anti-clockwise to release it from the bulbholder. If the socket contacts are dirty or corroded, they should be scraped clean and sprayed with electrical contact cleaner before the new bulb is installed.

7   Install the new bulb in the bulbholder by pressing it in and twisting

it clockwise. Press the bulbholder back into the headlight.

8   Check the operation of the sidelight.

US 1995 model, UK CBR900RR-R and RR-S
Refer to illustration 7.9 and 7.10

9   Unscrew the two sidelight lens retaining screws and remove the lens **(see illustration)**.

10   Pull the bulb out of the bulbholder **(see illustration)**.

11   Carefully push the new bulb into the bulbholder.

12   Check the lens seal before fitting the lens, and replace it if damaged. Fit the lens and install the retaining screws. Be careful not to overtighten the screws as the lens is easily damaged.

8   Headlight and sidelight - removal and installation

Headlight
Removal
Refer to illustration 8.2

1   Remove the upper fairing assembly as described in Chapter 8.

2   Unscrew the four headlight assembly retaining screws and remove the assembly from the fairing **(see illustration)**.

8.2 Unscrew the four screws (arrows) to release the headlight from the fairing - later model shown

9

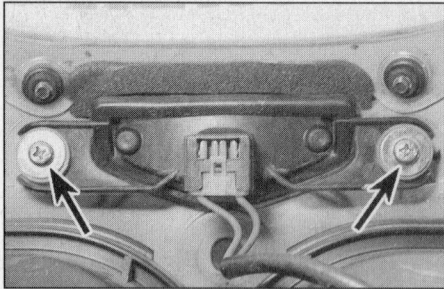

8.6 Unscrew the two screws (arrows) to release the sidelight from the fairing

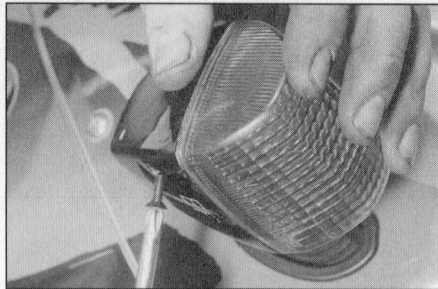

9.1 Unscrew the lens retaining screw from the base of the turn signal

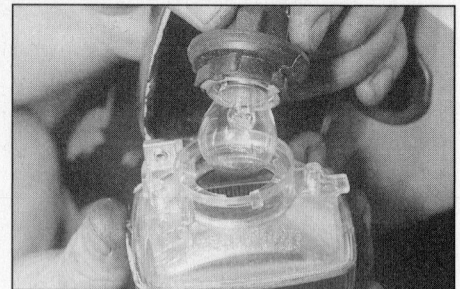

9.2 Release the bulbholder by twisting it 90° in either direction

Installation

3   Make sure the headlight assembly is correctly seated then install the four retaining screws and washers (where fitted) and tighten them securely.

4   Install the upper fairing as described in Chapter 8.

Sidelight - US 1995 model, UK CBR900RR-R and RR-S only

Removal

Refer to illustration 8.6

5   Remove the upper fairing assembly as described in Chapter 8.

6   Unscrew the two sidelight assembly retaining screws and remove the assembly from the fairing **(see illustration)**.

Installation

7   Make sure the sidelight assembly is correctly seated then install the two retaining screws and washers and tighten them securely.

8   Install the upper fairing as described in Chapter 8.

---

9   Turn signal, taillight and license plate bulbs - replacement

---

Turn signal bulbs

Refer to illustrations 9.1, 9.2 and 9.3

1   Unscrew the turn signal lens retaining screw from the base of the turn signal and withdraw the lens **(see illustration)**.

2   Twist the bulbholder 90° in either direction to release it from the lens **(see illustration)**.

3   Push the bulb into the holder and twist it counterclockwise (anti-clockwise) to remove it **(see illustration)**. Check the socket terminals for corrosion and clean them if necessary. Line up the pins of the new bulb with the slots in the socket, then push the bulb in and turn it clockwise until it locks into place.

Note: *It is a good idea to use a paper towel or dry cloth when handling the new bulb to prevent injury if the bulb should break and to increase bulb life.*

Note: *On US models, because the front turn signals double as running lights the offset pins can only be installed one way.*

4   Fit the bulbholder to the lens and twist the bulbholder clockwise until it is locked into place.

5   Install the lens back into the turn signal and securely tighten the retaining screw.

Taillight bulbs

Refer to illustration 9.7 and 9.8

6   Unlock the passenger seat using the ignition key and swing the seat up.

7   Twist the relevant bulbholder counterclockwise (anti-clockwise) to release it from the taillight unit **(see illustration)**.

8   Push the bulb into the holder and twist it counterclockwise (anti-clockwise) to remove it. Check the socket terminals for corrosion and clean them if necessary. Line up the pins of the new bulb with the slots in the socket, then push the bulb in and turn it clockwise until it locks into place **(see illustration)**. **Note:** *The pins on the bulb are offset so it can only be installed one way. It is a good idea to use a paper towel or dry cloth when handling the new bulb to prevent injury if the bulb should break and to increase bulb life.*

9   Install the bulbholder into the taillight unit and turn it clockwise until it is locked in place.

10   Swing the seat down into place and lock it.

License plate bulb - US models

11   Unbolt the license plate from the rear fender and ease it off the fender; if necessary disconnect the wiring under the passenger seat and pull it through the fender for improved access.

12   Unscrew the nuts from the lens holder studs and separate the license plate light components.

13   Push the bulb into the holder and twist it counterclockwise (anti-clockwise) to remove it. Check the socket terminals for corrosion and clean them if necessary. Line up the pins of the new bulb with the slots in the socket, then push the bulb in and turn it clockwise until it locks in place.

14   Install the license plate light components in a reverse of the removal sequence.

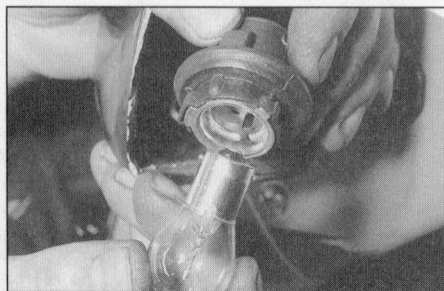

9.3 To release the bulb gently push it into the bulbholder and twist it counter-clockwise (anti-clockwise)

9.7 Twist the bulbholder counter-clockwise (anti-clockwise) to release it from the taillight unit

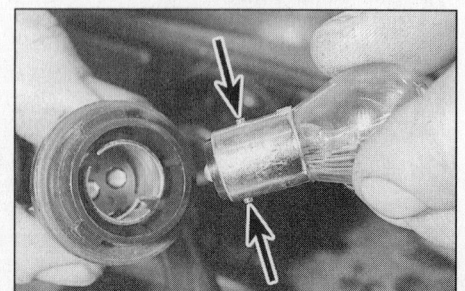

9.8 Note the pins of the taillight bulb are offset (arrows); make sure they are correctly aligned on installation

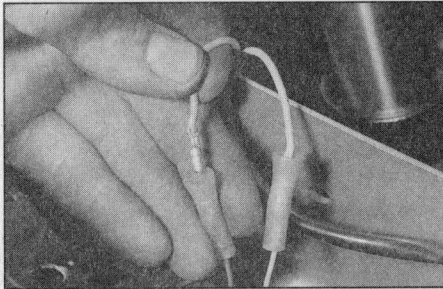
10.2 Disconnect the turn signal wiring at the connectors under the upper fairing inner cover

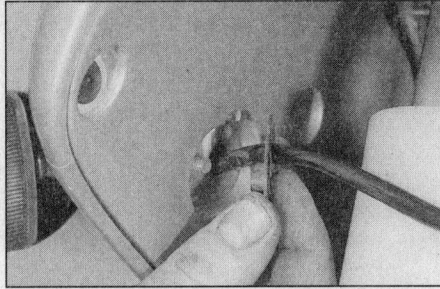
10.3 Unscrew the turn signal mounting nut and remove the collared backplate, noting how it fits

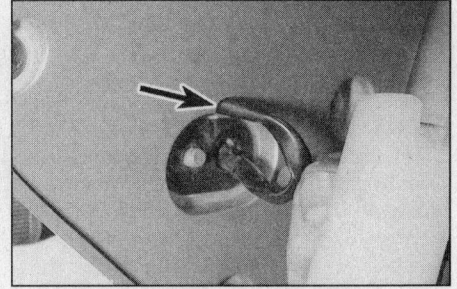
10.5 Feed the wiring back through the hole in the fairing (arrow)

10 Turn signal assemblies - removal and installation

Front

Removal

Refer to illustrations 10.2 and 10.3

1 Remove the upper fairing inner cover for the side you are working on (see Chapter 8).
2 Trace the turn signal wiring back from the turn signal and disconnect it at the connectors in the upper fairing **(see illustration).**
3 Pull the wiring through from the fairing to the turn signal mounting, noting its routing, then unscrew the turn signal mounting nut. Remove the collared backplate, noting how it fits, then remove the turn signal from the fairing **(see illustration).**

Installation

Refer to illustration 10.5

4 Install the turn signal in the fairing, then fit the collared backplate and the nut, tightening it securely.
5 Feed the wiring through the hole in the inner fairing panel, then reconnect it at the connectors and install the upper fairing inner cover (see Chapter 8) **(see illustration).**

Rear

Removal

6 Unlock the passenger seat using the ignition key and swing the seat up.
7 Trace the wiring back from the turn signal and disconnect it at the connectors.

11.3 On later models the turn signal relay is mounted at the front of the relay mounting bracket under the upper fairing right inner cover (arrow)

8 Unscrew the screw securing the turn signal assembly to the inside of the rear fender/mudguard, and remove it along with the collared backplate, noting how it fits.
9 Remove the turn signal from the fender/mudguard.

Installation

10 Install the turn signal in the rear fender/mudguard, then fit the collared backplate and the nut and tighten it securely.
11 Reconnect the wiring at the connectors, then swing the passenger seat down into place and lock it.

11 Turn signal circuit - check

Refer to illustration 11.3

1 The battery provides power for operation of the turn signal lights, so if they do not operate, always check the battery voltage first. Low battery voltage indicates either a faulty battery or a defective charging system. Refer to Section 3 for battery checks and Sections 29 and 30 for charging system tests. Also, check the fuses (see Section 5) and the switch (see Section 18).
2 Most turn signal problems are the result of a burned out bulb or corroded socket. This is especially true when the turn signals function properly in one direction, but fail to flash in the other direction. Check the bulbs and the sockets (see Section 9).
3 If the bulbs and sockets are good, check for power at the turn signal relay with the ignition "ON". The relay is mounted on the left side of the fairing mounting stay on early models, and at the front of the relay mounting bracket under the upper fairing left inner cover on later models **(see illustration)**. Refer to *wiring diagrams* at the end of the book to identify the power source terminal.
4 If power is present, check the wiring between the relay and the turn signal lights (see the *wiring diagrams* at the end of this Chapter).
5 If the wiring is good, replace the turn signal relay.

12 Brake light switches - check and replacement

Circuit check

1 Before checking any electrical circuit, check the bulb (Section 9) and fuses (see Section 5).
2 Using a test light connected to a good ground (earth), check for voltage at the brake light switch wiring connector (see Steps 8 and 9 for the rear brake light switch). If there's no voltage present, check the wire between the switch and the fusebox (see the *wiring diagrams* at the end of this Chapter).
3 If voltage is available, touch the probe of the test light to the other terminal of the switch, then pull the brake lever or depress the brake pedal. **Note:** *The wiring connector halves must be joined for this test and the probes inserted in the back of the connector.* If the test light doesn't light up, replace the switch.
4 If the test light does light, check the wiring between the switch and the brake lights (see the *wiring diagrams* at the end of this Chapter).

9

12.5 The front brake light switch (arrow) is mounted on the underside of the master cylinder

Switch replacement

Front brake lever switch

Refer to illustration 12.5

5 Remove the mounting screw and unplug the electrical connectors from the switch **(see illustration)**.

6 Detach the switch from the bottom of the front brake master cylinder.

7 Installation is the reverse of the removal procedure. The brake lever switch isn't adjustable.

Rear brake pedal switch

8 Remove the rider's seat as described in Chapter 8.

9 Trace the wiring back from the rear brake switch, located behind the rider's right footpeg bracket, and disconnect it at its wire connector. Work back along the wiring releasing it from any relevant clips and ties.

10 Unhook the spring from the switch and brake pedal and remove it from the bike.

11 Free the switch from the right footpeg bracket and remove it from the bike.

12 Install the switch by reversing the removal procedure, then adjust it by following the procedure described in Chapter 1.

13 Instrument cluster and speedometer cable - removal and installation

Instrument cluster

Removal

US 1993 and 1994 models, UK CBR900RR-N and RR-P

1 Unscrew the speedometer cable retaining ring from the rear of the instrument cluster and detach the cable.

2 Disconnect the two instrument cluster wiring connectors.

3 Unscrew the two instrument cluster mounting bolts and remove the collars.

4 Carefully maneuver the instrument cluster away from the bike, noting the alignment of the bosses on the cluster with the rubber grommets on the mounting bracket.

5 Inspect the cluster mounting dampers and mounting bracket grommets for signs of damage and replace if necessary.

US 1995 model, UK CBR900RR-R and RR-S

Refer to illustrations 13.6 and 13.7

6 Unscrew the three instrument cluster mounting bolts and remove the collars **(see illustration)**.

7 Carefully maneuver the instrument cluster out of position and disconnect the two cluster wiring connectors as they become accessible **(see illustration)**.

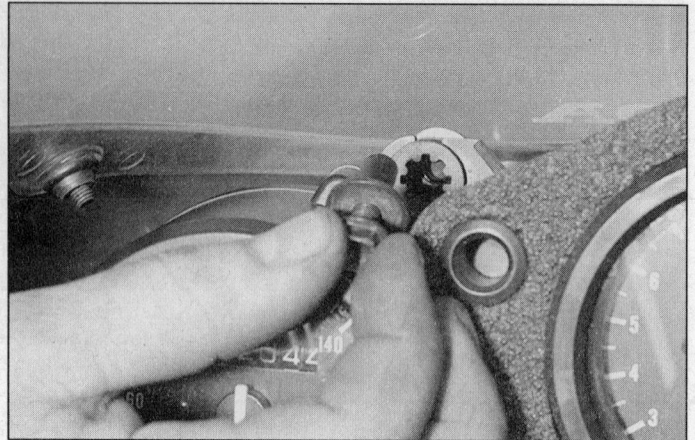

13.6 Unscrew the instrument cluster mounting bolts and remove the collars

8 Remove the instrument cluster from the bike. Inspect the cluster mounting dampers for signs of damage and replace if necessary.

Installation

US 1993 and 1994 models, UK CBR900RR-N and RR-P

9 Maneuver the instrument cluster into position, aligning the bosses on the cluster with the rubber grommets on the mounting bracket.

10 Fit the collars to the cluster. Install the mounting bolts and tighten them securely.

11 Connect the cluster wiring connectors.

12 Connect the speedometer cable and tighten the retaining ring securely.

US 1995 model, UK CBR900RR-R and RR-S

13 Maneuver the instrument cluster into position and connect the cluster wiring connectors.

14 Fit the collars to the cluster, then install the mounting bolts and tighten them securely

Speedometer cable - US 1993 and 1994 models, UK CBR900RR-N and RR-P only

Removal

15 Unscrew the speedometer cable retaining ring from the rear of the instrument cluster and detach the cable.

16 Unscrew the screw securing the lower end of the cable to the drive unit on the wheel and detach the cable.

17 Free the cable from its retaining guide on the left inner middle fairing panel and remove it from the bike, noting its correct routing.

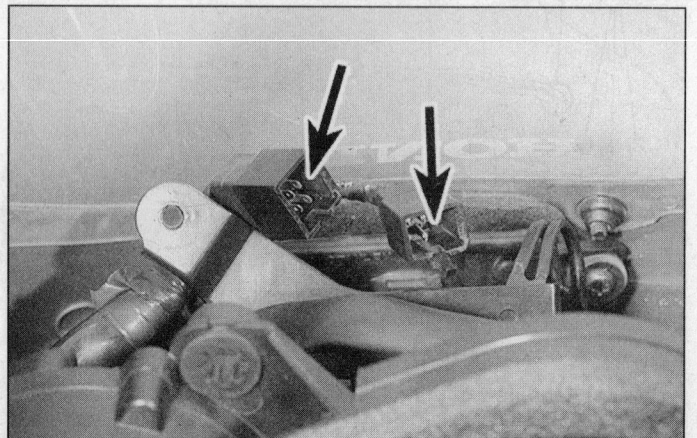

13.7 Maneuver the instrument cluster to gain access to the two wiring connectors and disconnect them

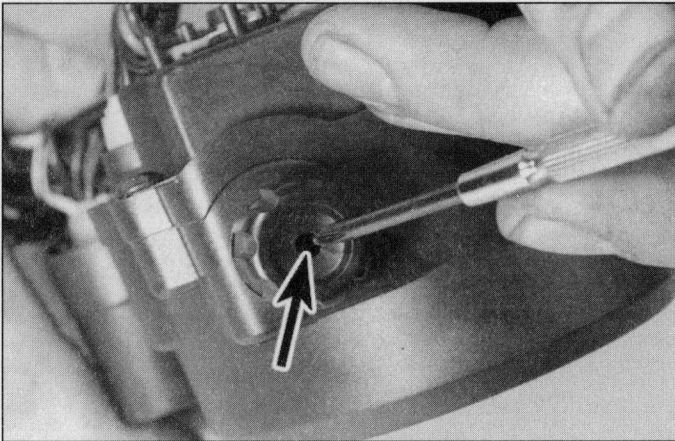

14.3 Unscrew the retaining screw from the center of the trip odometer knob (arrow)

Installation

18 Route the cable correctly and install it into the retaining guide on the left inner middle fairing panel.
19 Align the inner cable slot at the lower end of the cable with the drive gear dog on the front wheel and connect the cable to the drive. Install the cable retaining screw and tighten it securely.
20 Connect the cable upper end to the instrument cluster and tighten the retaining ring securely.
21 Check that the cable doesn't restrict steering movement or interfere with any other components.

14 Meters and gauges - check and replacement

Temperature gauge
Check
1 The temperature gauge check is described in Chapter 3.

Replacement
Refer to illustrations 14.3, 14.4a and 14.4b
2 Remove the instrument cluster from the bike (see Section 13).
3 Unscrew the retaining screw from the center of the trip odometer knob and remove the knob **(see illustration)**.

4 Unscrew the front cover retaining screws (early models have eight, later models have six) and remove the front cover. Note the position of the wiring clamp **(see illustrations)**.
5 Note the correct fitted position of the temperature gauge wires (the wiring color locations should be marked on the casing), then unscrew the wire retaining screws and detach the wires. The temperature gauge can now be removed from the casing.
6 Install the temperature gauge by reversing the removal sequence. Take care not to overtighten the retaining screws as the components are very fragile and can be easily damaged.

Tachometer
Check
7 Disconnect the large 9-pin wiring connector from the instrument cluster.
8 Remove the rider's seat as described in Chapter 8.
9 Disconnect the wiring connector from the ignition control module.
10 Using an ohmmeter or continuity tester, check for continuity between the yellow/green terminal on the instrument cluster 9-pin connector and the yellow/green terminal on the ignition control module connector. If no continuity is obtained, check the wiring between the two connectors (see the wiring diagrams at the end of this Chapter), and repair or replace the wiring as necessary. If continuity is obtained but the tachometer does not work, then it is faulty and must be replaced.

Replacement
11 Carry out the operations described above in Steps 2 through 4.
12 The tachometer on early models has one wire to it, whilst on later models there are two. Note the correct fitted position(s) of the tachometer wire(s) (the wiring color locations should be marked on the casing), then unscrew the wire retaining screws and detach the wires.
13 Unscrew the two tachometer retaining screws and lift the tachometer out of the casing.
14 Install the tachometer by reversing the removal sequence. Take care not to overtighten the retaining screws as the components are very fragile and can be easily damaged.

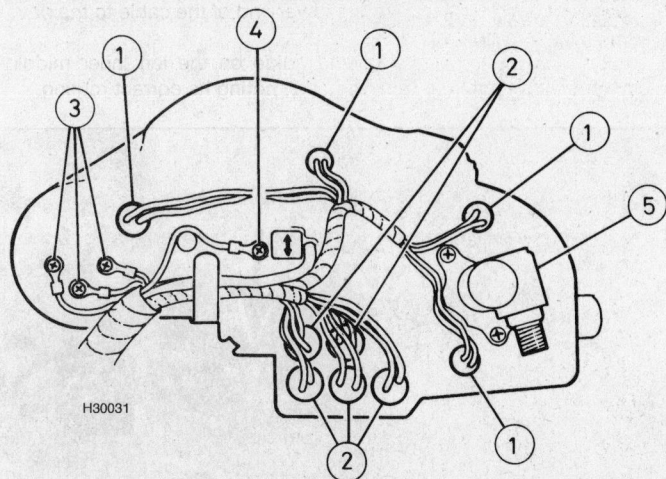

14.4a Rear view of instrument cluster - US 1993 and 1994 models, UK CBR900RR-N and RR-P

1 Instrument lights
2 Warning lights
3 Temperature gauge wires
4 Tachometer wire
5 Speedometer drive gearbox

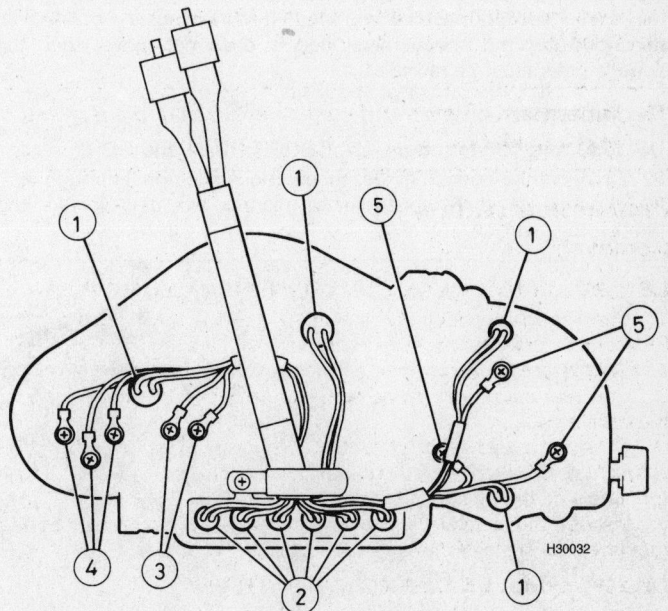

14.4b Rear view of instrument cluster - US 1995 model, UK CBR900RR-R and RR-S

1 Instrument lights
2 Warning lights
3 Tachometer wires
4 Temperature gauge wires
5 Speedometer wires

Speedometer
Check

US 1993 and 1994 models, UK CBR900RR-N and RR-P

15 Special instruments are required to properly check the operation of this meter. Take the machine to a Honda dealer or other qualified repair shop for diagnosis.

US 1995 model, UK CBR900RR-R and RR-S

16 If the system malfunctions, first check that the battery is fully charged and the terminals are clean, and that the wiring and fuses are in good condition.
17 Unscrew the two speed sensor bolts and remove the speed sensor from the front sprocket cover (Chapter 6, Section 17). Check the sensor for any signs of damage. If it does not rotate smoothly, it must be replaced. Otherwise replace the sensor and tighten the bolts.
18 To check the circuit, remove the fuel tank (Chapter 4), then trace the sensor wiring from the sensor to its connector on the right side of the frame. Using a voltmeter, connect the positive lead to the black/brown terminal on the connector, and the negative lead to the green/black terminal. With the ignition switched ON a reading of around 12 volts should be measured on the voltmeter. If no reading is obtained, check the wiring for continuity and that the connectors are free from corrosion and properly connected. If the wiring is faulty it must be repaired or replaced, and the connectors cleaned using electrical contact cleaner.
19 If the correct reading is obtained and the system is still faulty, remove the upper fairing (Chapter 8), and support the bike using an auxiliary stand or hoist so that the rear wheel is off the ground.
20 Using a voltmeter, connect the positive lead to the black/red terminal on the rear of the speedometer, and the negative lead to the green/black terminal. Turn the ignition switch ON, and with the bike in neutral, rotate the rear wheel as fast as possible by hand. If the sensor is operating correctly a fluctuating voltage reading between 0-5 volts should be measured on the voltmeter, the rate of fluctuation depending on the rate at which the rear wheel is turned.
21 If no reading is obtained, check the wiring between the speedometer terminals and the speed sensor for continuity and check that the connectors are free from corrosion and properly connected. If the wiring is faulty it must be repaired or replaced, and the connectors cleaned using electrical contact cleaner. If the wiring is not faulty, then the speed sensor must be replaced. If a reading is obtained as described but the speedometer needle does not move, then the speedometer must be replaced.

Replacement

US 1993 and 1994 models, UK CBR900RR-N and RR-P

22 Carry out the operations described above in Steps 2 through 4.
23 Unscrew the three speedometer gearbox mounting screws and remove the gearbox.

24 Unscrew the speedometer retaining screws and lift the speedometer out of the casing.
25 Install the speedometer by reversing the removal sequence. Take care not to overtighten the retaining screws as the components are very fragile and can be easily damaged.

US 1995 model, UK CBR900RR-R and RR-S

26 Carry out the operations described above in Steps 2 through 4.
27 Note the correct fitted position of the speedometer wires (the wiring color locations should be marked on the casing), then unscrew the wire retaining screws and detach the wires. The speedometer can now be removed from the casing.
28 Install the speedometer by reversing the removal sequence. Take care not to overtighten the retaining screws as the components are very fragile and can be easily damaged

---

15 Instrument and warning light bulbs - replacement

Refer to illustrations 15.2a and 15.2b
1 Remove the instrument cluster as described in Section 13.
2 Pull the relevant bulbholder out of the back of the cluster then pull the bulb out of its holder (see illustrations). If the socket contacts are dirty or corroded, they should be scraped clean and sprayed with electrical contact cleaner before a new bulb is installed.
3 Carefully push the new bulb into position, then push the bulbholder back into the rear of the cluster.
4 Install the instrument cluster as described in Section 13.

---

16 Oil pressure switch - check and replacement

Check

1 The oil pressure warning light should come on when the ignition (main) switch is turned ON and extinguish a few seconds after the engine is started. If the oil pressure light comes on whilst the engine is running, stop the engine immediately and carry out an oil pressure check as described in Chapter 2.
2 If the oil pressure warning light does not come on when the ignition is turned on, check the bulb (see Section 15) and fuses (see Section 5).
3 Remove the fairing lower section (or lower section right half only) as described in Chapter 8 to gain access to the oil pressure switch; the switch is screwed into the crankcase just below the end cover (see illustration 17.3 in Chapter 2). Peel back the rubber cover, then undo the retaining screw and detach the wiring connector from the switch. With the ignition switched ON, ground (earth) the wire on the crankcase and check that the warning light comes on. If the light comes on, the switch is defective and must be replaced.

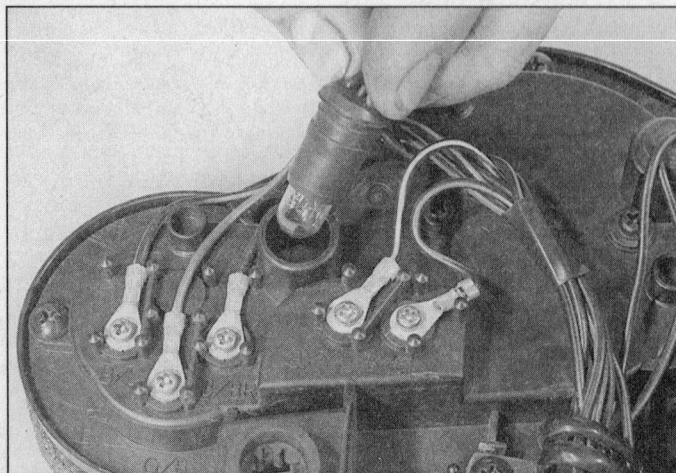

15.2a Pull the relevant bulbholder out of the casing . . .

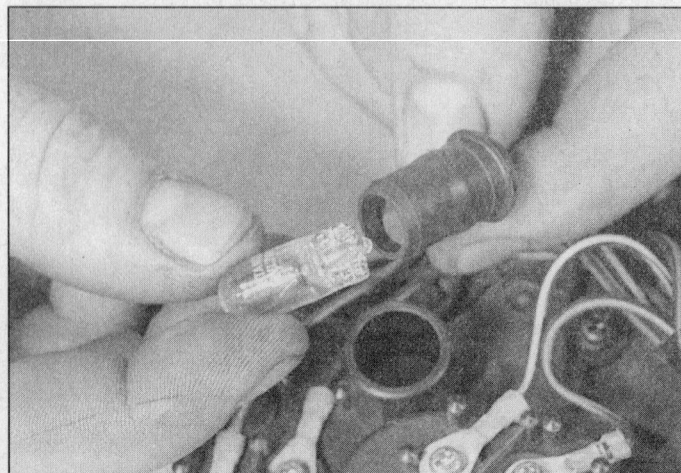

15.2b . . .and pull the bulb out of the bulbholder

4 If the light still does not come on, check for voltage at the wire terminal using a test light. If there is no voltage present, check the wire between the switch, the instrument cluster and fusebox for continuity (see the *wiring diagrams* at the end of this Chapter).

5 If the warning light comes on whilst the engine is running, yet the oil pressure is satisfactory, remove the wire from the oil pressure switch as described in Step 3. With the wire detached and the ignition switched ON the light should be out. If it is illuminated, the wire between the switch and instrument cluster must be grounded (earthed) at some point. If the wiring checks out OK, the switch must be assumed faulty and replaced.

Replacement

6 Detach the wire from the switch as described above in Step 3.
7 Unscrew the switch from the crankcase and swiftly plug the hole to prevent the escape of oil.
8 Make sure the switch threads are clean and dry and apply a thin coat of suitable sealant to the end 3 to 4 mm of its threads.
9 Remove the plug and screw the switch into the crankcase, tightening it securely.
10 Attach the wire, tightening its retaining screw securely, then seat the rubber cover over the switch.
11 Check the operation of the oil pressure warning light, then install the fairing section as described in Chapter 8.
12 Check the oil level (see Chapter 1) and top up if necessary.

17 Ignition (main) switch - check, removal and installation

Check

1 Remove the upper fairing right inner cover (see Chapter 8). Trace the ignition (main) switch wiring back from the base of the switch, and disconnect it at the connector.
2 Using an ohmmeter or a continuity test light, check the continuity of the terminal pairs (see the *wiring diagrams* at the end of this Chapter). Continuity should exist between the terminals connected by a solid line on the diagram when the switch is in the indicated position.
3 If the switch fails any of the tests, replace it.

Removal

Refer to illustration 17.5

4 Remove the upper fairing right inner cover (see Chapter 8). Trace the ignition (main) switch wiring back from the base of the switch, noting its routing and releasing it from any necessary retaining clips, and disconnect it at the connector.
5 Unscrew the two switch Torx bolts from the underside of the top triple clamp and remove the switch from the bike **(see illustration)**.

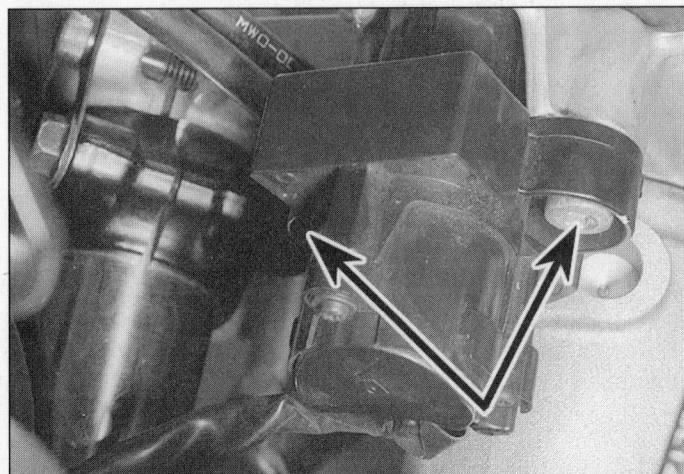

17.5 Unscrew the two Torx bolts (arrows) on the underside of the top triple clamp to release the ignition (main) switch

Installation

6 Maneuver the switch assembly into position making sure the wiring is correctly routed.
7 Thoroughly clean the switch bolts and apply a few drops of non-permanent locking compound to their threads. Install the bolts and tighten them to the specified torque setting.
8 Secure the switch wiring in position with any relevant clips or ties and reconnect it at the connector.

18 Handlebar switches - check

1 Generally speaking, the switches are reliable and trouble-free. Most troubles, when they do occur, are caused by dirty or corroded contacts, but wear and breakage of internal parts is a possibility that should not be overlooked. If breakage does occur, the entire switch and related wiring harness will have to be replaced with a new one, since individual parts are not usually available.
2 The switches can be checked for continuity using an ohmmeter or a test light. Always disconnect the battery negative cable, which will prevent the possibility of a short circuit, before making the checks.
3 Trace the wiring harness of the switch in question back to its connector(s) which are clipped to the fairing under the right inner cover. Unplug the relevant electrical connector(s).
4 Using the ohmmeter or test light, check for continuity between the terminals of the switch harness with the switch in the various positions (see the *wiring diagrams* at the end of this Chapter).
5 If the continuity check indicates a problem exists, remove the switch (Section 19) and spray the switch contacts with electrical contact cleaner. If they are accessible, the contacts can be scraped clean or polished with crocus cloth. If switch components are damaged or broken, it will be obvious when the switch is disassembled.

19 Handlebar switches - removal and installation

Left handlebar switch

Removal

Refer to illustration 19.3

1 Trace the wiring harness back from the switch to the wiring connectors and unplug the relevant electrical connector.
2 Work back along the harness, freeing it from all the relevant clips and ties, whilst noting its correct routing.
3 Unscrew the two switch retaining screws and remove the switch from the handlebar **(see illustration)**. Disconnect the two wires from the brake light switch.

19.3 Unscrew the two left handlebar switch retaining screws

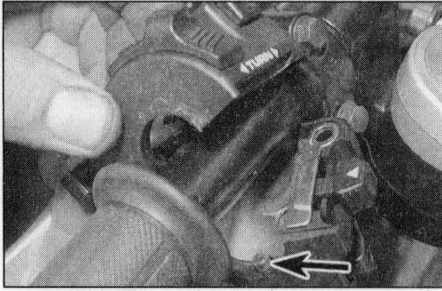

19.4 When installing the left switch, locate the peg on the switch (arrow) in the hole in the handlebar

19.6 Unscrew the two right handlebar switch retaining screws

19.7 When installing the right switch, locate the peg on the switch (arrow) in the hole in the handlebar

Installation

Refer to illustration 19.4

4   Installation is a reversal of the removal procedure, making sure that the locating peg on the lower half of the switch is correctly located in the hole in the handlebar **(see illustration)**.

Right handlebar switch

Removal

Refer to illustration 19.6

5   Disconnect the switch wiring connectors as described in Steps 1 and 2 above. Disconnect the two wires from the clutch switch.

6   Unscrew the two switch retaining screws and remove the switch from the handlebar **(see illustration)**.

Installation

Refer to illustration 19.7

7   Installation is a reversal of the removal procedure, making sure that the locating peg on the lower half of the switch is correctly located in the hole in the handlebar **(see illustration)**.

---

20   Neutral switch - check and replacement

Check

Refer to illustration 20.2

1   Before checking the electrical circuit, check the bulb (Section 15) and fuses (see Section 5).

2   Remove the fairing lower section (or lower section left half only) as described in Chapter 8. The switch is screwed into the left side of the crankcase, below the front sprocket cover **(see illustration)**.

3   Disconnect the wiring connector from the switch and shift the transmission into neutral.

4   With the wire detached and the ignition switched ON, the neutral light should be out. If not, the wire between the switch and instrument cluster must be grounded (earthed) at some point.

5   Ground (earth) the wire on the crankcase and check that the neutral light comes on. If the light does come on, the switch is defective.

6   If the light does not come on when the wire is grounded (earthed), check for voltage at the wire terminal using a test light. If there's no voltage present, check the wire between the switch, the instrument cluster and fusebox (see the *wiring diagrams* at the end of this Chapter).

Replacement

7   Remove the fairing lower section (or lower section left half only) as described in Chapter 8.

8   Disconnect the wiring connector from the switch **(see illustration 20.2)**.

9   Unscrew the switch and remove it from the crankcase. Recover the sealing washer and plug the switch opening to minimise oil loss whilst the switch is removed.

10   Clean the threads of the switch and fit a new sealing washer to it.

11   Remove the plug from the crankcase and install the switch. Tighten the switch to the torque setting specified at the beginning of the Chapter, then reconnect the wiring connector.

12   Check the operation of the neutral light.

13   Check the oil level (see Chapter 1) and top up if necessary.

---

21   Sidestand switch - check and replacement

Check

1   Before checking any electrical circuit, check the bulb (Section 15) and fuses (see Section 5).

2   To gain access to the sidestand switch wiring connector remove the rider's seat as described in Chapter 8. Trace the wiring from the switch to its connector and disconnect it.

3   Check the operation of the switch using an ohmmeter or continuity test light.

4   Set the meter to the ohms x 1 scale and connect the meter to the yellow/black and green wires on the switch side of the wiring connector. With the sidestand down (extended) there should be continuity between the terminals, and with the stand up there should be no continuity (infinite resistance).

5   Connect the meter to the green/white and green wires on the switch side of the connector. With the sidestand up there should be continuity between the terminals, and with the stand down there should be no continuity (infinite resistance).

6   If the switch does not perform as expected, it is defective and must be replaced.

7   If the switch is good, check the wiring between the wiring connector, instrument cluster and fusebox (see the *wiring diagrams* at the end of this book).

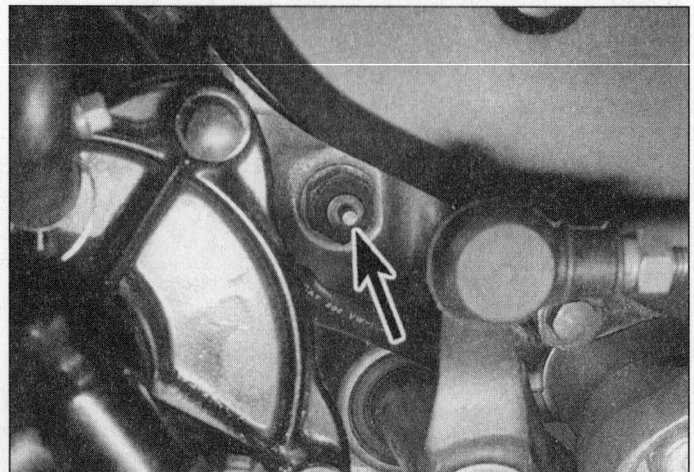

20.2 The neutral switch is located below the front sprocket cover on the left side of the engine (arrow)

Replacement

Refer to illustration 21.11

8   Disconnect the switch wiring connector as described in Step 2.

9   Work back along the switch wiring, freeing it from any relevant retaining clips and ties, noting its correct routing. Access to the wiring can be improved by removing the fairing lower section (or lower section left half only) as described in Chapter 8.

10   Unscrew the bolt securing the switch to the sidestand mounting bracket and remove the switch from the bike.

11   Fit the new switch to the rear of the sidestand bracket making sure the switch lug engages the hole in the sidestand, and the switch recess engages with the retaining pin **(see illustration)**. Install the switch retaining bolt and tighten it to the specified torque setting.

12   Make sure the wiring is correctly routed up to the connector and retained by all the necessary clips and ties.

13   Securely reconnect the wiring connector and check the operation of the switch and warning light as described in Chapter 1.

14   Install the rider's seat and, if removed, the fairing section as described in Chapter 8.

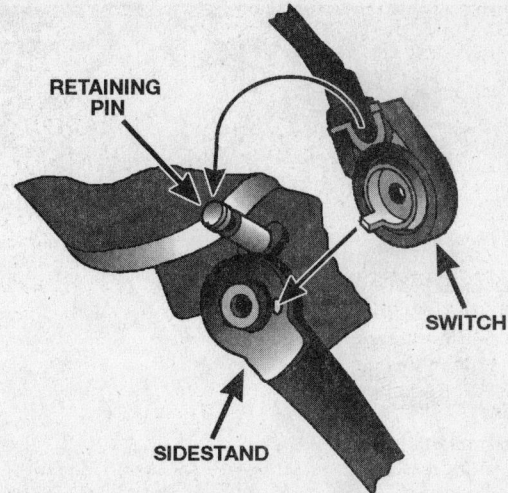

21.11 Make sure the sidestand switch lug engages the stand hole and the switch recess engages the retaining pin

22 Clutch diode - check and replacement

Refer to illustration 22.2

1   The clutch diode is plugged into the main wiring harness clipped to the right side of the frame. The diode is part of the starter safety circuit **(see illustration 1.1)** which prevents the starter motor operating whilst the transmission is in gear unless the clutch lever is pulled in. If the starter circuit is faulty, first check the fuses (see Section 5).

2   To gain access to the diode remove the fuel tank as described in Chapter 4. Remove the tape to free the diode from the wire harness, then unplug the diode **(see illustration)**.

3   Using an ohmmeter or a continuity test light, check for continuity between the terminals of the diode. Transpose the meter probes and check for continuity in the opposite direction. If the diode is serviceable there should be continuity in one direction (indicated by the arrow on the diode) and no continuity (infinite resistance) in the other. If not, the diode must be replaced.

4   If the diode is good, check the other components in the starter circuit (clutch switch, neutral switch and starter relay) as described in the relevant sections of this Chapter. If all components are good, check the wiring between the various components (see the *wiring diagrams* at the end of this book).

5   Plug the diode back into the harness and securely tape over it and around the surrounding harness wires. Install the fuel tank as described in Chapter 4.

23 Clutch switch - check and replacement

Check

Refer to illustration 23.2

1   The clutch switch is situated in the clutch lever mounting bracket. The switch is part of the starter safety circuit (see Section 1) which prevents the starter motor operating whilst the transmission is in gear unless the clutch lever is pulled in. If the starter circuit is faulty, first check the fuses (see Section 5).

2   To check the switch, disconnect the wiring connectors **(see illustration)**. Using an ohmmeter or a continuity test light, check for continuity between the terminals of the switch with the lever pulled into the handlebar, and no continuity (infinite resistance) with the lever released. If this is not so, the switch is faulty and must be replaced.

3   If the switch is good, check the other components in the starter circuit (clutch diode, neutral switch and starter relay) as described in the relevant sections of this Chapter. If all components are good, check the wiring between the various components (see the *wiring diagrams* at the end of this book).

Replacement

4   Disconnect the wiring connectors from the clutch lever switch.

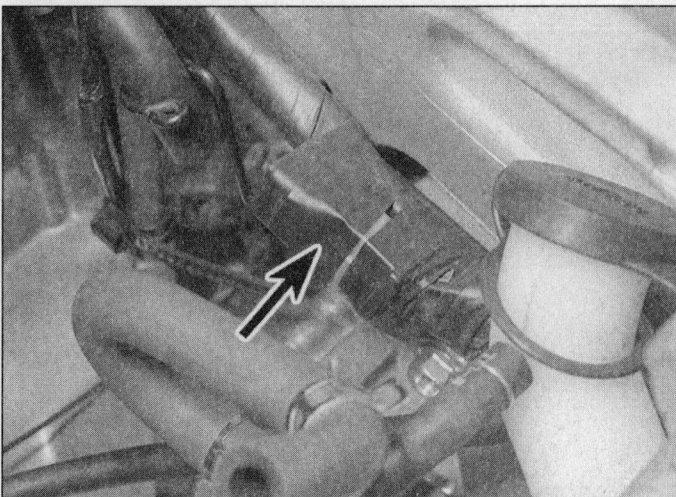

22.2 Clutch diode (arrow) is taped in wire harness

23.2 The clutch switch (arrow) is mounted on the clutch lever bracket

9

24.1 The horn is mounted to the underside of the lower triple clamp (arrow)

5   Slacken the upper adjuster to obtain maximum freeplay in the clutch cable then slacken and remove the clutch lever nut and pivot bolt.

6   Free the lever from the mounting bracket and remove the switch.

7   Install the new switch and fit the lever. Securely tighten the pivot bolt and nut then connect the switch wiring connectors.

8   Adjust the clutch cable as described in Chapter 1.

24 Horn - check and replacement

Check

Refer to illustration 24.1

1   The horn is mounted on the underside of the lower triple clamp **(see illustration)**.

2   Unplug the wiring connectors from the horn. Using two jumper wires, apply battery voltage directly to the terminals on the horn. If the horn sounds, check the switch (see Section 18) and the wiring between the switch and the horn (see the *wiring diagrams* at the end of this Chapter).

3   If the horn doesn't sound, replace it.

Replacement

4   Unplug the wiring connectors from the horn then unscrew the bolt

securing the horn to its mounting bracket and remove it from the bike.

5   Connect the wiring connectors to the new horn and securely tighten its retaining bolt.

25 Fuel pump - check, removal and installation

Warning: *Gasoline (petrol) is extremely flammable, so take extra precautions when you work on any part of the fuel system. Don't smoke or allow open flames or bare light bulbs near the work area, and don't work in a garage where a natural gas-type appliance (such as a water heater or clothes dryer) is present. If you spill any fuel on your skin, rinse it off immediately with soap and water. When you perform any kind of work on the fuel system, wear safety glasses and have a fire extinguisher suitable for a class B type fire (flammable liquids) on hand.*

Check

Refer to illustrations 25.1a, 25.1b and 25.13

1   The fuel pump is located under the fuel tank and to the left of the coolant resrvoir behind the engine. The fuel cut-off relay is located under the rider's seat behind the battery **(see illustrations)**.

2   The fuel pump is controlled through the fuel cut-off relay so that it runs whenever the ignition is switched ON and the ignition is operative (ie, only when the engine is turning over). As soon as the ignition is killed, the relay will cut off the fuel pump's electrical supply (so that there is no risk of fuel being sprayed out under pressure in the event of an accident).

3   It should be possible to hear or feel the fuel pump running whenever the engine is turning over - either place your ear close beside the pump or feel it with your fingertips. If you can't hear or feel anything, check the circuit fuse (see Section 5). If the fuse is good, check the pump and relay for loose or corroded connections or physical damage and rectify as necessary.

4   If the circuit is fine so far, switch the ignition OFF, unplug the relay's wiring connector and connect across the relay's black and black/blue terminals with a short length of insulated jumper wire. Switch the ignition ON; the pump should operate.

5   If the pump now works, either the relay or its wiring is at fault. Test the wiring as follows.

6   Check for full battery voltage at the relay's black terminal with the ignition switch ON. If there is no battery voltage, there is a fault in the circuit between the relay and the fuse - turn off the ignition, then trace and rectify the fault as outlined in Section 2; refer to the wiring diagrams at the end of this Chapter.

7   Disconnect the wire connectors from the relay, fuel pump and ignition control module. Using an ohmmeter, check for continuity between the yellow/blue wire on the relay's wire connector and the

25.1a The fuel pump is located under the fuel tank

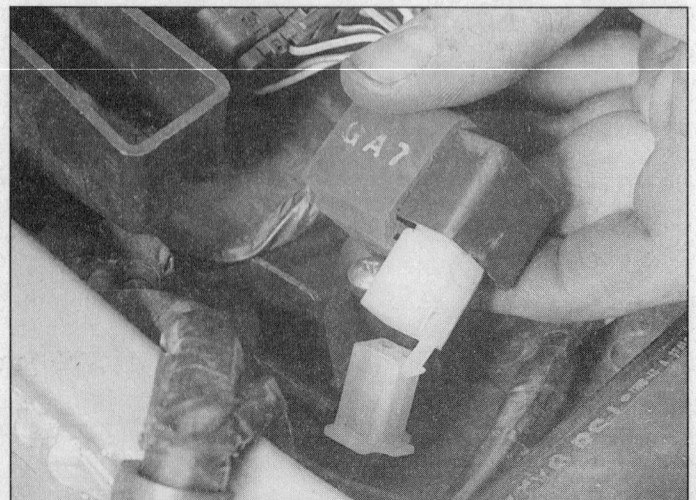

25.1b The fuel cut-off relay is located under the rider's seat

yellow/blue wire on the control module's wire connector. Similarly, check for continuity between the blue/yellow wire on the relay's wire connector and the blue/yellow wire on the control module's wire connector. Check for continuity between the black/blue wire on the relay's wire connector and the black/blue wire on the fuel pump's wire connector. Continuity should be indicated in all tests; if not, trace and rectify the fault as described in Section 2. Reconnect all wire connectors.

8   If the pump still does not work, trace the wiring from the pump and disconnect it at the black 2-pin wiring connector. Using a fully-charged 12 volt battery and two insulated jumper wires, connect the positive (+) terminal of the battery to the pump's black/blue terminal, and the negative (-) terminal of the battery to the pump's green terminal. The pump should operate. If the pump does not operate it must be replaced.

9   If the pump works and all the relevant wiring and connectors are good, then the relay is at fault. The only definitive test of the relay is to substitute one that is known to be good. If substitution does not cure the problem, bear in mind that the ignition control module could be faulty.

10   If the pump operates but is thought to be delivering an insufficient amount of fuel, first check that the fuel tank breather hose is unobstructed (except California models), that all fuel hoses are in good condition and not pinched or trapped. Check that the fuel filters in the fuel tank and fuel delivery hose are not blocked.

11   The fuel pump's output can be checked as follows: make sure the ignition switch is OFF. Remove the rider' s seat (see Chapter 8) and the fuel tank as described in Chapter 4.

12   Disconnect the fuel hose from the carburetor T-piece **(see illustration 6.4a in Chapter 4)**.

13   Trace the wiring from the fuel pump and disconnect it at the fuel cut-off relay's 4-pin connector **(see illustration 25.1b)**. Using a short length of insulated jumper wire, connect across the black and the black/blue wire terminals of the connector **(see illustration)**.

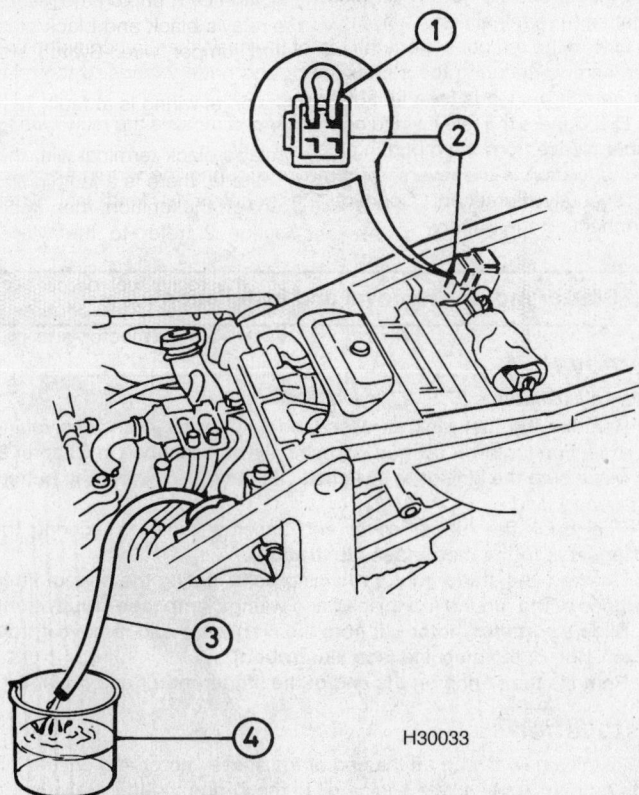

25.13 Checking the fuel pump flow rate

1 Jumper wire	3 Fuel outlet pipe
2 Relay 4-pin connector	4 Graduated beaker

14   Using a length of fuel hose from the fuel tank, or by using a dummy fuel tank, connect a fuel supply to the fuel filter. Place the fuel outlet hose end into a graduated beaker.

15   Turn the ignition switch ON and let fuel flow from the pump into the beaker for 5 seconds, then switch the ignition OFF.

16   Measure the amount of fuel that has flowed into the beaker, then multiply that amount by 12 to determine the fuel pump flow rate per minute. The minimum flow rate is 900 cc (30.4 US fl oz, 31.7 Imp fl oz) per minute. If the flow rate recorded is below the minimum required, then the fuel pump must be replaced.

Removal

Refer to illustration 25.19

17   Make sure both the ignition and the fuel tap are switched OFF. Remove the rider's seat as described in Chapter 8 and the fuel tank as described in Chapter 4.

18   Trace the wiring from the fuel pump and disconnect it at the black 2-pin connector.

19   Using a rag to mop up any spilled fuel, disconnect the two fuel hoses from the fuel pump, noting which pipe fits on which nozzle, then remove the pump with its rubber mounting sleeve from the mounting bracket, taking care not to snag the wiring **(see illustration)**.

20   To remove the fuel cut-off relay, disconnect its wiring connector and remove it from its mounting lug **(see illustration 25.1b)**.

Installation

21   Installation is a reverse of the removal procedure. Make sure the fuel hoses are correctly and securely fitted to the pump. Start the engine and look carefully for any signs of leaks at the pipe connections.

---

26   Starter relay - check and replacement

Check

1   If the starter circuit is faulty, first check the fuses (see Section 5). The starter relay is located behind the right side cover (see Chapter 8, if necessary).

2   With the ignition switch ON, the engine kill switch in RUN and the transmission in neutral, press the starter switch. The relay should click.

3   If the relay doesn't click, switch off the ignition and remove the relay as described below; test it as follows.

4   Set a multimeter to the ohms x 1 scale and connect it across the relay's starter motor and battery lead terminals. Using a fully-charged 12 volt battery and two insulated jumper wires, connect the positive (+) terminal of the battery to the yellow/red terminal of the relay, and the negative (-) terminal to the green/red terminal of the relay. At

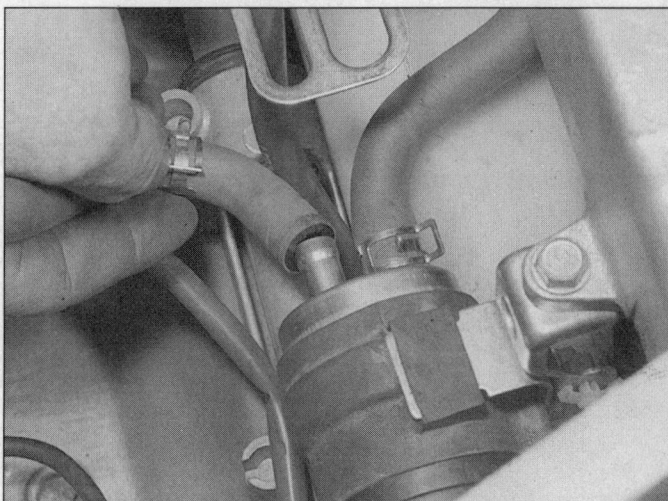

25.19 Disconnect the two fuel hoses from the pump, noting which pipe fits on which nozzle

26.7 Unscrew the two Allen screws to free the starter motor and battery leads from the starter relay

27.3 Peel back the rubber cover, then unscrew the nut and disconnect the starter cable from the motor

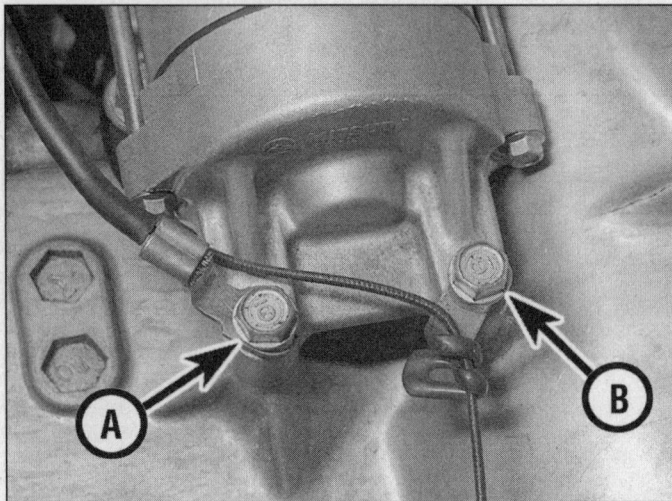

27.4 Unscrew the starter motor bolts, noting the fitted positions of the ground (earth) lead (A) and the wiring clip (B) . . .

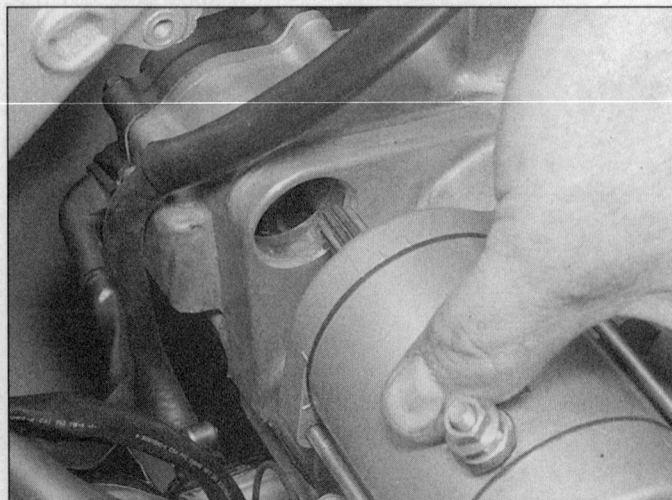

27.5 . . . and maneuver the starter motor out from the left side of the frame

this point the relay should click and the multimeter read 0 ohms (continuity). If this is the case the relay is serviceable and the fault lies in the starter switch circuit (check the clutch diode, clutch switch and neutral switch as described elsewhere in this Chapter); if the relay does not click when battery voltage is applied and indicates no continuity across its terminals, it is faulty and must be replaced.

Replacement

Refer to illustration 26.7

5   Remove the seat and the right side cover (see Chapter 8).
6   Disconnect the battery terminals, remembering to disconnect the negative (-) terminal first.
7   Unclip the red cover from the top of the relay, then unscrew the two Allen screws securing the starter motor and battery leads to the relay and detach the leads **(see illustration)**.
8   Disconnect the relay wiring connector and remove the relay with its rubber sleeve from the mounting lugs.
9   Installation is the reverse of removal ensuring the terminal screws are securely tightened. Connect the negative (-) lead last when reconnecting the battery.

27 Starter motor - removal and installation

Removal

Refer to illustrations 27.3, 27.4 and 27.5

1   Remove the fuel tank as described in Chapter 4 and the fairing lower section (or lower section left half only) as described in Chapter 8.
2   Make sure the ignition is switched OFF and disconnect the battery negative (-) lead.
3   Peel back the rubber cover and unscrew the nut securing the starter cable to the motor **(see illustration)**.
4   Unscrew the starter motor retaining bolts, noting the correct fitted positions of the ground (earth) lead and wiring clamp **(see illustration)**.
5   Slide the starter motor out from the crankcase and remove it from the left side of the machine **(see illustration)**.
6   Remove the O-ring on the end of the starter motor and replace it.

Installation

7   Install a new O-ring on the end of the starter motor and ensure it is in its groove; apply a little engine oil to the O-ring to aid installation.
8   Maneuver the motor into position and slide it into the crankcase.
9   Fit the retaining bolts, making sure the wiring clip is correctly fitted underneath the front bolt and the ground (earth) lead underneath the rear bolt, and tighten them securely.

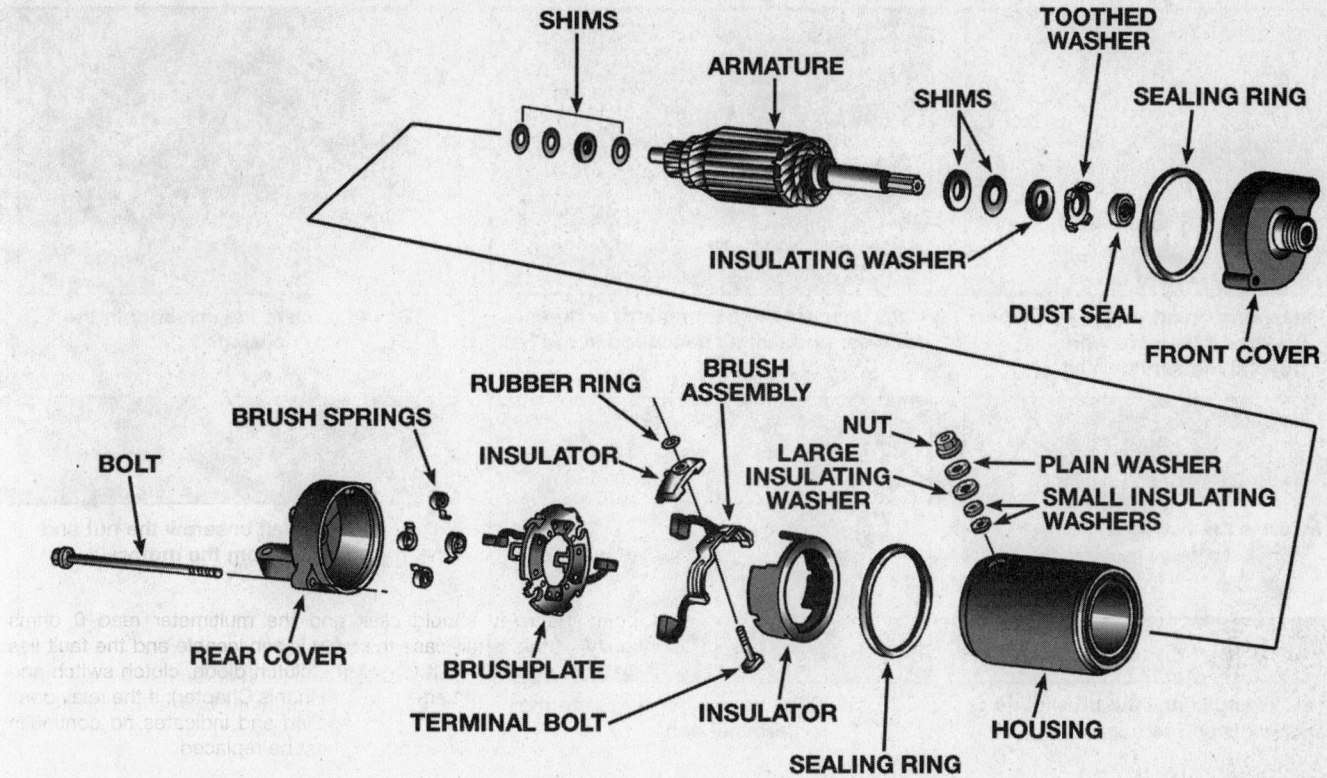

SHIMS ARMATURE SHIMS TOOTHED WASHER SEALING RING

INSULATING WASHER DUST SEAL FRONT COVER

BRUSH SPRINGS RUBBER RING BRUSH ASSEMBLY NUT PLAIN WASHER

BOLT INSULATOR LARGE INSULATING WASHER SMALL INSULATING WASHERS

REAR COVER BRUSHPLATE TERMINAL BOLT INSULATOR SEALING RING HOUSING

28.2a Exploded view of the starter motor

10 Connect the cable and securely tighten its retaining nut. Make sure the rubber cover is correctly seated over the terminal.
11 Connect the battery negative lead.
12 Install the fuel tank (see Chapter 4) and fairing section (Chapter 8).

---

28 Starter motor - disassembly, inspection and reassembly

Disassembly

Refer to illustrations 28.2a, 28.2b, 28.6 and 28.7
1 Remove the starter motor as described in Section 27.
2 Make alignment marks between the housing and end covers **(see illustrations)**.
3 Unscrew the two long bolts then remove the rear cover from the motor along with its sealing ring. Discard the sealing ring as a new one must be used for reassembly. Remove the shim(s) from the rear end of the armature noting their correct fitted positions.
4 Remove the front cover from the motor, together with its sealing ring. Discard the sealing ring as a new one must be used for

reassembly. Recover the toothed washer from the cover and slide off the insulating washer and shim(s) from the front end of the armature, noting their correct fitted locations.
5 Withdraw the armature from the housing.
6 Noting the correct fitted location of each washer, unscrew the nut from the terminal bolt and remove the plain washer, the various insulating washers and the rubber ring **(see illustration)**. Withdraw the terminal bolt and brushplate assembly from the housing and recover the insulator.
7 Lift the brush springs and slide the brushes out from their holders **(see illustration)**.

Inspection

Refer to illustrations 28.8 and 28.9
8 The parts of the starter motor that are most likely to require attention are the brushes. Measure the length of the brushes and compare the results to the brush length listed in this Chapter's Specifications **(see illustration)**. If any of the brushes are worn beyond the service limit, replace the brushplate assembly with a new one. If the brushes are not worn excessively, nor cracked, chipped, or otherwise damaged, they may be re-used.

28.2b Make alignment marks between the housing and the end covers before disassembly

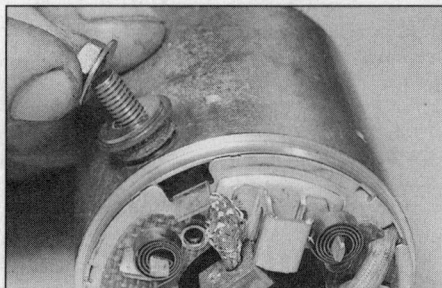

28.6 Unscrew the nut and remove the washers from the terminal bolt noting their correct fitted order

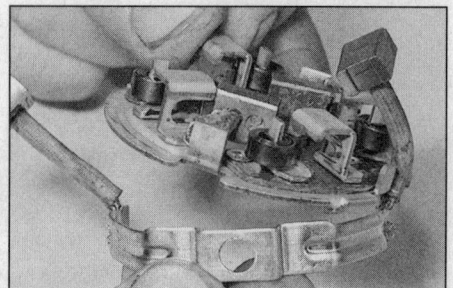

28.7 Lift the brush springs and slide the brushes out from their holders

9

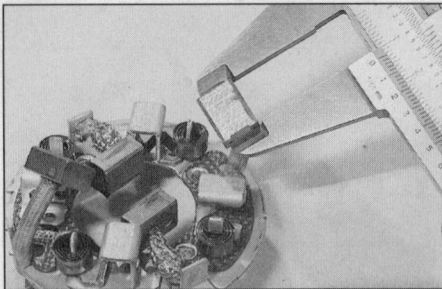

28.8 Measuring brush length - replace brushes if they are worn beyond the service limit

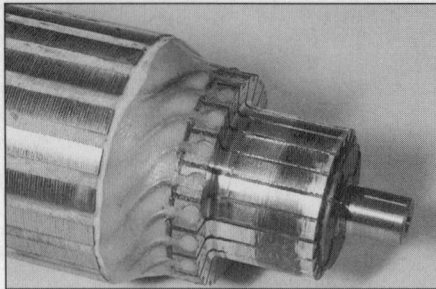

28.9 Inspect the commutator segments for wear and test as described in the text

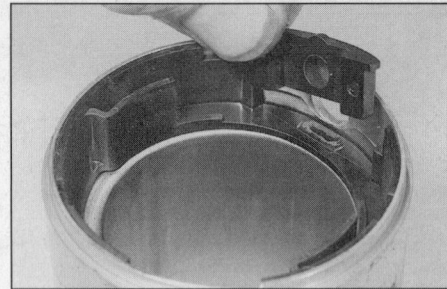

28.14a Locate the insulator in the housing . . .

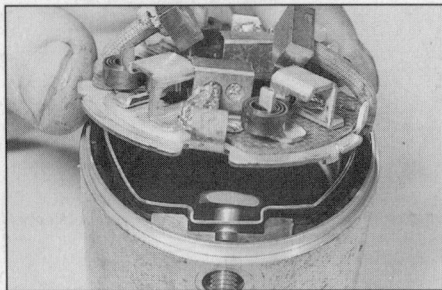

28.14b . . . and install the brushplate assembly and terminal bolt

28.15a Slide the rubber ring onto the terminal bolt . . .

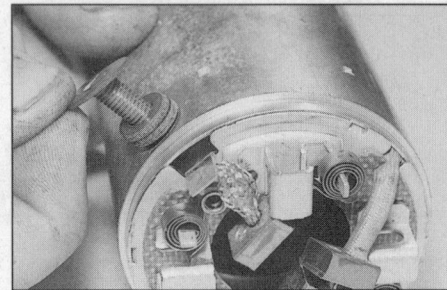

28.15b . . . followed by the insulating washers . . .

9   Inspect the commutator for scoring, scratches and discoloration **(see illustration)**. The commutator can be cleaned and polished with crocus cloth, but do not use sandpaper or emery paper. After cleaning, wipe away any residue with a cloth soaked in electrical system cleaner or denatured alcohol.

10   Using an ohmmeter or a continuity test light, check for continuity between the commutator bars. Continuity should exist between each bar and all of the others. Also, check for continuity between the commutator bars and the armature shaft. There should be no continuity between the commutator and the shaft. If the checks indicate otherwise, the armature is defective.

11   Check the starter pinion gear for worn, cracked, chipped and broken teeth. If the gear is damaged or worn, replace the starter motor.

12   Inspect the insulating washers and front cover dust seal for signs of damage and replace if necessary.

Reassembly

Refer to illustrations 28.14a, 28.14b, 28.15a, 28.15b, 28.15c, 28.16, 28.17a, 28.17b, 28.18, 28.19a, 28.19b, 28.20, 28.21, 28.22 and 28.23

13   Lift the brush springs and slide all the brushes back into position in their holders.

14   Fit the insulator to the housing and install the brushplate. Insert the terminal bolt through the brushplate and housing **(see illustrations)**.

15   Slide the rubber ring and small insulating washer(s) onto the bolt, followed by the large insulating washer(s) and the plain washer. Fit the nut to the terminal bolt and tighten it securely **(see illustrations)**.

16   Locate the brushplate assembly in the housing making sure its tab is correctly located in the housing slot **(see illustration)**.

17   Insert the armature in the front of the housing and locate the brushes on the commutator bars. Check that each brush is securely pressed against the commutator by its spring and is free to move easily in its holder **(see illustrations)**.

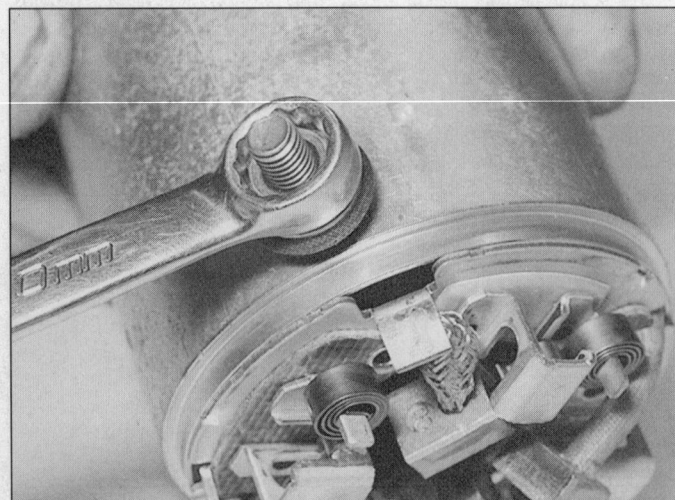

28.15c . . . and plain washer, fit the nut and tighten it securely

28.16 Make sure brushplate tab (arrow) is correctly located in the housing slot

28.17a Insert the armature from the front of the housing . . .

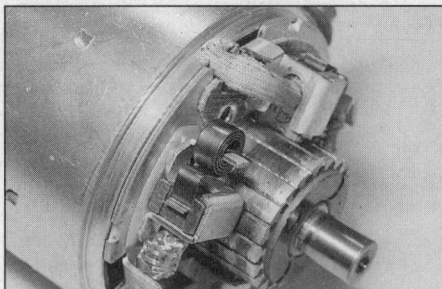

28.17b . . . and locate the brushes on the commutator

28.18 Fit the toothed washer to the front cover so that its teeth engage with the cover ribs

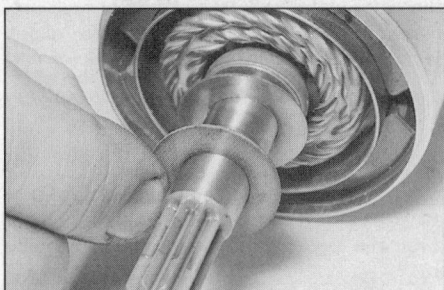

28.19a Install the shims and washers on the front of the armature making sure they are fitted in the correct order

28.19b Fit the sealing ring (arrow) and install the front cover

28.20 Fit the rear shims to the armature in the order noted on removal

18 Fit the toothed washer to the front cover so that its teeth are correctly located with the cover ribs **(see illustration)**. Apply a smear of grease to the cover dust seal lip.

19 Slide the shim(s) onto the front end of the armature shaft then fit the insulating washer. Fit a new sealing ring to the housing and carefully slide the front cover into position, aligning the marks made on removal **(see illustrations)**.

20 Fit the shims to the rear of the armature shaft **(see illustration)**.

21 Ensure the brushplate inner tab is correctly located in the housing slot and fit a new sealing ring to the housing **(see illustration)**.

22 Align the rear cover groove with the brushplate outer tab and install the cover **(see illustration)**.

23 Check the marks made on removal are correctly aligned then fit the long bolts and tighten them securely **(see illustration)**.

24 Install the starter motor as described in Section 27.

29 Charging system testing - general information and precautions

1 If the charging system is suspect, the system as a whole should be checked first, followed by testing of the individual components (the alternator stator coils and the voltage regulator/rectifier). **Note:** *Before*

beginning the checks, make sure the battery is fully charged and that all system connections are clean and tight.

2 Checking the output of the charging system and the performance of the various components within the charging system requires the use of a multimeter (with voltage, current and resistance checking facilities).

3 When making the checks, follow the procedures carefully to prevent incorrect connections or short circuits, as irreparable damage to electrical system components may result if short circuits occur.

4 If a multimeter is not available, the job of checking the charging system should be left to a Honda dealer.

30 Charging system - leakage and output test

1 If the charging system of the machine is thought to be faulty, remove the seat (see Chapter 8) and unhook the battery retaining strap. Perform the following checks.

Leakage test

Refer to illustration 30.3

2 Turn the ignition switch OFF and disconnect the lead from the battery negative (-) terminal.

28.21 Make sure that the brushplate tab is correctly located in the housing (arrow) . . .

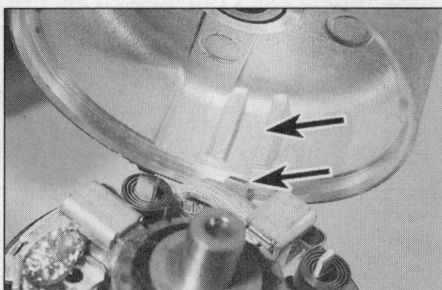

28.22 . . . and install the rear cover, aligning its groove with the brushplate outer tab (arrows)

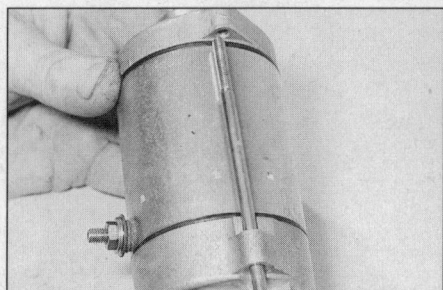

28.23 Align the marks made on removal and install the starter motor bolts

9

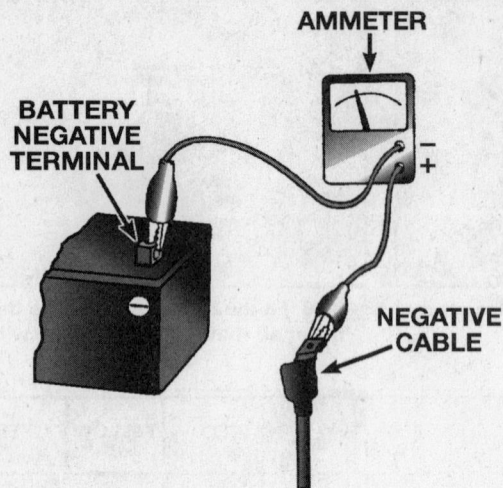

30.3 Checking the charging system leakage rate, connect the meter as shown

3   Set the multimeter to the mA (milli Amps) function and connect its negative (-) probe to the battery negative (-) terminal, and positive (+) probe to the disconnected negative (-) lead **(see illustration)**. With the meter connected like this the reading should not exceed 0.1 mA.
4   If the reading exceeds the specified amount it is likely that there is a short circuit in the wiring. Thoroughly check the wiring between the various components (see the *wiring diagrams* at the end of this book).
5   If the reading is below the specified amount, the leakage rate is satisfactory. Disconnect the meter and connect the negative (-) lead to the battery, tightening it securely, Check the alternator output as described below.

Output test

6   Start the engine and warm it up to normal operating temperature.
7   Allow the engine to idle and connect a multimeter set to the 0-20 volts scale (voltmeter) across the terminals of the battery (positive (+) lead to battery positive (+) terminal, negative (-) lead to battery negative (-) terminal). Slowly increase the engine speed to 5000 rpm and note the reading obtained. At this speed the voltage should be 13.0 to 15.5 volts. If the voltage is below this it will be necessary to check the alternator and regulator as described in the following Sections. **Note:** *Occasionally the condition may arise where the charging voltage is excessive. This condition is almost certainly due to a faulty regulator/rectifier which should be tested as described in Section 32.*

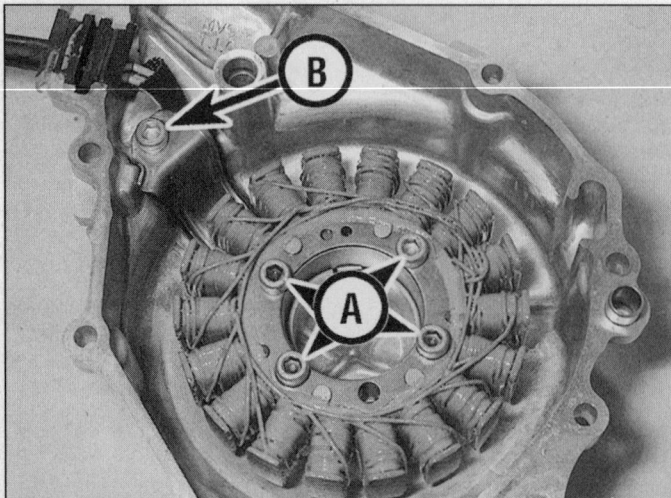

31.4 Alternator stator coil retaining bolts (A) and wiring clip bolt (B)

31 Alternator stator coils - check and replacement

Check

1   Remove the rider's seat as described in Chapter 8 to gain access to the wiring connectors located in the rubber boot.
2   Disconnect the 3-pin block connector containing the yellow wires. Using a multimeter set to the ohms x 1 (ohmmeter) scale measure the resistance between each of the yellow wires on the alternator side of the connector, taking a total of three readings, then check for continuity between each terminal and ground (earth). If the stator coil windings are in good condition there should be no continuity (infinite resistance) between any of the terminals and ground (earth) and the three readings should be within the range shown in the Specifications at the start of this Chapter. If not, the alternator stator coil assembly is at fault and should be replaced. **Note:** *Before condemning the stator coils, check the fault is not due to damaged wiring between the connector and coils.*

Replacement

Refer to illustration 31.4
3   Remove the left crankcase end cover as described in Section 21 of Chapter 2. Discard the gasket as a new one must be used.
4   Unscrew the bolt and remove the wiring retaining clip from inside the cover **(see illustration)**.
5   Undo the four bolts and separate the stator coil assembly from the cover.
6   Remove all trace of sealant from the wiring grommet and apply a smear of fresh sealant to the grommet.
7   Clean the crankcase and bolt threads and apply a few drops of non-permanent locking compound to the stator retaining bolt threads. Fit the stator coil assembly to the cover and tighten the retaining bolts to the torque setting specified at the beginning of the Chapter.
8   Ensure the grommet is correctly seated in the casing then install the retaining clip and securely tighten its retaining screw.
9   Using a new gasket fit the cover to the engine as described in Section 21 of Chapter 2.

32 Regulator/rectifier unit - check and replacement

Check

Refer to illustrations 32.1 and 32.6
1   Remove the right side cover (see Chapter 8) and disconnect the wiring connector from the regulator/rectifier unit **(see illustration)**.

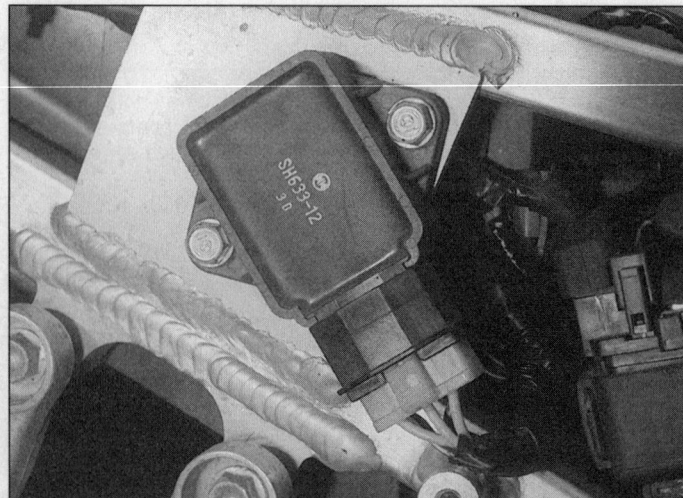

32.1 The regulator/rectifier unit is secured to the frame behind the right side cover by two bolts

2   Connect the negative (-) probe of the multimeter to a suitable ground (earth) point then switch the ignition switch ON and carry out the following checks.

3   Set the multimeter to the 0-20 dc volts setting then connect the meter positive (+) probe to the red/white terminal of the wiring connector and check for voltage. Full battery voltage should be present. Switch the ignition switch OFF.

4   Switch the multimeter to the resistance (ohms) scale. Check for continuity between the green terminal of the wiring connector and ground (earth) then check the resistance between the two yellow terminals of the wiring connector. There should be continuity between the green terminal and ground (earth) and a resistance reading of 0.1 to 0.3 ohms should be obtained between the two yellow terminals of the wiring connector.

5   If the above checks do not provide the expected results check the wiring between the battery, regulator/rectifier and alternator (see the *wiring diagrams* at the end of this book).

6   If the wiring checks out, the regulator/rectifier unit is probably faulty. Using a multimeter set to the appropriate resistance scale check the resistance between the various terminals of the regulator/rectifier **(see illustration)**. If the readings do not compare closely with those shown in the accompanying table, the regulator/rectifier unit can be considered faulty. **Note:** *The use of certain multimeters could lead to false readings being obtained. Therefore, if the above check shows the regulator/rectifier unit to be faulty take the unit to a Honda dealer for confirmation of its condition before replacing it.*

Replacement

7   Remove the right side cover (see Chapter 8).

8   Disconnect the wiring connector from the rectifier/regulator unit then unscrew the two bolts and remove the unit **(see illustration 32.1)**.

9   On installation, tighten the retaining bolts securely and connect the wiring connector. Install the side cover (see Chapter 8).

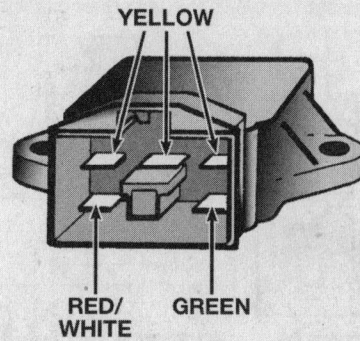

UNIT: KΩ

− \ +	RED/ WHITE	YELLOW 1	YELLOW 2	YELLOW 3	GREEN
RED/WHITE		∞	∞	∞	∞
YELLOW 1	0.5-10		∞	∞	∞
YELLOW 2	0.5-10	∞		∞	∞
YELLOW 3	0.5-10	∞	∞		∞
GREEN	1-20	0.5-10	0.5-10	0.5-10	

32.6 Regulator/rectifier unit terminal identification and resistance readings

Wiring diagrams overleaf

9

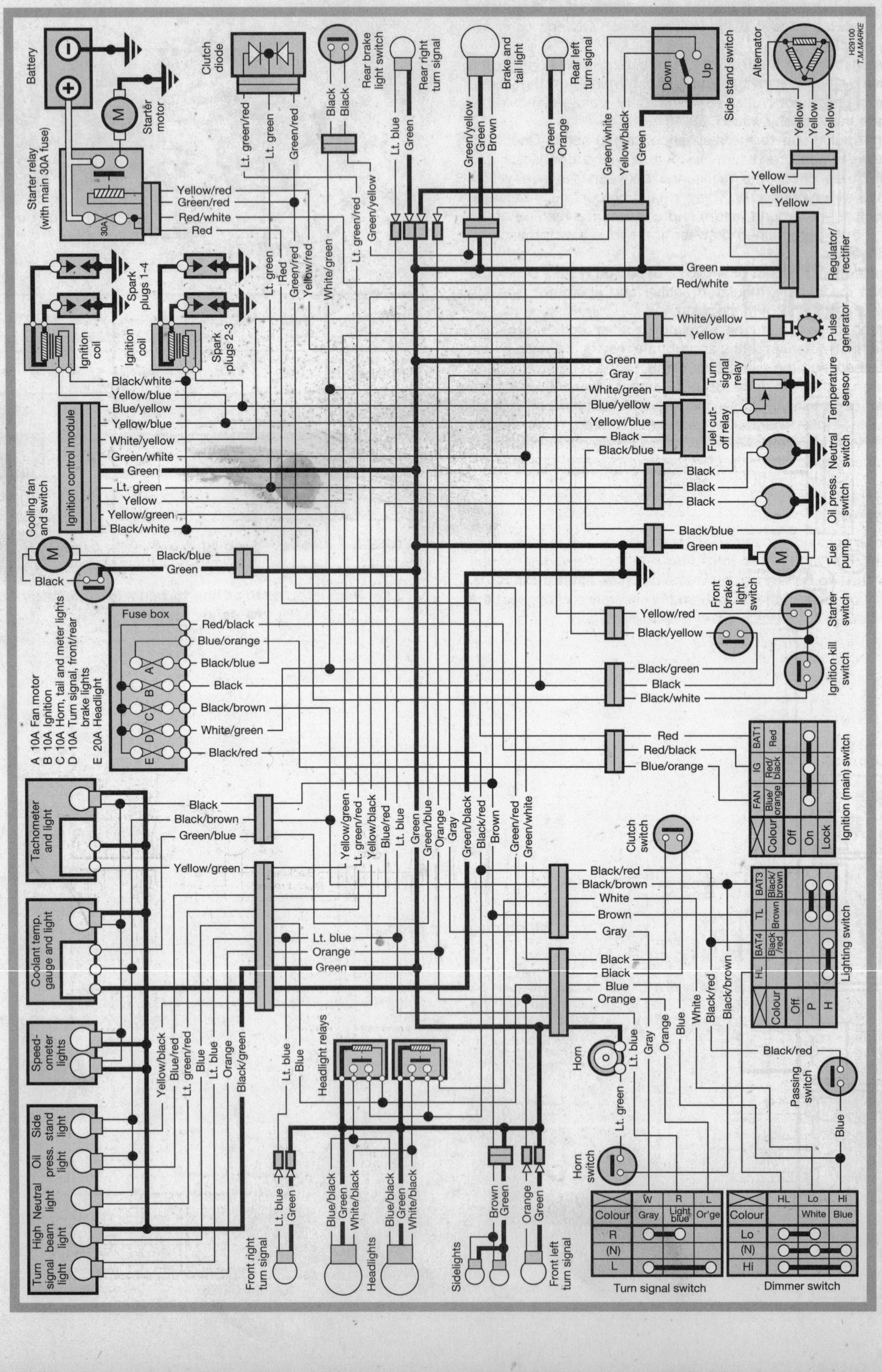

Battery

Starter motor

Starter relay (with main 30A fuse)

30A

Clutch diode

Rear brake light switch

Rear right turn signal

Brake and tail light

Rear left turn signal

Down Up

Side stand switch

Alternator

H29100 T.M.MARKE

Yellow
Yellow
Yellow

Regulator/ rectifier

Green
Red/white

Yellow/red
Green/red
Red/white
Red

Lt. green/red
Lt. green
Green/red

Lt. green
Red
Green/red
Yellow/red
White/green

Lt. green/red

Green/yellow

Lt. blue
Green

Green/yellow
Green
Brown

Green
Orange

Green/white
Yellow/black
Green

Spark plugs 1-4

Ignition coil

Spark plugs 2-3

Ignition coil

Black/white
Yellow/blue
Blue/yellow
Yellow/blue
White/yellow
Green/white
Green
Lt. green
Yellow
Yellow/green
Black/white

Ignition control module

Cooling fan and switch

Black/blue
Green

Black

M

White/yellow
Yellow

Pulse generator

Green
Gray
White/green
Blue/yellow
Yellow/blue
Black
Black/blue

Turn signal relay

Fuel cut-off relay

Temperature sensor

Black
Black
Black

Neutral switch

Oil press. switch

Black/blue
Green

Fuel pump
M

Front brake light switch

Yellow/red
Black/yellow

Starter switch

Black/green
Black
Black/white

Ignition kill switch

Fuse box

A 10A Fan motor
B 10A Ignition
C 10A Horn, tail and meter lights
D 10A Turn signal, front/rear brake lights
E 20A Headlight

Red/black
Blue/orange
Black/blue
Black
Black/brown
White/green
Black/red

E D C B A

Red
Red/black
Blue/orange

BAT1 Red
IG Red/ black
FAN Blue/ orange
Colour On Lock

Ignition (main) switch

Tachometer and light

Black
Black/brown
Green/blue
Yellow/green

Yellow/green
Lt. green/red
Yellow/black
Blue/red
Lt. blue
Green/blue
Orange
Gray
Green/black
Brown
Green/red
Green/white

Black/red
Black/brown
White
Brown
Gray
Black
Black
Blue
Orange

Clutch switch

BAT3 Brown
TL Brown/ brown
BAT4 Black/ red
HL Black/brown
Colour P H

Lighting switch

Coolant temp. gauge and light

Lt. blue
Orange
Green

Speed-ometer lights

Yellow/black
Blue/red
Lt. green/red
Blue
Lt. blue
Orange
Black/green

Lt. blue
Blue

Headlight relays

Horn

Lt. blue
Gray
Orange
Blue

White
Black/red
Black/brown

Black/red

Passing switch

Blue

Oil press. light

Side stand light

Neutral light

High beam light

Turn signal light

Lt. blue
Green

Blue/black
Green
White/black

Headlights

Blue/black
Green
White/black

Brown
Green

Orange
Green

Front right turn signal

Sidelights

Front left turn signal

Lt. green

Horn switch

Colour	W Gray	R Light blue	L Or'ge
R			
(N)			
L			

Turn signal switch

Colour	HL	Lo White	Hi Blue
Lo			
(N)			
Hi			

Dimmer switch

Wiring diagram - UK CBR900RR-N and CBR900RR-P

Battery
Starter motor
Starter relay (with main 30A fuse)
30A
Clutch diode
Rear brake light switch
Rear right brake light
Rear right turn signal
Brake and tail lights
Rear left turn signal
Side stand switch
Alternator
Regulator/rectifier
Pulse generator
Turn signal relay
Fuel cut-off relay
Temperature sensor
Neutral switch
Oil press. switch
Fuel pump
Front brake light switch
Starter switch
Ignition kill switch
Ignition (main) switch
Clutch switch
Lighting switch
Passing switch
Dimmer switch
Turn signal switch
Horn
Horn switch
Front left turn signal
Sidelight
Headlights
Headlight relays
Front right turn signal
Oil press. light
Side stand light
Neutral light
High beam light
Right turn signal light
Left turn signal light
Coolant temp. gauge and light
Speedometer and lights
Tachometer and light
Fuse box
Speed sensor
Cooling fan and switch
Ignition control module
Ignition coil
Spark plugs 1-4
Spark plugs 2-3

A 10A Fan motor
B 10A Ignition
C 10A Horn, tail and meter lights
D 10A Turn signal, front/rear brake lights
E 20A Headlight

Spark plugs 1-4

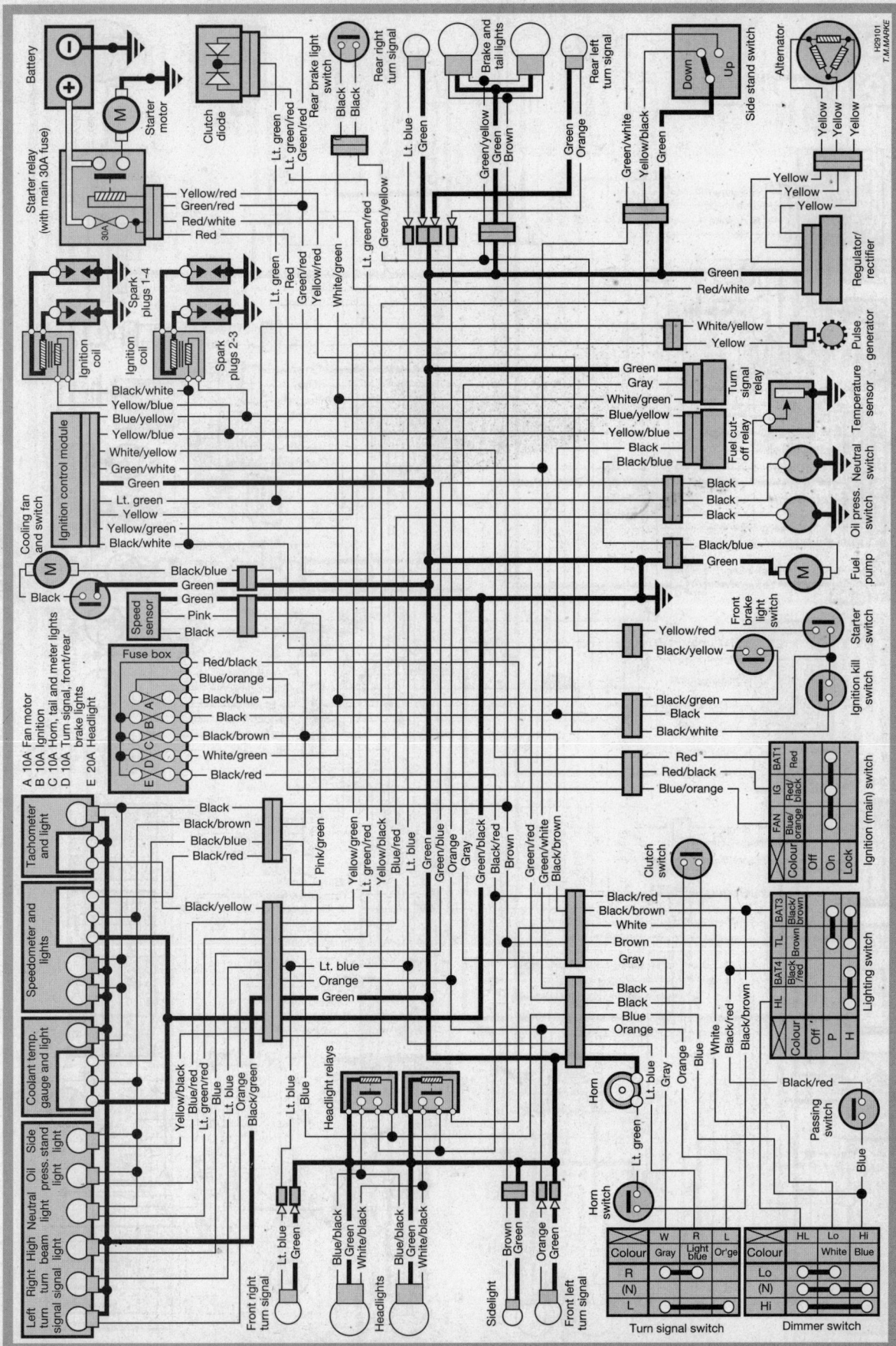

Wiring diagram - UK CBR900RR-R and CBR900RR-S

9

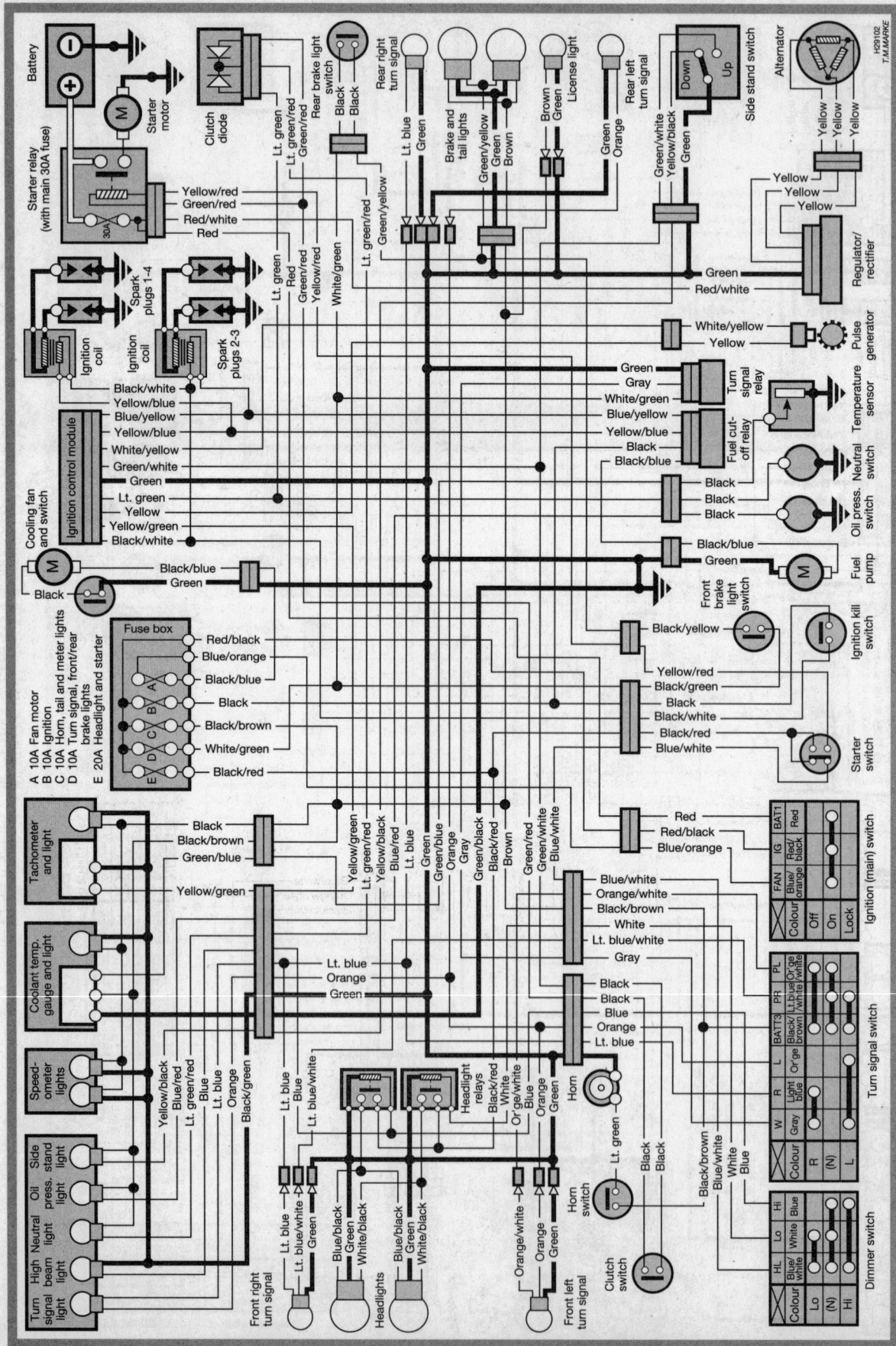

Wiring diagram - US 1993 and 1994 models

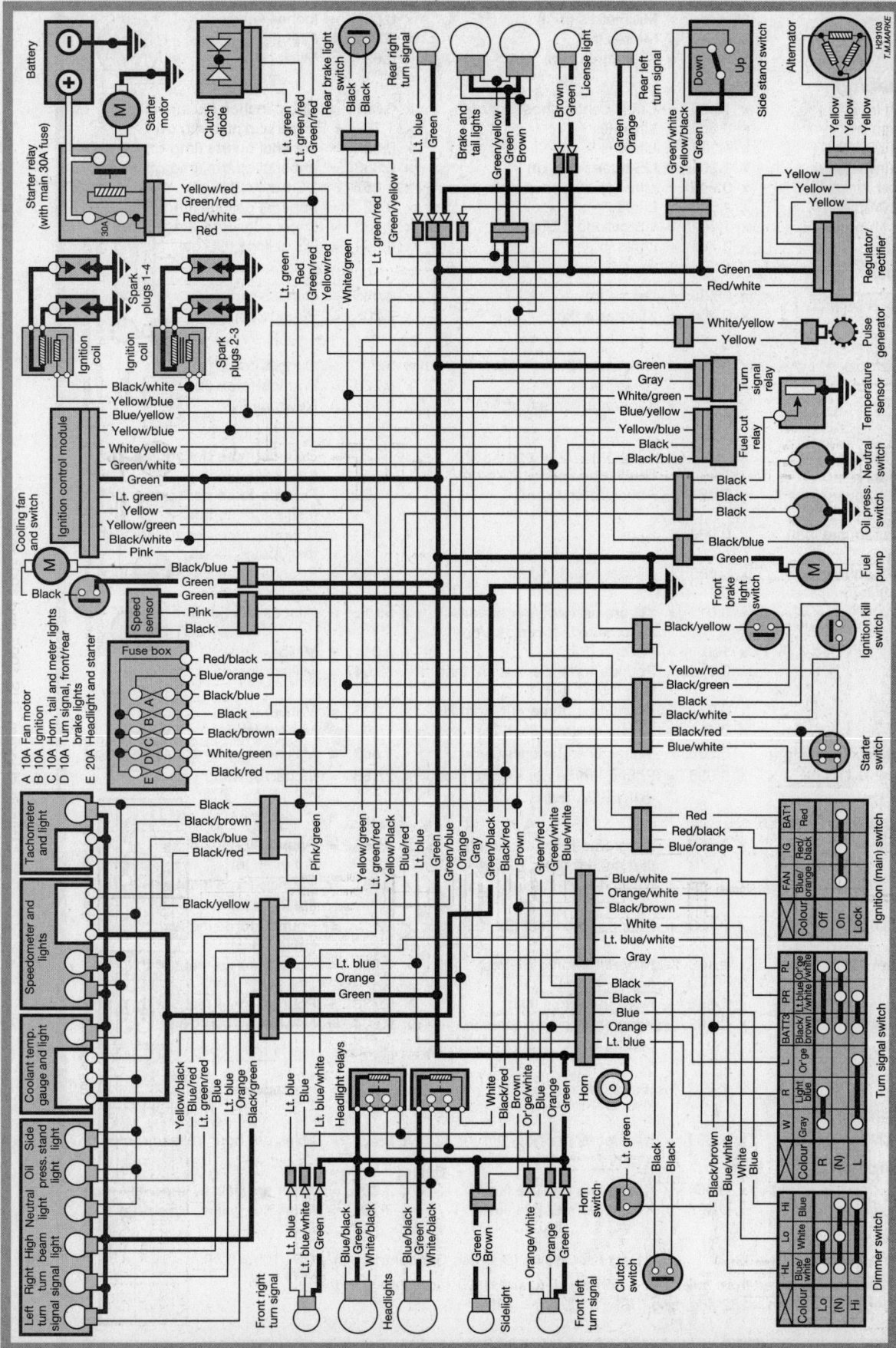

Wiring diagram - US 1995 model

9

Conversion factors

Length (distance)

Inches (in)	25.4	= Millimetres (mm)	x 0.0394	=	Inches (in)
Feet (ft)	0.305	= Metres (m)	x 3.281	=	Feet (ft)
Miles	1.609	= Kilometres (km)	x 0.621	=	Miles

Volume (capacity)

Cubic inches (cu in; in³)	x 16.387	= Cubic centimetres (cc; cm³)	x 0.061	=	Cubic inches (cu in; in³)
Imperial pints (Imp pt)	x 0.568	= Litres (l)	x 1.76	=	Imperial pints (Imp pt)
Imperial quarts (Imp qt)	x 1.137	= Litres (l)	x 0.88	=	Imperial quarts (Imp qt)
Imperial quarts (Imp qt)	x 1.201	= US quarts (US qt)	x 0.833	=	Imperial quarts (Imp qt)
US quarts (US qt)	x 0.946	= Litres (l)	x 1.057	=	US quarts (US qt)
Imperial gallons (Imp gal)	x 4.546	= Litres (l)	x 0.22	=	Imperial gallons (Imp gal)
Imperial gallons (Imp gal)	x 1.201	= US gallons (US gal)	x 0.833	=	Imperial gallons (Imp gal)
US gallons (US gal)	x 3.785	= Litres (l)	x 0.264	=	US gallons (US gal)

Mass (weight)

Ounces (oz)	x 28.35	= Grams (g)	x 0.035	=	Ounces (oz)
Pounds (lb)	x 0.454	= Kilograms (kg)	x 2.205	=	Pounds (lb)

Force

Ounces-force (ozf; oz)	x 0.278	= Newtons (N)	x 3.6	=	Ounces-force (ozf; oz)
Pounds-force (lbf; lb)	x 4.448	= Newtons (N)	x 0.225	=	Pounds-force (lbf; lb)
Newtons (N)	x 0.1	= Kilograms-force (kgf; kg)	x 9.81	=	Newtons (N)

Pressure

Pounds-force per square inch (psi; lbf/in²; lb/in²)	x 0.070	= Kilograms-force per square centimetre (kgf/cm²; kg/cm²)	x 14.223	=	Pounds-force per square inch (psi; lbf/in²; lb/in²)
Pounds-force per square inch (psi; lbf/in²; lb/in²)	x 0.068	= Atmospheres (atm)	x 14.696	=	Pounds-force per square inch (psi; lbf/in²; lb/in²)
Pounds-force per square inch (psi; lbf/in²; lb/in²)	x 0.069	= Bars	x 14.5	=	Pounds-force per square inch (psi; lbf/in²; lb/in²)
Pounds-force per square inch (psi; lbf/in²; lb/in²)	x 6.895	= Kilopascals (kPa)	x 0.145	=	Pounds-force per square inch (psi; lbf/in²; lb/in²)
Kilopascals (kPa)	x 0.01	= Kilograms-force per square centimetre (kgf/cm²; kg/cm²)	x 98.1	=	Kilopascals (kPa)
Millibar (mbar)	x 100	= Pascals (Pa)	x 0.01	=	Millibar (mbar)
Millibar (mbar)	x 0.0145	= Pounds-force per square inch (psi; lbf/in²; lb/in²)	x 68.947	=	Millibar (mbar)
Millibar (mbar)	x 0.75	= Millimetres of mercury (mmHg)	x 1.333	=	Millibar (mbar)
Millibar (mbar)	x 0.401	= Inches of water (inH₂O)	x 2.491	=	Millibar (mbar)
Millimetres of mercury (mmHg)	x 0.535	= Inches of water (inH₂O)	x 1.868	=	Millimetres of mercury (mmHg)
Inches of water (inH₂O)	x 0.036	= Pounds-force per square inch (psi; lbf/in²; lb/in²)	x 27.68	=	Inches of water (inH₂O)

Torque (moment of force)

Pounds-force inches (lbf in; lb in)	x 1.152	= Kilograms-force centimetre (kgf cm; kg cm)	x 0.868	=	Pounds-force inches (lbf in; lb in)
Pounds-force inches (lbf in; lb in)	x 0.113	= Newton metres (Nm)	x 8.85	=	Pounds-force inches (lbf in; lb in)
Pounds-force inches (lbf in; lb in)	x 0.083	= Pounds-force feet (lbf ft; lb ft)	x 12	=	Pounds-force inches (lbf in; lb in)
Pounds-force feet (lbf ft; lb ft)	x 0.138	= Kilograms-force metres (kgf m; kg m)	x 7.233	=	Pounds-force feet (lbf ft; lb ft)
Pounds-force feet (lbf ft; lb ft)	x 1.356	= Newton metres (Nm)	x 0.738	=	Pounds-force feet (lbf ft; lb ft)
Newton metres (Nm)	x 0.102	= Kilograms-force metres (kgf m; kg m)	x 9.804	=	Newton metres (Nm)

Power

Horsepower (hp)	x 745.7	= Watts (W)	x 0.0013	=	Horsepower (hp)

Velocity (speed)

Miles per hour (miles/hr; mph)	x 1.609	= Kilometres per hour (km/hr; kph)	x 0.621	=	Miles per hour (miles/hr; mph)

Fuel consumption*

Miles per gallon, Imperial (mpg)	x 0.354	= Kilometres per litre (km/l)	x 2.825	=	Miles per gallon, Imperial (mpg)
Miles per gallon, US (mpg)	x 0.425	= Kilometres per litre (km/l)	x 2.352	=	Miles per gallon, US (mpg)

Temperature

Degrees Fahrenheit = (°C x 1.8) + 32   Degrees Celsius (Degrees Centigrade; °C) = (°F - 32) x 0.56

It is common practice to convert from miles per gallon (mpg) to litres/100 kilometres (l/100km), where mpg (Imperial) x l/100 km = 282 and mpg (US) x l/100 km = 235

Index